COMPETITION LAW'S INNOVATION FACTOR

In recent years, market definition has come under attack as an analytical tool of competition law. Scholars have increasingly questioned its usefulness and feasibility. That criticism is thrown into sharp relief in dynamic, innovation-driven markets that do not correspond to the static markets on which the concept of the relevant market was modelled. This book explores that controversy from a comparative legal perspective, taking into account both EU competition law and US antitrust law. It examines the manifold ways in which courts and competition authorities in the EU and US have factored innovation-related considerations into market delineation, covering: innovative product markets, product differentiation, future markets, issues going beyond market definition proper – such as innovation competition, innovation markets and potential competition –, intellectual property rights, innovative aftermarkets and multi-sided platforms. The book finds that, going forward, the role of market definition in dynamic contexts needs to focus on its function of market characterisation rather than on the assessment of market power.

Volume 23 in the series Hart Studies in Competition Law

Hart Studies in Competition Law

Fairness in Antitrust: Protecting the Strong from the Weak
Adi Ayal

European Merger Remedies: Law and Policy
Dorte Hoeg

Media Ownership and Control: Law, Economics and Policy in an Indian and International Context
Suzanne Rab and Alison Sprague

The Interface between Competition and the Internal Market: Market Separation under Article 102 TFEU
Vasiliki Brisimi

Anti-Cartel Enforcement in a Contemporary Age: Leniency Religion
Edited by Caron Beaton-Wells and Christopher Tran

Public Procurement and the EU Competition Rules
Albert Sánchez Graells

The Concept of Abuse in EU Competition Law: Law and Economic Approaches
Pınar Akman

The Competitive Effects of Minority Shareholdings: Legal and Economic Issues
Panagiotis Fotis and Nikolaos Zevgolis

The More Economic Approach to EU Antitrust Law
Anne C Witt

Private Power, Online Information Flows and EU Law
Angela Daly

The Role of Competitors in the Enforcement of State Aid Law
Fernando Pastor-Merchante

The Legality of Bailouts and Buy Nationals: International Trade Law in a Crisis
Kamala Dawar

A Critical Account of Article 106(2) TFEU: Government Failure in Public Service Provision
Jarleth Burke

Dawn Raids Under Challenge
Helene Andersson

A Framework for European Competition Law: Co-ordinated Diversity
Christopher Townley

Evidence Standards in EU Competition Enforcement: The EU Approach
Andriani Kalintiri

The Metaphysics of Market Power: The Zero-sum Competition and Market Manipulation Model
George Raitt

Competition Law's Innovation Factor

The Relevant Market in Dynamic Contexts in the EU and the US

Viktoria H S E Robertson

•HART•

OXFORD • LONDON • NEW YORK • NEW DELHI • SYDNEY

HART PUBLISHING
Bloomsbury Publishing Plc
Kemp House, Chawley Park, Cumnor Hill, Oxford, OX2 9PH, UK
1385 Broadway, New York, NY 10018, USA
29 Earlsfort Terrace, Dublin 2, Ireland

First published in Great Britain 2020

Reprinted 2021

A catalogue record for this book is available from the British Library.

Library of Congress Cataloging-in-Publication data

Names: Robertson, Viktoria, author.

Title: Competition law's innovation factor : the relevant market in dynamic contexts in the EU and the US / Viktoria Robertson.

Description: Oxford, UK ; New York, NY : Hart, 2020 | Series: Hart studies in competition law; volume 23 | Includes bibliographical references and index.

Identifiers: LCCN 2019042599 (print) | LCCN 2019042600 (ebook) | ISBN 9781509931897 (hardback) | ISBN 9781509931903 (Epub)

Subjects: LCSH: Antitrust law—United States. | Antitrust law—European Union countries. | Technological innovations—Law and legislation—United States. | Technological innovations—Law and legislation—European Union countries. | Markets—Law and legislation—United States. | Markets—Law and legislation—European Union countries.

Classification: LCC K3842 .R63 2020 (print) | LCC K3842 (ebook) | DDC 343.2407/21—dc23

LC record available at https://lccn.loc.gov/2019042599

LC ebook record available at https://lccn.loc.gov/2019042600

ISBN: HB: 978-1-50993-189-7
ePDF: 978-1-50993-191-0
ePub: 978-1-50993-190-3

Typeset by Compuscript Ltd, Shannon

Printed and bound in Great Britain by TJ International Ltd, Padstow, Cornwall

To find out more about our authors and books visit www.hartpublishing.co.uk. Here you will find extracts, author information, details of forthcoming events and the option to sign up for our newsletters.

PREFACE

The delineation of the relevant product market has led to numerous debates within the antitrust community. There is a palpable tension between academics who argue in favour of abolishing market definition, those who want to retain it as an analytical tool but see a need to increase market definition's legal certainty, competition authorities that provide practical guidance on market definition while at the same time being policy-makers, and courts that do not accept an antitrust case unless it is founded on a solid market delineation. With innovation considerations high on the antitrust radar, it is clear that dynamic market contexts inevitably add to that tension. When I first embarked on the research that would lead to this study in the autumn of 2014, I was intrigued by these multi-faceted debates. I believed that an in-depth analysis of market delineation in innovative market environments was needed in order to gain a more thorough understanding of this first vital step in antitrust analysis. Owing to the often-global dimension of innovative markets, it seemed necessary for this research to rely upon a comparative analysis. The present study is the fruit of this work. It states the law and literature as it stood on 1 May 2019. Developments occurring thereafter could only be incorporated on rare occasions.

Over the course of the past five years, I have had the opportunity to discuss my work with a number of outstanding colleagues. I am greatly indebted to them for often thought-provoking, sometimes challenging and always rewarding discussions. In particular, I wish to thank Rachel Brandenburger, John Cartwright, David Evans, Hans Friederiszick, Michal Gal, Stefan Holzweber, Pablo Ibáñez Colomo, Maria Ioannidou, Wolfgang Kerber, Jiří Kindl, Mark Lemley, Phil Malone, Sandra Marco Colino, Paco Marcos, Giorgio Monti, Julian Nowag, Sofia Pais, Carlos Ragazzo, Alan Riley, Aidan Robertson, Barry Rodger, Sam Simon, Danny Sokol, Miguel Sousa Ferro, Stefan Storr, Maurice Stucke, Eric Talley, Florence Thépot, Mark Williams and Angela Zhang. I am particularly grateful to three competition law academics whose support was instrumental in finishing this book, and whose dedication to the frontiers of competition law is inspiring: Mike Carrier, Bill Kovacic and Heike Schweitzer.

I am indebted to Tomislav Borić, who has been a source of invaluable encouragement and inspiration for my academic career. When I first knocked on his door in 2006 to interview for the position of student assistant, I could not have imagined the academic journey this would lead to. It was his unbounded enthusiasm for comparative law that first motivated me to embark on such a comprehensive comparative legal study, and I am very glad I did.

To Ariel Ezrachi, I am very grateful for hosting me at the Oxford Centre for Competition Law and Policy (CCLP) throughout the academic year of 2016/17 and for regularly welcoming me back. It was in the wonderful working environment of the CCLP that the main research for this study was completed. I truly appreciate our many discussions, the helpful suggestions he made concerning my research on market delineation, as well as the time Ariel and his family made to make us feel at home in Oxford.

I would like to thank Simon Whittaker for making my return to St John's College Oxford during the academic year of 2016/17 such an enjoyable one, enabling me to touch base with my student roots at St John's – but in the even more comfortable and stimulating environment of the Senior Common Room.

When my research started to take shape, I had the opportunity to discuss my initial findings at a number of conferences. I want to thank all the participants who discussed ideas with me at the Stanford International Junior Faculty Forum in October 2015, the CLaSF Workshop on digital markets in April 2016, the ASCOLA conference in Leiden in June and July 2016, the Comparative Law Discussion Group at Oxford University in May 2017, the Antitrust Enforcement Symposium at Oxford University in June 2017, the ASCOLA conference in Stockholm in June 2017 and the ABA/NYU Next Generation of Antitrust Scholars Conference in January 2018.

I am particularly grateful to the research institutions that enabled me to conduct this study. The University of Graz, my first alma mater and employer, has been tremendously supportive of my research, as has the Institute of Corporate and International Commercial Law at the University of Graz at which I am currently based. Not only did they considerably stock up the competition law library, but they also generously supported my many longer and shorter research stays abroad. Sanja and Susanne provided invaluable administrative support. I am also grateful for the generous research funding that I received from the Heinrich Graf Hardegg Foundation. A comparative work such as the present requires considerable travelling and I would like to thank a number of research institutions that hosted me for various periods of time over the past five years: the CCLP and the Institute of European and Comparative Law at my second alma mater, Oxford University (thank you, Jenny!), the College of Europe in Bruges, the European University Institute in Florence, the Max Planck Institute for Comparative and International Private Law in Hamburg, Stanford Law School and Rutgers Law School at Camden, New Jersey.

I am grateful to Sinéad Moloney and the entire team at Hart Publishing for their editorial support at every step along the way. Two anonymous reviewers took the time to make thoughtful suggestions, which were greatly appreciated.

Throughout my research, I had the good fortune to be surrounded by a group of marvellous friends who provided encouragement and much-needed distraction, both of which I appreciated in equal measure. Finally, I thank my wonderful family for their love, patience and support throughout this research journey: my parents, my brother and his gorgeous family, my parents-in-law and my sister-in-law, all of whom were central to the completion of this study. And, above all, my amazing husband Rafael and our brilliant daughters, who are thrilled that I am publishing this study with an imprint of Bloomsbury, the home of Harry Potter.

VHSER
Oxford, 1 July 2019

CONTENTS

PART II
THE INNOVATION FACTOR IN MARKET DELINEATION UNDER EU COMPETITION AND US ANTITRUST LAW

PART III
RECONCEPTUALISING THE LEGAL FRAMEWORK FOR DELINEATING ANTITRUST MARKETS IN DYNAMIC CONTEXTS

TABLE OF CASES

European Union

Court of Justice of the European Union

Advocates General before the Court of Justice of the European Union

General Court of the European Union

European Commission

United States of America

US Supreme Court

US Federal Courts of Appeal

US Federal District Courts

Germany

Italy

Sweden

United Kingdom

TABLE OF LEGISLATION AND SOFT LAW INSTRUMENTS

European Union

Legislation

Soft Law Instruments

United States of America

Legislation

Soft Law Instruments

Other Jurisdictions

Austria

Germany

Japan

South Korea

PART I

Market Definition and Innovation

1

Introduction

There is no subject in antitrust law more confusing than market definition.[1]

The importance of market delineation for antitrust can easily be likened to the importance of the tennis court in tennis: for competition law purposes, the relevant market stakes out the playing field upon which competition law provides the rules of the game, and competition law judges act as referees. The competitive constraints that the market players face are used as the playing field's demarcation lines. If market behaviour affects issues that lie outside of the playing field, the line umpire signals an 'out' and competition law does not apply.

In order to decide whether a ball was inside or outside the court, tennis provides for certain procedures. The same is true of competition law, where the judge must determine whether behaviour occurred on a relevant market or not. On clay courts, the chair umpire may carry out ball mark inspections.[2] Similarly, some competition law judges directly delineate the market by relying on the evidence brought before them. On a hard court, chair umpires may rely on hawk-eye or other line calling assistance. In competition law, judges may rely on economic experts to illuminate questions of fact pertaining to market definition. What unites tennis and competition law is that in tennis, '[t]he referee is the final authority on all questions of tennis law',[3] including on deciding whether the ball was in or out. In competition law, it is the judge who is the final authority on all questions of competition law, including on deciding whether market behaviour took place on the relevant market or outside of it.

Different types of courts – be they clay, carpet, grass or hard courts – require different skills for line calls, just like different market environments in competition law do. In antitrust, this is particularly true of highly dynamic markets in which innovation plays a decisive role. When the playing field is innovative, it is frequently uncertain where the demarcation lines should be. The line umpires are at a loss. This is the question addressed in the present volume. While tennis has

[1] *US Healthcare v Healthsource*, 986 F2d 589, 598 (1st Cir 1993). In this study, the terms 'market delineation' and 'market definition' are used interchangeably, as are the terms 'competition law' and 'antitrust law'.

[2] International Tennis Federation, 'ITF Rules of Tennis' (2012) Appendix V.

[3] ibid.

achieved unification on line calls through the International Tennis Federation's Rules, transnational competition law lacks harmonisation and thus encounters the added complexity of different jurisdictions relying on (sometimes only slightly) different rules for market delineation, even where they relate to the same innovative market – one and the same game, so to speak.

The present study turns to two antitrust jurisdictions that are globally significant, namely the US and the EU. It compares their market definition frameworks as applied in dynamic contexts, thereby encountering a multitude of legal and economic approaches. After disassembling and comparatively re-organising both frameworks, consideration is given to the question of whether and how convergence of all or some parts of these frameworks could increase legal certainty and more accurately depict innovation to the benefit of consumers across the globe.

I. Innovation and the Relevant Market: The Issues at Stake

The relevant market is one of the most complex and contentious legal concepts in competition law, particularly in innovative market environments. While market definition merely serves as an analytical tool for antitrust purposes,[4] the analytical tools we employ may well determine the outcome. Drawing on economic insights, market definition attempts to determine substitutability amongst products, both from a demand and from a supply perspective.[5] It positions the legally relevant market within actual economic activity, while at the same time setting out which area of economic life is legally relevant for the antitrust assessment. As such, it provides the analytical framework for any antitrust analysis,[6] perhaps even representing antitrust's 'analytical core'.[7]

Courts, antitrust authorities and scholars alike have repeatedly emphasised the unparalleled importance of market definition for antitrust cases.[8] The reason for this importance is, quite simply put, that a company's market behaviour cannot properly be legally assessed in the abstract, but only with reference to the

[4] MR Baye, 'Market Definition and Unilateral Competitive Effects in Online Retail Markets' (2008) 4 *Journal of Competition Law & Economics* 639, 652.

[5] ABA Section of Antitrust Law, *Market Power Handbook: Competition Law and Economic Foundations*, 2nd edn (Chicago, ABA Publishing, 2012) 63.

[6] This wording is based on M Blaschczok, *Kartellrecht in zweiseitigen Wirtschaftszweigen: Eine Untersuchung vor dem Hintergrund der ökonomischen Forschung zu 'two-sided markets'* (Baden-Baden, Nomos, 2015) 55.

[7] DA Crane, 'Market Power without Market Definition' (2014) 90 *Notre Dame Law Review* 31, 33.

[8] R Pitofsky, 'New Definitions of Relevant Market and the Assault on Antitrust' (1990) 90 *Columbia Law Review* 1805, 1807; *Eastman Kodak v Image Technical Services*, 504 US 451, 469 fn 15 (1992); European Commission, Notice on the definition of relevant market for the purposes of Community competition law [1997] OJ C372/5 (EU Market Definition Notice 1997) para 4.

market(s) in which the company is active:[9] the relevant market(s). The soundness of antitrust market definition has a direct impact on the quality of antitrust law as such.[10] While some regard market definition and the ensuing market power analysis as necessarily conflated,[11] the present research focuses on market definition as a separate analytical step, thus essentially ascribing a standalone value to market definition. This focus can again be understood against the metaphor of line calls in tennis: where competition law is not even sure what the relevant market is, it cannot properly judge whether the game that is being played by market participants is pro- or anti-competitive. In order to do so, it must first set out to delineate and thoroughly understand that playing field.

Market definition is the foundation that subsequent antitrust analysis builds upon. Relying on market analysis presupposes two steps: the identification of the companies that are catering to a particular customer demand, and based thereupon the analysis of competition on that market.[12] Depending on the competition conditions prevailing in the relevant market, antitrust authorities may or may not have competition concerns. The position of a company in the relevant market can have a plethora of consequences under antitrust law, including the application of stricter antitrust rules if a company is found to enjoy significant market power. Therefore, it is of crucial importance to define the relevant antitrust market in a predictable, coherent way. However, market definition as such is 'hardly an exact science',[13] facing companies with legal uncertainty in antitrust matters. This problem is exacerbated in highly dynamic market environments, as market definition gives but a snapshot of economic reality.

In economics, it is believed that, in the long run, dynamic competition – or innovation – can generate greater consumer welfare than static competition.[14] This insight has become widely accepted in antitrust,[15] even if its consequences for antitrust law are not yet fully understood. As a first step, however, it is clear

[9] D Cameron, MA Glick and D Mangum, 'Comments on Articles in the Kaplow Special Issue' (2012) 57 *Antitrust Bulletin* 957, 960; L Peeperkorn and V Verouden, 'Market Definition' in J Faull and A Nikpay (eds), *The EU Law of Competition*, 3rd edn (Oxford, Oxford University Press, 2014) § 1.134.

[10] RJ van den Bergh and A Giannaccari, 'L'approcio più economico nel diritto comunitario della concorrenza: Il più è troppo o non (ancora) abbastanza?' (2014) XVI *Mercato concorrenza regole* 393, 425.

[11] For instance, see L Kaplow, 'Why (Ever) Define Markets?' (2010) 124 *Harvard Law Review* 438.

[12] MB Coate and JH Fischer, 'Is Market Definition Still Needed after All These Years' (2014) 2 *Journal of Antitrust Enforcement* 422, 428.

[13] D Waelbroeck, 'Vertical Agreements: 4 Years of Liberalisation by Regulation N. 2790/99 after 40 Years of Legal (Block) Regulation' in H Ullrich (ed), *The Evolution of European Competition Law: Whose Regulation, Which Competition?* (Cheltenham, Edward Elgar, 2006) 87.

[14] See JA Schumpeter, *Capitalism, Socialism and Democracy*, 5th edn (London, Allen & Unwin, 1976) 84 f; JG Sidak and DJ Teece, 'Dynamic Competition in Antitrust Law' (2009) 5 *Journal of Competition Law & Economics* 581, 600.

[15] O Kolstad, 'Competition Law and Intellectual Property Rights: Outline of an Economics-Based Approach' in J Drexl (ed), *Research Handbook on Intellectual Property and Competition Law* (Cheltenham, Edward Elgar, 2008) 4; HJ Hovenkamp, 'Antitrust and Innovation: Where We are and Where We Should Be Going' (2011) 77 *Antitrust Law Journal* 749, 751.

that the proper use of market definition in innovative environments is a necessary precondition for applying antitrust law to the long-term benefit of consumers.[16] There is widespread concern that traditional antitrust tools may focus too much on static market conditions, price and homogeneous products, thus leading to a competition analysis that cannot properly take into account product diversification, future products and innovation competition.[17] In dynamic market environments, innovation regularly overthrows the current market order and pushes the limits of the relevant market. This aspect of innovative markets needs to be analysed with legal certainty in mind.

Innovation as understood in the present context refers to new or significantly improved products or processes.[18] This includes the digital markets that are currently at the forefront of many policy discussions both in the EU and in the US,[19] but is much broader than that. Any industry that exhibits significant dynamic characteristics is included in this definition, covering innovation in production, in service provision and extending to self-driving cars and the pharmaceutical industry. The dynamic characteristics of innovative industries face market definition with a number of important challenges,[20] as the analytical framework for market definition was developed in static rather than in dynamic environments. These challenges revolve around the fast-moving nature of dynamic industries, the importance of potential competition and intellectual property rights in innovative markets, innovation rather than price as the most important parameter of competition in these markets, multi-sided platforms, innovative and proprietary aftermarkets, and network effects that are at play in dynamic environments. A rich body of literature cautions that the dynamic characteristics of innovative markets must be taken into account in antitrust law.[21] Such an innovation-conscious approach necessarily needs to begin with market definition.

In the EU, it is understood that the delineation of a relevant market is a crucial aspect of competition analysis.[22] In its case law, the Court of Justice of the European

[16] International Competition Network, 'Competition Enforcement and Consumer Welfare – Setting the Agenda' (May 2011) 19 f.

[17] J Drexl, 'Anticompetitive Stumbling Stones on the Way to a Cleaner World: Protecting Competition in Innovation without a Market' (2012) 8 *Journal of Competition Law & Economics* 507, 508; MA Lemley and MP McKenna, 'Is Pepsi Really a Substitute for Coke? Market Definition in Antitrust and IP' (2012) 100 *Georgetown Law Journal* 2055, 2058.

[18] See ch 3, section I.

[19] For instance, see European Commission, Conference on 'Shaping Competition Policy in the Era of Digitisation' (Brussels, 17 January 2019); Federal Trade Commission, 'Hearings on Competition and Consumer Protection in the 21st Century' (2018–19), www.ftc.gov/policy/hearings-competition-consumer-protection.

[20] RC Lind and P Muysert, 'Innovation and Competition Policy: Challenges for the New Millennium' (2003) 24 *European Competition Law Review* 87, 88.

[21] See JG Sidak and DJ Teece, 'Rewriting the Horizontal Merger Guidelines in the Name of Dynamic Competition' (2009) 16 *George Mason Law Review* 885, 894; GA Manne and JD Wright, 'Innovation and the Limits of Antitrust' (2010) 6 *Journal of Competition Law & Economics* 153; J Galloway, 'Driving Innovation: A Case for Targeted Competition Policy in Dynamic Markets' (2011) 34 *World Competition* 73.

[22] EU Market Definition Notice 1997 (n 8) para 4.

Union (CJEU) has confirmed that the product and geographical dimensions of the relevant market must be established for antitrust analysis,[23] and that particular regard must be had to demand and supply substitutability.[24] The General Court regards market definition as the first analytical step for any case in which an infringement of Article 102 of the Treaty on the Functioning of the European Union (TFEU)[25] is postulated. This is so because the abuse of a dominant position is conditional upon the existence of such a dominant position, and finding a dominant position on a relevant market in turn presupposes market definition.[26] The European Commission has issued several binding block exemption regulations for certain categories of agreements that are not subject to scrutiny under Article 101 TFEU – provided certain market share thresholds are not surpassed.[27] These market share thresholds require the calculation of market shares on a previously defined relevant market.

In 2015, Margrethe Vestager, the Commissioner for Competition, reminded the antitrust community of how vital market definition is to EU competition law,[28] as it represents the first step in virtually any European competition law case. The European Commission as the EU's leading competition authority has published numerous (non-binding) Notices and Guidelines that elaborate on that authority's approach to market definition. This soft law represents the Commission's interpretation of the Treaty provisions on competition law and the case law relating to them. Although the Commission's understanding is subject to the EU Courts' interpretation,[29] the General Court has held that 'in so far as the definition of the product market involves complex economic assessments on the part of the Commission, it is subject to only limited review by the [EU] judicature'.[30] This results in the Commission enjoying considerable autonomy and thus weight in questions of market definition.

The Commission is acutely aware of the need to assess relevant markets in their particular context. Amongst other things, this is evidenced by its Guidelines on Horizontal Co-operation, which contain separate sections dealing with market definition for specific types of agreements. For research and development (R&D) agreements, for instance, the Commission urges that market definition might need

[23] Case 27/76 *United Brands v Commission* EU:C:1978:22, para 10. Together, the CJEU and the General Court (previously the Court of First Instance) are referred to as EU Courts.

[24] Case 85/76 *Hoffmann-La Roche v Commission* EU:C:1979:36, para 28; Case 6/72 *Europemballage and Continental Can v Commission* EU:C:1973:22, para 33.

[25] Consolidated Version of the Treaty on the Functioning of the European Union [2016] OJ C202/47 (TFEU). The following uses the current appellation of the TFEU articles, even when discussing events that occurred before the renumbering by the Treaty of Lisbon in 2009.

[26] Case T-62/98 *Volkswagen v Commission* EU:T:2000:180, para 230.

[27] On these, see ch 2, section II.B.

[28] M Vestager, 'Thoughts on Merger Reform and Market Definition' (Studienvereinigung Kartellrecht, Brussels, 12 March 2015).

[29] EU Market Definition Notice 1997 (n 8) para 6.

[30] Case T-301/04 *Clearstream v Commission* EU:T:2009:317, para 47; Case T-201/04 *Microsoft v Commission* EU:T:2007:289, para 482.

to include existing product markets, existing technology markets and also the agreement's influence on competition in innovation.[31] In its Technology Transfer Block Exemption Regulation, it distinguishes separate product and technology markets that, combined with the relevant geographical markets, make up the relevant market for the purposes of that regulation.[32] This paves the way for relying on innovation-specific considerations in delineating antitrust markets.

In the *Microsoft/Skype* case, both the Commission (2011) and the General Court (2013) were prepared to accept that market shares only have a limited value in the face of rapid innovation.[33] The underlying issue, of course, was that market boundaries may inadvertently shift in highly dynamic markets such as the one at issue in that case. The Commission has recognised the two-sided nature of some markets in decisions relating to online advertising and internet search engines.[34] The Commission also dealt with market definition in innovative digital market environments in its three *Google* cases, which led to record fines.[35] The question whether intellectual property rights (IPRs) should help delineate antitrust markets has also been considered in a number of cases.[36]

In the US, the US Supreme Court places considerable weight on market definition, holding that the relevant market decides the case in most antitrust proceedings.[37] In several landmark rulings,[38] the Supreme Court has set out how antitrust markets should be defined based on demand and supply substitutability. It remains to be established to what extent market definition differs when applied under section 1 or 2 of the Sherman Act or under section 7 of the Clayton Act.[39]

[31] European Commission, Guidelines on the applicability of Article 101 of the Treaty on the Functioning of the European Union to horizontal co-operation agreements [2011] OJ C11/1, paras 112 f.

[32] Commission Regulation (EU) 316/2014 on the application of Article 101(3) of the Treaty on the Functioning of the European Union to categories of technology transfer agreements [2014] OJ L93/17 (TTBER), art 1(j)–(m).

[33] *Microsoft/Skype* (Case COMP/M.6281) Commission Decision of 7 October 2011, paras 78, 99; Case T-79/12 *Cisco Systems & Messagenet v Commission* EU:T:2013:635, para 69.

[34] *Google/DoubleClick* (Case COMP/M.4731) Commission Decision of 11 March 2008 [2008] OJ C184/10, paras 20, 290; *Microsoft/Yahoo! Search Business* (Case COMP/M.5727) Commission Decision of 18 February 2010, paras 47, 100.

[35] *Google Search (Shopping)* (Case AT.39740) Commission Decision of 27 June 2017 [2018] OJ C9/11, currently on appeal as Case T-612/17 *Google and Alphabet v Commission* [2017] OJ C369/37; *Google Android* (Case AT.40099) Commission Decision of 18 July 2018, currently on appeal as Case T-604/18 *Google and Alphabet v Commission* [2018] OJ C445/21; *Google Search (AdSense)* (Case AT.40411) Commission Decision of 20 March 2019, currently on appeal as Case T-334/19 *Google and Alphabet v Commission* [2019] OJ C255/46.

[36] See ch 6.

[37] *Eastman Kodak v Image Technical Services*, 504 US 451, 469 fn 15 (1992).

[38] On these, see ch 2, section I.A.

[39] Sherman Antitrust Act (1890), 15 USC §§ 1–7, as amended; Clayton Antitrust Act (1914), 15 USC §§ 12–27, as amended. See, eg, JO von Kalinowski, 'Market Definition under Section 2: The Applicability of Clayton Act Section 7 Analysis' (1978) 10 *Southwestern University Law Review* 95; L Griggs, 'A Teleological Approach to Market Definition – Has it Led to Single Product Market Definition?' (2002) 4 *University of Notre Dame Australia Law Review* 77.

Recently, the US antitrust agencies have put less emphasis on market definition. In their joint 2010 Horizontal Merger Guidelines, the US Department of Justice (DoJ) and the Federal Trade Commission (FTC) point out that market definition is no longer the necessary starting point when analysing a proposed merger.[40] In addition, they consider that '[r]elevant markets need not have precise metes and bounds'[41] and are thought of as an approximation to the reality of the market, while in the EU, market shares will often decide on the applicability of certain antitrust rules, thus presupposing a rather precise definition of the relevant market.

Despite their view that market definition in innovation-intensive industries does not require any analysis out of the ordinary,[42] the US antitrust agencies have taken market definition in innovative markets seriously. Several of their policy documents relate to the issue of how to address innovation in antitrust market definition. In 1995, the DoJ and the FTC issued Antitrust Guidelines for the Licensing of Intellectual Property (IP Guidelines), in which they state that innovation and consumer welfare are the common goals of the intellectual property and antitrust laws.[43] This assertion was maintained in the 2017 update to the IP Guidelines.[44] In the context of antitrust analysis, the agencies will not equate ownership of an IPR with the possession of market power.[45] This conclusion is today strongly supported by case law, which has found that a patent does not in itself confer market power.[46] Instead, market power will need to be proved with reference to the relevant market at issue, which might or might not be delineated by IPRs.

The agencies' IP Guidelines introduced the concept of R&D markets (previously innovation markets).[47] The R&D market concept was – and to some extent still is – vividly discussed in the literature.[48] Such an R&D market 'consists of the assets comprising research and development related to the identification of a commercializable product, or directed to particular new or improved goods or processes, and the close substitutes for that research and development'.[49] The R&D market is distinct from the technology market,

[40] US Department of Justice and Federal Trade Commission, Horizontal Merger Guidelines (19 August 2010) (US Horizontal Merger Guidelines 2010) § 4.

[41] ibid.

[42] United States in OECD (ed), *Policy Roundtable: Market Definition* (2012) DAF/COMP(2012)19, 331.

[43] US Department of Justice and Federal Trade Commission, Antitrust Guidelines for the Licensing of Intellectual Property (6 April 1995) (US IP Guidelines 1995) § 1.0.

[44] US Department of Justice and Federal Trade Commission, Antitrust Guidelines for the Licensing of Intellectual Property (14 January 2017) (US IP Guidelines 2017) § 1.0.

[45] ibid §§ 2.0, 2.2.

[46] *CSU and Others v Xerox Corp*, 203 F3d 1322, 1325 (Fed Cir 2000); *Illinois Tool Works and Others v Independent Ink*, 547 US 28, 31 (2006).

[47] US IP Guidelines 1995 (n 43) § 3.2.3; US IP Guidelines 2017 (n 44) § 3.2.3.

[48] See ch 5, section II.

[49] US IP Guidelines 2017 (n 44) § 3.2.3.

especially where it relates to products that do not yet exist. Importantly, the agencies will only rely on an R&D market where the special assets or characteristics of a company are determinative for that company's ability to engage in the R&D in question.[50] The agencies have acknowledged that part of the criticism directed at the innovation market concept was based on the lack of insight as to how market structure and innovation are interrelated.[51] The agencies' Competitor Collaboration Guidelines of 2000 define three possible markets that can be looked at when assessing horizontal agreements: product, technology and innovation markets.[52] Developing case law already demonstrates how the agencies[53] and the courts[54] have tried to tackle the delineation of innovation-intensive markets.

The arbitrary nature of market definition from an economics point of view has been criticised for many decades.[55] The realisation that various forms of innovation are essential for consumer welfare[56] – and indeed for social welfare at large – has only heightened this problem. Conceptually, the relevant market as it applies in competition law is in dire need of an in-depth inquiry, especially for highly dynamic markets.[57] In addition, the last few years have seen an important debate questioning whether there is any need for market definition in antitrust at all.[58] In the context of rapid innovation and IPRs, this debate can have an important impact on market delineation in these market environments. The present study ties these different strands together: it deconstructs and reconceptualises the legal analytical framework for delineating innovative markets from a comparative perspective encompassing EU competition and US antitrust law, thereby providing an impetus for overhauling said legal framework with the characteristics of innovative markets in mind.

[50] ibid.

[51] United States in OECD (ed), *Policy Roundtable: Merger Review in Emerging High Innovation Markets* (2002) DAFFE/COMP(2002)20, 149.

[52] Federal Trade Commission and US Department of Justice, Antitrust Guidelines for Collaborations among Competitors (April 2000) (US CC Guidelines 2000) §§ 3.32(a) to (c).

[53] eg, *Sensormatic Electronics*, 119 FTC 520 (1995); *Upjohn and Others*, 121 FTC 44 (1996); Federal Trade Commission, 'Statement Concerning *Google/DoubleClick*, Case 071-0170' (20 December 2007); *Motorola Mobility & Google*, 156 FTC 147 (2013).

[54] eg, *Realcomp II Ltd v FTC*, Case 9320 (ALJ) (10 December 2009).

[55] EH Chamberlin, 'Product Heterogeneity and Public Policy' (1950) 40 *American Economic Review* 85, 86; GE Hale and RD Hale, *Market Power: Size and Shape under the Sherman Act* (Boston, Little, Brown & Comp, 1958) 111.

[56] Consumer welfare is here broadly understood as the promotion of consumer interests, which includes price, service, quality, choice and innovation, rather than the more technical understanding of consumer welfare in the Chicagoan sense; see also E Buttigieg, *Competition Law: Safeguarding the Consumer Interest: A Comparative Analysis of US Antitrust Law and EC Competition Law* (Alphen aan den Rijn, Kluwer Law International, 2009) 1.

[57] R Podszun, 'The Arbitrariness of Market Definition and an Evolutionary Concept of Markets' (2016) 61 *Antitrust Bulletin* 121, 123.

[58] See Kaplow, 'Why (Ever) Define Markets?' (n 11); and ch 2, section IV.

II. The Parameters of this Study

A. The Relevant Market as a Legal Concept of Competition Law

The present study is based on the premise that the relevant market is a legal concept which is employed in competition law.[59] Relevant antitrust markets are legal 'constructs to facilitate analysis of particular alleged conduct rather than facts in their own right'.[60] The relevant market as understood by competition lawyers is not identical to the market in a business sense[61] nor to the market as it is generally understood in economics.[62] This is also the reason why managers' insights into demand substitution might be of great value for antitrust market definition, but their views as to what they regard as the market might not.[63] When incorporating the economic concept of a relevant market into competition law, this concept becomes a legal concept that relates to the same core, but at the same time takes on distinct conceptions that satisfy the requirements of the law.[64] Within competition law, this concept fulfils a number of functions, as will be discussed in Chapter 2.

In analysing a legal concept such as the relevant market, it is important to bear in mind the context within which it operates.[65] When 'rationally reconstruct[ing]' the relevant market concept, one converts that concept into a more concrete form while retaining its actual meaning.[66] In doing so, one is necessarily influenced by two factors: by what one assumes the concept's original meaning to be and by what one believes the concept's meaning ought to be. The present volume positions the legal concept of the relevant market within the innovation context, thus understanding innovation as the inevitable backdrop of which competition law should be mindful. Well-established principles of competition law are reconsidered in the dynamic light of innovation, while at the same time recognising the legal and cultural limitations that are embodied in the law.

[59] See VHSE Robertson, 'The Relevant Market in Competition Law: A Legal Concept' (2019) 7 *Journal of Antitrust Enforcement* 158.

[60] RL Smith, 'Defining and Proving Markets and Market Power' in J Duns, A Duke and BJ Sweeney (eds), *Comparative Competition Law* (Cheltenham, Edward Elgar, 2015) 32.

[61] EU Market Definition Notice 1997 (n 8) para 3.

[62] PA Geroski, 'Thinking Creatively about Markets' (1998) 16 *International Journal of Industrial Organization* 677, 678; R Podszun and B Franz, 'Was ist ein Markt? – Unentgeltliche Leistungsbeziehungen im Kartellrecht' (2015) 3 *Neue Zeitschrift für Kartellrecht* 121, 125.

[63] See also JB Baker, 'Market Definition: An Analytical Overview' (2007) 74 *Antitrust Law Journal* 129, 139; C Caffarra and M Walker, 'An Exploration into the Use of Economics before Courts in Europe' (2010) 1 *Journal of European Competition Law & Practice* 158, 159.

[64] See Robertson, 'Legal Concept' (n 59) 164.

[65] See already Å Frändberg, 'An Essay on Legal Concept Formation' in JC Hage and D von der Pfordten (eds), *Concepts in Law* (Berlin, Springer, 2009) 15.

[66] T Spaak, 'Explicating the Concept of Legal Competence' in JC Hage and D von der Pfordten (eds), *Concepts in Law* (Berlin, Springer, 2009) 69.

B. On the Relationship between Law and Economics in Competition Law

Antitrust law is based on a certain understanding of how markets work and how they contribute to overall social or consumer welfare – an understanding that itself was greatly shaped by economics. This is one of the reasons why the question to what extent economic analysis is and should be used in competition law remains one of the major recurring themes in comparative competition law.[67] The present study acknowledges that the foundations of competition law are often rooted in economic thinking, making economics a natural sister discipline for competition lawyers. At the same time, however, market definition can be regarded as a troubled 'marriage of the economic and legal disciplines' with divorce just around the corner.[68] The increasing reliance of competition law on economics is often met with considerable scepticism, both because industrial economics itself is divided about the theoretical foundations of antitrust economics and because an economic approach may prove too complex to be justiciable.[69] The law must prescribe the scope for economic analysis in a competition law case,[70] while that economic analysis cannot replace the legal analysis.[71] Where competition law (also) pursues certain non-economic goals, economics will often not provide the insights that the law requires.[72] When economics reaches its limits, it is the law that must take over.[73]

While this study recognises the importance of economics for the relevant market concept, it does not employ the efficiency-based methodology of law and economics. Law and economics tries to predict the reactions both by individuals and by firms to the laws under scrutiny.[74] It sees law as a tool for steering human behaviour in order to attain certain policy objectives, most importantly efficiency.[75] As dynamic efficiency cannot be as readily measured as allocative or

[67] D Geradin, 'Competition Law' in JM Smits (ed), *Elgar Encyclopedia of Comparative Law*, 2nd edn (Cheltenham, Edward Elgar, 2012) 210.

[68] ML Glassman, 'Market Definition as a Practical Matter' (1980) 49 *Antitrust Law Journal* 1155, 1155.

[69] K Heyer, 'A World of Uncertainty: Economics and the Globalization of Antitrust' (2005) 72 *Antitrust Law Journal* 375, 379; ILO Schmidt, 'The Suitability of the More Economic Approach for Competition Policy: Dynamic vs Static Efficiency' (2007) 28 *European Competition Law Review* 408, 408–10.

[70] Y Katsoulacos, S Avdashevaa and S Golovanova, 'Legal Standards and the Role of Economics in Competition Law Enforcement' (2017) 12 *European Competition Journal* 277, 278.

[71] Case T-1/89 *Rhône-Poulenc v Commission* EU:T:1991:38, Opinion of AG Vesterdorf, 957.

[72] RJ van den Bergh, 'The More Economic Approach in European Competition Law: Is More Too Much or Not Enough?' in M Kovač and A-S Vandenberghe (eds), *Economic Evidence in EU Competition Law* (Cambridge, Intersentia, 2016) 14.

[73] RH Pate, 'The Common Law Approach and Improving Standards for Analyzing Single Firm Conduct' (Thirtieth Annual Conference on International Antitrust Law and Policy, New York, 23 October 2003) 27 f.

[74] R Cooter and T Ulen, *Law & Economics*, 6th edn (Boston, Pearson, 2012) 3 f.

[75] ibid 9; U Kischel, *Comparative Law* (A Hammel (trans), Oxford, Oxford University Press, 2019) § 3 para 56.

productive efficiencies, both of which are more static in nature, innovation is not only hard to predict, but also difficult to measure retrospectively.[76] In addition, the relevant market is not a fully-fledged legal rule which could be analysed from a law and economics point of view – it is but an analytical tool, albeit a very consequential one.

From a comparative perspective, there was a noticeable push towards a more economics-based approach to competition law in general and market delineation in particular within the European Commission in the 1990s,[77] while economics has for many decades played an important role in US market definition and continues to do so.[78] Only recently, the US Supreme Court conceded that courts may re-interpret the antitrust laws in the light of evolving insights from economics,[79] thus paving the way for an economics-minded understanding of antitrust law. In Europe, it remains more controversial than in the US whether economics should fill antitrust concepts with meaning.[80]

Broaching the theme of convergence, which occupies a special place in comparative competition law, there is a widespread belief that economic insights shared across jurisdictions may provide an important impetus for convergence between competition law systems.[81] However, this expectation is generally based on the premise that economic insights are uniform,[82] which arguably they are not.[83] The question then turns on which economic school ought to drive this envisaged convergence.

As was stated above, the relevant market concept as relied upon in antitrust shares its roots with economics, but has a specific legal conception.[84] The innate differences between antitrust law on the one hand and antitrust economics on the

[76] JW Markham, 'The Joint Effect of Antitrust and Patent Laws upon Innovation' (1966) 56 *American Economic Review* 291, 291; Schmidt, 'More Economic Approach' (n 69) 408; MA Schilling, 'Towards Dynamic Efficiency: Innovation and its Implications for Antitrust' (2015) 60 *Antitrust Bulletin* 191, 192.

[77] See M Monti, 'Market Definition as a Cornerstone of EU Competition Policy' (SPEECH/01/439, Helsinki, 5 October 2001); T Ackermann, 'European Competition Law' in K Riesenhuber (ed), *European Legal Method* (Cambridge, Intersentia, 2017) § 20.22.

[78] GR Hall, 'Market Definition and Antitrust Policy' (1963) 20 *Washington & Lee Law Review* 47, 47; HJ Hovenkamp, 'The Reckoning of Post-Chicago Antitrust' in A Cucinotta, R Pardolesi and RJ van den Bergh (eds), *Post-Chicago Developments in Antitrust Law* (Cheltenham, Edward Elgar, 2002) 1 f.

[79] *Kimble v Marvel Entertainment*, 576 US ___ (2015), 135 SCt 2401, 2412 f (2015).

[80] See P Ibáñez Colomo, 'Beyond the "More Economics-Based Approach": A Legal Perspective on Article 102 TFEU Case Law' (2016) 53 *CML Rev* 709, 711.

[81] L-H Röller, 'Antitrust Economics: Catalyst for Convergence' (George Mason Law Review Symposium, Washington, DC, 6 October 2004); J Vickers, 'Competition Law and Economics: A Mid-Atlantic Viewpoint' (2007) 3 *European Competition Journal* 1; A Devlin and M Jacobs, 'Antitrust Divergence and the Limits of Economics' (2010) 104 *Northwestern University Law Review* 253, 256, 262; DJ Gerber, 'Global Competition Law Convergence: Potential Roles for Economics' in T Eisenberg and G Ramello (eds), *Comparative Law and Economics* (Cheltenham, Edward Elgar, 2016) 206, 214.

[82] A Ezrachi, 'Sponge' (2017) 5 *Journal of Antitrust Enforcement* 49, 60.

[83] DJ Gerber, 'Competition Law and the Institutional Embeddedness of Economics' in J Drexl, L Idot and J Monéger (eds), *Economic Theory and Competition Law* (Cheltenham, Edward Elgar, 2009) 24.

[84] DF Turner, 'The Role of the "Market Concept" in Antitrust Law' (1980) 49 *Antitrust Law Journal* 1145, 1147; Robertson, 'Legal Concept' (n 59) 164.

other cannot be ignored: economics can provide a very informative basis from which to venture into an antitrust discussion, and it can provide models that try to explain complex economic realities and attempt to predict future economic conduct. Law, on the other hand, needs to provide legally binding rules upon which all those subjected to them can rely in order to evaluate the legality of their conduct, their rights and obligations. As such, the law must provide a coherent legal test for defining antitrust markets that antitrust analysis can build upon. While the law may very well learn from the insights of antitrust economics, including behavioural economics, and can use this discipline when applying its legal concepts to new economic phenomena (such as platform markets), its primary task remains the provision of normative guidance. Once reliable normative guidance is established, economic expertise can serve to elucidate questions of fact as the relevant market is delineated in practice, for instance, through expert opinions commissioned by the parties to an antitrust lawsuit or by the court itself. Therefore, while economics can inform and help us to elucidate legal principles, it is ultimately the law – or the judge or authority applying the law – which must make the normative decision. It is this normative guidance on market definition that the present study is concerned with. While it draws on insights from economics as well as from the economic analysis of the law, it adopts a distinctly legal approach when reconceptualising the legal framework for market delineation in an innovation context.

III. The Course of this Study

The study proceeds in three parts. In Part I, this introduction is followed by Chapter 2, which sets out the aims and purposes of market definition in EU competition and US antitrust law. It investigates the relevant market concept under US antitrust law, highlighting the substantive tests that the US Supreme Court has developed in this respect. It then turns to the functions of market definition under the three main US antitrust provisions: sections 1 and 2 of the Sherman Act and section 7 of the Clayton Act. It also analyses agency guidance on market definition, particularly under the Horizontal Merger Guidelines 2010. Thereafter, it moves on to the relevant market concept under EU competition law, discussing the substantive tests under the EU Courts' case law and the emerging role of the General Court, as well as the vague concept of sufficient interchangeability. It then considers the functions of market definition under the three main EU competition law provisions: Articles 101 and 102 TFEU and the EU Merger Regulation.[85] Commission guidance on market definition is discussed in some detail. Subsequently, the chapter takes a first comparative look at the functions of the relevant market in these two jurisdictions, finding that market definition can

[85] Council Regulation (EC) 139/2004 on the control of concentrations between undertakings [2004] OJ L24/1 (EU Merger Regulation).

be ascribed two essential functions: that of providing a basis for assessing market power in a given case, and functions outside of market power assessment that can be summarised as market characterisation. As will be seen, these functions are not mutually exclusive, but can complement each other. While the market power function makes available insights for a competition law assessment based on a structural approach, the market characterisation function provides the necessary market context in order to understand and apply the competition theory of harm and an analysis of anti-competitive effects.

The debate on whether antitrust can and should continue to rely on market definition is also detailed, referencing amongst others the scholarship of Louis Kaplow and Richard Markovits. The validity of the arguments presented by these scholars is critically discussed. While these authors tend to focus on market definition's market power function, they often neglect its role for market characterisation. Their arguments cannot automatically be applied to this second function of market definition. A European perspective on that debate is developed, focusing on market definition as a legal concept rather than an economic tool.

Chapter 2 is a comparative inquiry into antitrust law as it developed and now stands in the area of market definition. It is conducted as a comparison of concepts or a *Konzeptvergleich*, relying on an inverted functional approach to comparative law that strives to understand the functions that the legal concept of the relevant market fulfils in EU and US antitrust law.[86] Instead of focusing on a socio-legal problem, as the functional method of comparison would,[87] the comparison of concepts focuses on a legal concept that is present in both jurisdictions under scrutiny, striving to uncover its respective content and putting it into a comparative perspective. The study therefore examines to what extent the legal concept of the relevant antitrust market fulfils the same or different functions in the EU and US, and whether in its application it is subject to the same or different interpretation(s). While both the EU and US rely on the same terminology, it cannot be ruled out that we are dealing with a linguistic *faux ami*.[88]

Chapter 3 centres on the framework of innovation. This aspect of the research is problem-oriented, containing both descriptive and evaluative aspects. It defines the notion of innovation as understood in the present context and assesses the specific challenges that innovative markets pose for antitrust market definition.

[86] See Kischel (n 75) § 3, paras 166, 190.

[87] On that method, see M Rheinstein, 'Teaching Comparative Law' (1938) 5 *University of Chicago Law Review* 615, 617 ff; E Rabel, 'Die Fachgebiete des Kaiser-Wilhelm-Instituts für ausländisches und internationales Privatrecht (gegründet 1926) (1900–1935)' in M Planck (ed), *25 Jahre Kaiser Wilhelm-Gesellschaft zur Förderung der Wissenschaften*, vol III (Berlin, Springer, 1937) 82; K Zweigert and H Kötz, *Einführung in die Rechtsvergleichung auf dem Gebiete des Privatrechts*, 3rd edn (Tübingen, Mohr Siebeck, 1996) 33; R Michaels, 'The Functional Method of Comparative Law' in M Reimann and R Zimmermann (eds), *The Oxford Handbook of Comparative Law*, 2nd edn (Oxford, Oxford University Press, 2019) 345.

[88] On language as one of the major pitfalls of comparative law, see P de Cruz, *Comparative Law in a Changing World*, 3rd edn (New York, Routledge-Cavendish, 2008) 220.

It is seen how the features of innovative market environments cannot easily be accommodated by conventional antitrust market definition as outlined in Chapter 2. Several points of view on incorporating innovation considerations into market definition are then discussed, including the Schumpeter v Arrow debate, views from competition policy and the issue of error costs associated with market definition.

Part II of this volume is dedicated to the innovation factor in antitrust market delineation. It uncovers the many ways in which antitrust authorities, courts and legal scholarship in the EU and the US have met the various challenges of dynamic markets when defining antitrust markets, especially in the last two decades. This analysis is made against the background of the general legal frameworks for antitrust market definition currently in place in both the EU and US. Data analysed for market definition in innovative markets includes competition authority decisions and court judgments as well as antitrust law scholarship and policy documents. The different solutions that have been adopted regarding antitrust market definition in the face of innovation on both sides of the Atlantic are juxtaposed and their respective strengths and weaknesses, convergences and divergences are analysed. Importantly, the solutions found in the two jurisdictions are scrutinised against the background of their functional role in antitrust, and it is discussed how well they depict innovative markets as a basis for further antitrust scrutiny.

To set the scene, Chapter 4 looks at innovative product markets, discussing the innovative nature of certain markets and the application of conventional substitutability tests to them. Product differentiation – for instance, in the form of the online/offline paradigm or concerning the functionalities of a product – is highlighted as an area of concern, as are interoperability issues that might lead to the delineation of narrow markets. Current markets are closely connected to future markets, and the question is addressed at what point innovation becomes a question of tomorrow's markets rather than today's – and how these can meaningfully be conceptualised for antitrust market definition.

Chapter 5 looks at ways to capture antitrust markets before they are properly established, in particular through potential competition considerations. The US-inspired R&D market concept is also discussed, showing that the EU never fully embraced this approach. The focus then moves to innovation competition as a dimension that is situated outside of market definition proper, but that is highly relevant to the present study as it takes up where market definition leaves off. Chapter 6 deals with the role of IPRs in delineating relevant antitrust markets, showing how IPRs are no longer seen as synonymous with market power, at least in theory. It also highlights a string of cases in which the relevant market was delineated along the lines of an IPR, particularly in the case of successful IPRs that convey a strong brand image. Technology markets are discussed as a useful way of conceptualising licensing markets. Next, innovative aftermarkets (Chapter 7) are looked at, showing how these profitable markets are often sealed off by innovators, particularly with the help of IPRs. Multi-sided platforms (Chapter 8) are then looked at in some detail, with a particular emphasis on digital platform markets.

Here, a remarkable clash between traditional and more dynamic approaches to market definition can be observed. By tracing the case law, it is seen that courts demonstrate a growing awareness of current economics literature. Chapter 9 is a foray into industrial organisation, and cautiously suggests to what limited extent standard economic tests such as the hypothetical monopolist test or concentration levels are applicable in innovative markets. It also briefly outlines issues of geographical market definition in innovative markets.

A dialectical comparison ensures that findings arrived at in Part II enter into a dialogue, presenting comparative conclusions in a straightforward manner.[89] This methodology understands comparative legal research as a hermeneutic, dynamic process that continuously puts previous findings into a new perspective.[90] It lends itself both to analytical comparison and to law reform proposals building on the dialectical comparison. It follows an integrative approach both in its research design and in presenting and discussing the research findings.[91] The dialectical comparison allows for a new analysis of traditional antitrust market definition in the light of innovation considerations by giving the two jurisdictions at the heart of this study the space to interact, while creating as few artificial barriers in the form of separate jurisdictional chapters as possible.

Part III ties the findings of the previous chapters together and provides a comprehensive yet succinct picture for reconceptualising the legal framework(s) for defining antitrust markets in innovative environments. The reconceptualisation in Chapter 10 builds on the rich comparative law analysis in Part II and adds a law reform approach to the comparative analysis with a view to improving market definition *de lege ferenda*.[92] The chapter develops two sets of model guidance for delineating innovative markets, each resting on a typology that is based on the functions of market definition uncovered in Chapter 2. The first set of guidelines focuses on the function of market definition that informs market power assessments and is more traditional and static in nature, while valuing innovation as one of the goals of competition law. The second set of model guidance focuses on the market characterisation role of market definition, is more dynamic in nature and has innovation as one of its central themes. In substantive terms, the guidance draws on the many insights gained in Part II and closely follows the structure of that part of the book. The two different approaches to market delineation in innovative markets are then discussed with a view to establishing how well they depict innovation, and which consequences each approach might entail for the subsequent substantive antitrust assessment.

[89] eg, A Tschentscher, 'Dialektische Rechtsvergleichung – Zur Methode der Komparistik im öffentlichen Recht' (2007) 62 *JuristenZeitung* 807.

[90] ibid 815; J Husa, *A New Introduction to Comparative Law* (Oxford, Hart Publishing, 2015) 96.

[91] Tschentscher (n 89) 807, 809, 815; G Samuel, *An Introduction to Comparative Law Theory and Method* (Oxford, Hart Publishing, 2014) 92; Kischel (n 75) § 3, para 53.

[92] On comparative law as a basis for law reform, see already L-J Constantinesco, *Rechtsvergleichung, Band II: Die rechtsvergleichende Methode* (Cologne, Carl Heymanns, 1972) 371–73.

Chapter 11 reflects on market definition's role in dynamic contexts and argues that a reconceptualisation of market definition as proposed here makes it unnecessary to do away with the concept altogether. The issue of legal culture is touched upon, which might not readily allow for the adoption of (parts of) the guidelines in certain jurisdictions. Discussion is given to what degree market definition in its current state is failing innovation through its focus on static price competition, and it is suggested that this bias may be remedied – at least to a certain degree – by relying on a more innovation-sensitive market definition framework.

2

The Functions of the Relevant Market in EU Competition and US Antitrust Law

Market definition 'translates an economic reality … into the realm of the law'[1] and thereby provides the very basis for competition law analysis. Rather than being an aim in itself,[2] it fulfils a number of functions in antitrust law. In order to discuss if and how market definition needs to adapt to dynamic market contexts, an insight into how relevant product markets are commonly delineated in the EU and the US is necessary. In the following, the intention is to recall the most important milestones in the development of antitrust market definition with a view to understanding the aims and purposes of this analytical step. After analysing the functions of market definition under US antitrust and EU competition law, we take a first comparative look at the functions of market definition in these two jurisdictions. Then we turn to the broader debate on the continued usefulness of market definition in antitrust law, adding a European view on market definition as a legal concept.

I. The Relevant Product Market under US Antitrust Law

In the eyes of the US Supreme Court, 'market definition generally determines the result of [any antitrust] case'.[3] Antitrust policy must also necessarily reflect on the market it is being applied to.[4] However, the US antitrust agencies have recently put less emphasis on market definition. This approach has not (yet) been validated by the courts.

[1] R Podszun, 'The Arbitrariness of Market Definition and an Evolutionary Concept of Markets' (2016) 61 *Antitrust Bulletin* 121, 129.

[2] See, eg, RG Harris and TM Jorde, 'Antitrust Market Definition: An Integrated Approach' (1984) 72 *California Law Review* 3, 4; MR Baye, 'Market Definition and Unilateral Competitive Effects in Online Retail Markets' (2008) 4 *Journal of Competition Law & Economics* 639, 652.

[3] *Eastman Kodak v Image Technical Services*, 504 US 451, 469 fn 15 (1992). See also JB Baker, 'Market Definition: An Analytical Overview' (2007) 74 *Antitrust Law Journal* 129, 129.

[4] *Continental TV v GTE Sylvania*, 433 US 36, 53 fn 21 (1977).

In the following sections, we trace the legal tests for market definition as they developed in the US courts against the background of the main US antitrust provisions, with a view to understanding the functions that market definition fulfils under US law.

A. Substantive Tests in the Case Law

Anti-competitive behaviour always relates to a certain relevant product market, a fact that is also reflected in the legal provisions of US antitrust law. As neither of these provisions gives any guidance on how to delineate this relevant market, the courts had to make sense of market definition. US courts have in turn been accused of relying on an 'intuitive approach' when defining markets rather than on a proper analytical framework.[5]

The first ever reported US case to use the term 'relevant market' within US antitrust law was the 1948 merger decision of *Columbia Steel*.[6] A few years later, a District Court found that while 'section 7 [of the Clayton Act] does not contain the word "market" [i]t is clear ... that "line of commerce" signifies a product market'.[7] In *Grinnell* (1966), the Supreme Court clarified that a line of commerce within the meaning of section 7 of the Clayton Act and a part of trade or commerce within the meaning of section 2 of the Sherman Act referred to one and the same concept of antitrust market.[8] Since then, market definition under the various antitrust provisions has converged.[9] Nevertheless, there remains some debate on whether markets can be delineated in the same way under different antitrust provisions.[10] In particular, the question of whether the hypothetical monopolist test can be relied upon in non-merger cases poses itself.[11] In the present context, and while acknowledging that market definition is highly fact-specific, it is argued that legal certainty requires that the methodology for delineating relevant antitrust markets should be the same under all antitrust provisions.

[5] Harris and Jorde, 'Integrated Approach' (n 2) 7 (direct quote); DF Turner, 'The Role of the "Market Concept" in Antitrust Law' (1980) 49 *Antitrust Law Journal* 1145, 1150.

[6] *United States v Columbia Steel Co*, 334 US 495, 508, 519, 520, 527 (1948).

[7] *United States v Bethlehem Steel Corporation*, 168 FSupp 576, 587 f (SDNY 1958) (continuing that '"section of the country" refers to a geographic market').

[8] *United States v Grinnell Corp*, 384 US 563, 573 (1966).

[9] *Clean Conversion Technologies v CleanTech Biofuels*, Case 12-cv-239-L (JMA) (SD Cal 20 August 2012) 7 (containing further references); GJ Werden, 'The History of Antitrust Market Delineation' (1992) 76 *Marquette Law Review* 123, 166.

[10] MA Glick, DJ Cameron and DG Mangum, 'Importing the Merger Guidelines Market Test in Section 2 Cases: Potential Benefits and Limitations' (1997) 42 *Antitrust Bulletin* 121; GJ Werden, 'Market Delineation under the Merger Guidelines: Monopoly Cases and Alternative Approaches' (2000) 16 *Review of Industrial Organization* 211.

[11] LJ White, 'Market Power and Market Definition in Monopolization Cases' in ABA Section of Antitrust Law (ed), *Issues in Competition Law and Policy* (Chicago, ABA Publishing, 2008) 914, 924; JD Ratliff and DL Rubinfeld, 'Online Advertising: Defining Relevant Markets' (2010) 6 *Journal of Competition Law & Economics* 653, 671.

i. Product Interchangeability: Between Cross-price Elasticity of Demand and Product Characteristics

Product interchangeability became one of the first cornerstones of antitrust market definition in the US. Although not explicitly referring to the concept of an antitrust market, already as early as in 1916, Judge Hand held that a product market comprised all those products that were functionally exchangeable with each other, based on consumer behaviour or production cost.[12] This concept of product interchangeability was soon incorporated into market definition. In *Times-Picayune Publishing* (1953), a case under the Sherman Act, the Supreme Court referred to product substitutability and the necessity to draw market boundaries. It also referred to cross-elasticities of demand in order to delineate markets.[13] Cross-price elasticity remains central to market definition to this day.[14]

In *Cellophane* (1956), the Supreme Court started from the premise that in order to determine a product market, one needed to analyse the product's characteristics and use, as well as demand substitutability.[15] It held that '[t]he "market" … is composed of products that have reasonable interchangeability for the purposes for which they are produced – price, use and qualities considered'.[16] Supplementing this qualitative assessment, interchangeability was then expressed as the cross-elasticity of demand between the product under scrutiny – cellophane – and other wrapping materials, such as aluminium foil.[17] Linking back to section 2 of the Sherman Act, the Supreme Court found that 'commodities reasonably interchangeable by consumers for the same purposes make up that "part of the trade or commerce" monopolization of which may be illegal'.[18] The Court therefore relied on two benchmarks in its substantive market analysis: a qualitative analysis of the products' interchangeability and a quantitative measurement of cross-price elasticity of demand.[19] Ever since *Cellophane*, the courts have applied that market definition test by combining 'economic theory, data, and pragmatic judgment'.[20]

In another case involving du Pont, *du Pont-General Motors* (1957), the Supreme Court relied on the products' 'sufficient peculiar characteristics and uses' to hold that they constituted a line of commerce under section 7 of the Clayton Act.[21] Being based on product interchangeability as perceived by customers, this test

[12] *United States v Corn Products Refining Co*, 234 Fed 964 (SDNY 1916); appeal dismissed in *United States v Corn Products Refining Co*, 249 US 621 (1918).

[13] *Times-Picayune Publishing Co v United States*, 345 US 594, 612 fn 13 (1953).

[14] RA Posner, *Antitrust Law*, 2nd edn (Chicago, University of Chicago Press, 2001) 148.

[15] *United States v EI du Pont de Nemours & Co (Cellophane)*, 351 US 377, 393 (1956).

[16] ibid 404 (direct quote); U Schwalbe and F Maier-Rigaud, 'Background Note' in OECD (ed), *Policy Roundtable: Market Definition* (2012) DAF/COMP(2012)19, 30 fn 26.

[17] *United States v EI du Pont de Nemours & Co (Cellophane)*, 351 US 377, 380 (1956).

[18] ibid 395 (direct quote), 404.

[19] See H Nevo, *Definition of the Relevant Market: (Lack of) Harmony between Industrial Economics and Competition Law* (Cambridge, Intersentia, 2015) 62.

[20] CL Sagers, *Antitrust*, 2nd edn (New York, Wolters Kluwer, 2014) 74.

[21] *United States v EI du Pont de Nemours & Co (du Pont-General Motors)*, 353 US 586, 593 f (1957).

does not relate to cross-price elasticity of demand. Within the span of a year, the Supreme Court therefore chose different ways to approach the question of how to establish product interchangeability: through cross-price elasticity of demand combined with a qualitative assessment of the product's price, use and qualities, or simply through a product's peculiar characteristics and uses.

ii. Supply-Side Substitutability

Although demand substitutability is the main focus of the market definition inquiry, *Columbia Steel* (1948) was a reminder that supply-side substitutability also matters. While the government had sought to limit the market to certain types of rolled steel products, the Supreme Court insisted that supply-side substitution had to be taken into account. As rolled steel producers could use their production facilities for all types of rolled steel products, this wider market constituted the relevant market.[22] Thus, supply-side substitution effectively enlarged the market, leading to the finding that the merger would not significantly reduce competition on that wider market.

However, the Supreme Court was not always consistent in its approach to supply-side substitution. In *Alcoa* (1964), it replaced the lower court's market delineation with a (sub-)market definition that entirely rested on demand substitutability: aluminium and copper conductors were used for different purposes, and there was a price differential between the two. This, according to the Supreme Court majority, warranted a relevant market that only included aluminium conductors.[23] The three dissenting judges emphasised that supply substitutability should have played a more prominent role and would have led to an entirely different antitrust market, as there was 'complete manufacturing interchangeability between copper and aluminum'.[24] Such a market definition would have been in line with *Columbia Steel*.

iii. Brown Shoe: *Sub-markets*

The Supreme Court case law on antitrust market definition from 1962 to 1982 attracted serious criticism.[25] The case of *Brown Shoe* (1962) has been referred to as the 'low point'[26] of US market definition and showcased how narrowly relevant antitrust markets can be delineated.[27] In that case, the Supreme Court tried

[22] *United States v Columbia Steel Co*, 334 US 495, 510 (1948).

[23] *United States v Aluminum Co of America (Alcoa)*, 377 US 271, 277 (1964).

[24] ibid 283–85 (direct quote at 285).

[25] Werden, 'History' (n 9) 185 fn 417; P Areeda and HJ Hovenkamp, *Antitrust Law: An Analysis of Antitrust Principles and Their Application*, 4th edn (Alphen aan den Rijn, Wolters Kluwer, 2017) § 533.

[26] G Niels, H Jenkins and J Kavanagh, *Economics for Competition Lawyers*, 2nd edn (Oxford, Oxford University Press, 2016) § 2.04.

[27] DD Martin, 'The Brown Shoe Case and the New Antimerger Policy' (1963) 53 *American Economic Review* 340, 348.

to reconcile the various formulas for defining the relevant antitrust market that it had previously relied upon.[28] As Congress had not given any guidance on the particular methodology for antitrust market definition,[29] the Supreme Court set out to shed light on this important issue. It held that a product market could be delineated based on 'reasonable interchangeability of use or the cross-elasticity of demand between the product itself and substitutes for it'.[30] In a footnote to this statement, the Court added that the cross-elasticity of supply could also constitute a relevant feature.[31] Yet, *Brown Shoe* did not succeed in bringing forward a proper analytical framework for market definition.[32] Justice Harlan, concurring, criticised the fact that the majority had not sufficiently taken supply substitutability into account.[33] Relying on *Cellophane* – which had arguably never advanced such a notion – the Court went on to add that sub-markets could exist within antitrust markets. These sub-markets could be determined by relying on a number of indicia: industry or public recognition of a sub-market as a separate economic entity; the product's peculiar characteristics and uses; unique production facilities; distinct customers; distinct prices; sensitivity (ie, of the sales)[34] to price changes; and specialised vendors.[35] These seven factors have been influential across the spectrum of antitrust cases. Nevertheless, many of them have been criticised,[36] particularly in relation to their applicability to dynamic markets.[37]

Some courts have interpreted the sub-market concept so as to reconcile it with previous case law and economic theory, for instance, holding that *Brown Shoe*'s sub-market concept is merely a misnomer, as this 'analysis simply clarifies whether two products are in fact "reasonable" substitutes and are therefore part of the same market'.[38] The indicia have also been regarded as practical guidelines for lower courts, not as a rejection of more sophisticated economic analysis.[39] Despite the controversy surrounding this case, it continues to be a leading precedent on market definition that is widely relied upon.[40]

[28] L Maisel, 'Submarkets in Merger and Monopolization Cases' (1983) 72 *Georgetown Law Journal* 39, 42 f.

[29] *Brown Shoe v United States*, 370 US 294, 320 f (1962).

[30] ibid 325.

[31] ibid 325 fn 42.

[32] MB Coate and JJ Simons, 'In Defense of Market Definition' (2012) 57 *Antitrust Bulletin* 667, 673; Turner (n 5) 1151.

[33] *Brown Shoe v United States*, 370 US 294, 367 (1962).

[34] See C Pleatsikas and DJ Teece, 'The Analysis of Market Definition and Market Power in the Context of Rapid Innovation' (2001) 19 *International Journal of Industrial Organization* 665, 673.

[35] *Brown Shoe v United States*, 370 US 294, 325 (1962).

[36] Glick, Cameron and Mangum (n 10) 129.

[37] See ch 4, section II.E.

[38] *Geneva Pharmaceuticals Technology v Barr Laboratories*, 386 F3d 485, 496 (2d Cir 2004).

[39] *Reifert v South Central Wisconsin MLS*, 450 F3d 312, 320 (7th Cir 2006).

[40] *United States v Anthem and Cigna*, 236 FSupp3d 171, 197 (DDC 2017); *United States v Aetna/Humana*, 240 FSupp3d 1, 23 ff (DDC 2017); HJ Hovenkamp, 'Markets in Merger Analysis' (2012) 57 *Antitrust Bulletin* 887, 894 f, 900.

iv. Case Law after Brown Shoe: *Making Sense of Market Definition*

Following its much-criticised sub-market concept, the Supreme Court relied on a cluster market concept in *Philadelphia National Bank* (1963), essentially holding that commercial banking consisted of a cluster of products and services that could be regarded as a line of commerce and thus constituted the relevant product market under section 7 of the Clayton Act.[41] While this new concept was not warmly received, the European Court of Justice accepted a similar market delineation many years later in *Lombard Club*.[42] *Philadelphia National Bank* was also important because it denied the possibility of taking efficiencies into account that occur outside of the relevant market.[43]

A number of cases since *Brown Shoe* have attempted to consolidate that case with earlier case law on market definition. In the *Continental Can* merger case (1964), for instance, the Supreme Court referred to *Brown Shoe* and held that '[c]oncededly these guidelines offer no precise formula for judgment'.[44] In *Continental Can*, it was found that the can industry and the glass container industry were relevant lines of commerce, while the Department of Justice (DoJ) had also claimed that almost a dozen end-use markets (eg, containers for the beer industry, the health industry and so on) should be delineated.[45] Noting the tension between *Times-Picayune* and *Cellophane*, the Supreme Court held that when 'defining the product market between these terminal extremes, we must recognize meaningful competition where it is found to exist'.[46] It noted considerable competition between glass and metal containers for use in different industries and concluded that for the purposes of section 7 of the Clayton Act, the relevant market consisted of both glass and metal containers for any end uses in which the two competed.[47]

In *Grinnell* (1966), the Supreme Court emphasised that a relevant market 'may be of such a character that substitute products must also be considered, as customers may turn to them if there is a slight increase in the price of the main product'.[48] Conceptually, this analysis pointed towards a version of the hypothetical monopolist test. The products at issue in *Grinnell* were alarm services for properties that were connected with a 24-hour manned central station that would send a guard by the property in the case of an alarm.[49] As in *Philadelphia National Bank*, these were held to constitute a cluster market.[50] The Supreme Court – perhaps wrongly – concluded that such central station property protection services

[41] *United States v Philadelphia National Bank*, 374 US 321, 356 (1963).

[42] Joined Cases C-125, C-133, C-135 and C-137/07 P *Raiffeisen Zentralbank Österreich and Others v Commission (Lombard Club)* EU:C:2009:576. See section II.A.iv below.

[43] *United States v Philadelphia National Bank*, 374 US 321, 371 (1963).

[44] *United States v Continental Can Co*, 378 US 441, 449 (1964).

[45] ibid 447.

[46] ibid 449.

[47] ibid 450–53, 455, 457.

[48] *United States v Grinnell Corp*, 384 US 563, 571 (1966).

[49] ibid 566 f.

[50] ibid 573.

were not interchangeable with other property protection services because 'for many customers, only central station protection will do'; nevertheless, the Court acknowledged that some customers might be willing to switch to other protection services.[51] In defining the market so narrowly, it possibly overestimated Grinnell's market power.[52] What should have mattered in a further step was the magnitude of customers willing to turn to alternative protection services and the question whether those customers' switching would have made a price increase unprofitable.

In another case from 1966, the Supreme Court held that the geographical market did not have to be delineated 'by metes and bounds'.[53] A few years later, the Court clarified that while relevant antitrust markets cannot always 'be defined with scientific precision', the government did have to define the market it was relating to with at least 'rough approximation'.[54] This discussion on the precision of market definition is relevant for both the product and the geographical dimensions of the market, and continues to this day.

The Supreme Court's antitrust cases until the late 1960s were adjudicated in an era in which the structure-conduct-performance (SCP) paradigm of the Harvard School was influential, a factor that can be sensed in many of the judgments just discussed.[55] The landmark cases on market definition from this time continue to be relied upon today, thus carrying a structural type of antitrust analysis into the present day.[56]

Having revisited the substantive legal tests for market definition under US antitrust law, we now turn to the three most important substantive provisions in US antitrust law – sections 1 and 2 of the Sherman Act as well as section 7 of the Clayton Act – in order to appraise the specific functions that market definition plays under each of them.

B. Section 1 of the Sherman Act and Market Definition

Section 1 of the Sherman Act states that every contract or conspiracy that restrains trade among the US or with third countries is illegal. US case law has established that per se antitrust offences, such as price cartels, are always considered to be

[51] ibid 574.

[52] R Pitofsky, 'New Definitions of Relevant Market and the Assault on Antitrust' (1990) 90 *Columbia Law Review* 1805, 1816 f.

[53] *United States v Pabst Brewing*, 384 US 546, 549 (1966).

[54] *United States v Connecticut National Bank*, 418 US 656, 669 (1974).

[55] Hovenkamp, 'Markets in Merger Analysis' (n 40) 889 f. On the main tenets of the Harvard School, see DB Audretsch, WJ Baumol and AE Burke, 'Competition Policy in Dynamic Markets' (2001) 19 *International Journal of Industrial Organization* 613, 615; WE Kovacic, 'The Intellectual DNA of Modern US Competition Law for Dominant Firm Conduct: The Chicago/Harvard Double Helix' [2007] *Columbia Business Law Review* 1; A Devlin, *Fundamental Principles of Law and Economics* (New York, Routledge, 2015) 306.

[56] See, for instance, the references to *Brown Shoe*, *Cellophane* and *Times-Picayune* in *United States v American Express*, Case 10-CV-4496 (NGG) (RER) (EDNY 19 February 2015) § III.

anti-competitive. In per se cases, the offence's effects on the relevant market do not need to be shown, and the relevant market may not necessarily need to be defined in order to establish antitrust liability.[57] However, even for per se offences, one will often need to rely on a relevant market if one wants to assess whether a company has entered into a cartel with a competitor, ie, a company operating on the same relevant market, or is engaging in vertical price-fixing with a non-competitor.[58] In addition, the Supreme Court has held on numerous occasions that tying constitutes a per se offence only where the supplier exerts market power over the tying product market,[59] thus requiring an analysis of the relevant market both for the finding of two separate markets and for market power analysis.

Outside of per se offences, company behaviour is assessed under a rule of reason. Market definition is widely relied upon in US antitrust law when assessing possible rule of reason offences under section 1 of the Sherman Act, ie, behaviour that is only deemed anti-competitive if it has anti-competitive effects on the market. According to the Supreme Court, market definition and the subsequent analysis of market power are the basis for assessing whether an agreement can affect competition.[60] Vertical restraints need to be assessed with regard to the conditions of competition on the relevant product market.[61] As the Supreme Court held in *American Express* (2018), in vertical cases 'courts usually cannot properly apply the rule of reason without an accurate definition of the relevant market'.[62]

Within rule of reason analysis, the Supreme Court has developed a quick look approach that does not require the demonstration of actual anti-competitive effects on the market.[63] As the quick look analysis is not based on a structural analysis, no relevant antitrust market needs to be established.[64] However, the Supreme Court has also warned that its case law under section 1 of the Sherman Act could not simply be categorised as 'per se', 'quick look' and 'rule of reason' – it was more nuanced than this.[65] This calls into question the seemingly reduced role of market

[57] United States in OECD (ed), *Policy Roundtable: Market Definition* (2012) DAF/COMP(2012)19, 321 f; *FTC v Superior Court Trial Lawyers Association*, 493 US 411 (1990).

[58] MA Lemley and MP McKenna, 'Is Pepsi Really a Substitute for Coke? Market Definition in Antitrust and IP' (2012) 100 *Georgetown Law Journal* 2055, 2077.

[59] *Jefferson Parish Hospital District v Hyde*, 466 US 2, 13 f (1984) (containing further references).

[60] *FTC v Indiana Federation of Dentists*, 476 US 447, 460 (1986) (but proof of actual effects may make market definition unnecessary). See also *Eastern Food Services v Pontifical Catholic University Services Association*, 357 F3d 1, 5 (1st Cir 2004).

[61] *Continental TV v GTE Sylvania*, 433 US 36, 45 (1977).

[62] *Ohio v American Express*, 585 US ___ (2018), 138 SCt 2274, 2285 (2018). Taking issue with this general statement as regards vertical restraints, see the dissenting opinion in ibid 2297. For criticism of the horizontal/vertical distinction, see C Ritter, 'Antitrust in Two-Sided Markets: Looking at the US Supreme Court's *Amex* Case from an EU Perspective' (2019) 10 *Journal of European Competition Law & Practice* 172, 173 f.

[63] *California Dental Association v FTC*, 526 US 765, 770 (1999).

[64] JA Keyte and NR Stoll, 'Markets? We Don't Need No Stinking Markets! The FTC and Market Definition' (2004) 49 *Antitrust Bulletin* 593, 600.

[65] *California Dental Association v FTC*, 526 US 765, 779 (1999).

definition. In *California Dental*, for instance, the Supreme Court held that there was a need for a more in-depth look at the market under scrutiny.[66]

US courts regard the delineation of a relevant market as an inherent part of the legal test under section 1 of the Sherman Act.[67] Market definition is useful in '§ 1 allegations because a market definition provides the context against which to measure the competitive effects of an agreement'.[68] In *Bell Atlantic v Twombly* (2007), the US Supreme Court referred to the specific antitrust market when characterising the aim of the alleged anti-competitive conduct, namely to keep competitors out of the market.[69]

In the past, the Supreme Court has agreed with lower courts that analysing effects on the relevant market *can* be substituted by proof of actual detrimental effects.[70] Where past behaviour is concerned, direct evidence of harm to competition may be more convincing than a competition analysis that is based on market definition.[71] Similarly, the Supreme Court has been willing to find market power based on considerations outside of market definition. In *Actavis* (2013), it was called upon to assess a reverse payment settlement under section 1 of the Sherman Act. The patent owner had paid considerable sums to keep a potential competitor out of its market.[72] In that case, the Supreme Court considered that the mere fact that a reverse payment was agreed upon might suggest that the patent owner derived market power from the patent at issue.[73]

C. The Relevant Market in Monopolisation Cases

Section 2 of the Sherman Act prohibits the monopolisation of trade and the attempt to monopolise among the US or with third countries. One of the two prerequisites for finding that a company has infringed the monopolisation aspect of section 2 of the Sherman Act is that the company in question had monopoly power on the relevant market or attempted to obtain such.[74] The Supreme Court has underlined the importance of market definition in this context, holding that '[w]ithout a definition of that market, there is no way to measure [a company's] ability to lessen or

[66] ibid.

[67] *Golden Bridge v Motorola*, 547 F3d 266, 271 (5th Cir 2008); *Online Travel Company (OTC) Hotel Booking Antitrust Litigation*, 997 FSupp2d 526, 534 (ND Tex 2014).

[68] *Geneva Pharmaceuticals Technology v Barr Laboratories*, 383 F3d 485, 496 (2d Cir 2004).

[69] *Bell Atlantic v Twombly*, 550 US 544, 564 f (2007).

[70] *FTC v Indiana Federation of Dentists*, 476 US 447, 460 f (1986).

[71] JD Richards, 'Is Market Definition Necessary in Sherman Act Cases When Anticompetitive Effects Can Be Shown with Direct Evidence?' (2012) 26 *Antitrust* 53, especially 57.

[72] *FTC v Actavis*, 570 US ___ (2013), 133 SCt 2223 (2013).

[73] ibid 2236.

[74] *United States v Grinnell Corp*, 384 US 563, 570 (1966); RD Blair and CK Carruthers, 'The Economics of Monopoly Power in Antitrust' in KN Hylton (ed), *Antitrust Law and Economics* (Cheltenham, Edward Elgar, 2010) 64.

destroy competition'.[75] In *Spectrum Sports* (1993), the Supreme Court insisted that in order to claim that a defendant had attempted to monopolise, the plaintiff had to 'prove a dangerous probability of actual monopolization, which has generally required a definition of the relevant market and examination of market power'.[76] The (attempted) monopolisation element of section 2 of the Sherman Act therefore usually calls for the relevant market to be delineated.

Under section 2 of the Sherman Act, market definition serves not only as a tool to establish monopoly power, but also as a context for the behaviour leading to (attempted) monopolisation. The perspective of market definition under section 2 is oriented towards the past, assessing whether a specific market has been monopolised through anti-competitive behaviour or whether monopolisation has been attempted through anti-competitive means.

Several cases have shed light on the Supreme Court's reliance on market definition under section 2 of the Sherman Act in the new millennium. In *Trinko* (2004), a refusal-to-deal case, the Court emphasised that an infringement of section 2 of the Sherman Act required the finding of 'monopoly power in the relevant market'.[77] In *Weyerhaeuser* (2007), which concerned the attempted monopolisation of a market through predatory bidding for a key input to that market, two markets were concerned: the input-purchasing market, on which plaintiff and defendant were competitors, and the market for the end product.[78] It was the interplay between those markets that proved central to the case. Finally, *Pacific Bell* (2009) concerned a margin squeeze on the digital subscriber line (DSL) market for internet connectivity. This economic market consisted of two relevant antitrust markets – the (upstream) wholesale and the (downstream) retail levels – and the case rested on the finding that these were two distinct markets and that the defendant had used market power on the wholesale market to squeeze margins on the retail market.[79]

Contrary to European case law, there are no market share levels at which courts presume monopoly power for the purposes of section 2 of the Sherman Act. Some courts have required market shares significantly larger than 55 per cent, while others have found that market shares below 70 per cent would not lead to a finding of monopoly power.[80] In another case, market shares between 80 and 95 per cent were considered high enough to indicate monopoly power.[81] This

[75] *Walker Process Equipment v Food Machinery*, 382 US 172, 177 (1965). But see AM Stein and BJ Brett, 'Market Definition and Market Power in Antitrust Cases: An Empirical Primer on When, Why and How' (1979) 24 *New York Law School Law Review* 639, 643 fn 12.

[76] *Spectrum Sports v McQuillan*, 506 US 447, 455 (1993).

[77] *Verizon Communications v Trinko*, 540 US 398, 407 (2004).

[78] *Weyerhaeuser v Ross-Simmons Hardwood Lumber*, 549 US 312, 315, 321 (2007) (while the input market was the purchasing market for red alder saw logs, the output market was that of finished lumber).

[79] Ultimately, the Supreme Court found that no case for a margin squeeze could be made. *Pacific Bell v Linkline*, 555 US 438, 442, 449 (2009).

[80] United States, *Market Definition* (n 57) 324 f fn 22.

[81] *Eastman Kodak v Image Technical Services*, 504 US 451, 481 (1992).

reflects the insight that market characteristics play a significant role in the determination of monopoly power, and static market shares may not do justice to a specific market's features.

In monopolisation and merger cases, market definition serves 'to identify all of the important restraints on the ability of the defendant to exercise market power'.[82] One way of establishing these restraints is the hypothetical monopolist test (see section I.F.i below). However, this test faces considerable constraints in the context of monopolisation, which have become known as the cellophane fallacy associated with the *Cellophane* case discussed above. In that case, the Supreme Court relied on too broad a relevant market when it assumed that cellophane was, from the consumers' point of view, interchangeable with other wrapping materials. The Court's market definition was based on prices that the dominant undertaking had already considerably raised above market price ('monopoly price'), which in turn influenced consumer response to a further increase.[83] Instead, cross-elasticities of demand should have been measured by applying prices that resembled competitive price levels or but-for prices.[84] As the market was too broadly defined, no anti-competitive monopolisation was established. Despite the cellophane fallacy, the hypothetical monopolist test may serve as conceptual guidance for market definition in section 2 cases. However, where a company already enjoys monopoly power, it cannot be applied,[85] or at least not in a straightforward manner.

D. US Merger Control and Market Definition

Section 7 of the Clayton Act prohibits mergers that would substantially lessen competition or create a monopoly on a specific product and geographical market.[86] Market definition has long been held to be 'absolutely indispensable'[87] in merger cases.[88] In terms of substance, many merger cases rely on reasonable interchangeability rather than on cross-elasticity of demand. This principle may take consumer preference into account, but more often relies on alternatives available to consumers.[89]

Based on the legislative history of the Clayton Act, a market defined in the context of a merger case may differ slightly from how the same market would be

[82] Glick, Cameron and Mangum (n 10) 122.
[83] WM Landes and RA Posner, 'Market Power in Antitrust Cases' (1981) 94 *Harvard Law Review* 937, 960 f.
[84] Werden, 'History' (n 9) 139.
[85] Werden, 'Market Definition Under the Merger Guidelines' (n 10) 214, 216.
[86] *United States v Philadelphia National Bank*, 374 US 321, 356 (1963).
[87] Turner (n 5) 1146.
[88] *United States v El du Pont de Nemours & Co (du Pont-General Motors)*, 353 US 586, 593 (1957); *Brown Shoe v United States*, 370 US 294, 324 (1962); *United States v Marine Bancorporation*, 418 US 602, 618 (1974).
[89] JA Keyte and KB Schwartz, '"Tally-Ho!": UPP and the 2010 Horizontal Merger Guidelines' (2011) 77 *Antitrust Law Journal* 587, 600, 604, 607 f.

defined in the case of (attempted) monopolisation.[90] This is due to the fact that merger analysis requires a prospective antitrust assessment that takes future developments into account – most importantly the merger itself – while sections 1 and 2 of the Sherman Act necessarily assess past behaviour. Nevertheless, the Supreme Court in *Grinnell* aligned market definition under section 2 of the Sherman Act and section 7 of the Clayton Act, holding that there was 'no reason to differentiate between "line" of commerce in the context of the Clayton Act and "part" of commerce for purposes of the Sherman Act'.[91]

During the 1960s, it was criticised that the courts employed market definition with the aim of barring mergers,[92] a sentiment that Justice Stewart voiced in his dissent in *Von's Grocery*.[93] This is an ongoing concern, as the more recent case of *FTC v Whole Foods Market* (2007) and the commentary that followed it showed.[94] The instrumentalisation of the analytical step of market definition has long been an issue and can only be countered by straightforward guidelines that – to the extent possible – do not leave considerable room for interpretation when being applied.

In *Continental Can* (1964), the Supreme Court highlighted that 'the purpose of delineating a line of commerce is to provide an adequate basis for measuring the effects of a given acquisition'.[95] It also held that a merger and its possible negative impact on competition had to be assessed functionally, while bearing in mind the relevant antitrust market and its structure, as well as the market's development in the past and its likely development in the future.[96] By considerably broadening the relevant market in that case, as discussed above, the Supreme Court found that the merger may not raise competition concerns.[97]

Phoebe Putney (2013) was the first merger case to be decided by the Supreme Court in over three decades.[98] It concerned a two-to-one hospital merger that the FTC held to violate section 5 of the Federal Trade Commission Act (FTCA)[99] and section 7 of the Clayton Act by creating a monopoly, and thereby substantially lessening competition on the market for acute-care hospital services.[100] The market definition was crucial to the competition analysis, as the finding that a monopoly was created through the merger rested on it.

[90] See on this JND, 'Product Market Definition under the Sherman and Clayton Acts' (1962) 110 *University of Pennsylvania Law Review* 861.

[91] *United States v Grinnell Corp*, 384 US 563, 573 (1966); on *Grinnell*, see above, section I.A.iv.

[92] JE Lopatka, 'Market Definition?' (2011) 39 *Review of Industrial Organization* 69, 74.

[93] *United States v Von's Grocery*, 384 US 270, 301 (1966).

[94] *FTC v Whole Foods Market*, 502 FSupp2d 1 (DDC 2007); FB Cross and RL Miller, *The Legal Environment of Business: Text and Cases – Ethical, Regulatory, Global, and E-Commerce Issues*, 7th edn (Mason, Cengage, 2009) 674 (newer editions no longer include this passage).

[95] *United States v Continental Can Co*, 378 US 441, 457 (1964).

[96] ibid 458.

[97] ibid 465 f.

[98] *FTC v Phoebe Putney Health System*, 568 US 216 (2013).

[99] Federal Trade Commission Act (1914), 15 USC §§ 41–58, as amended.

[100] *FTC v Phoebe Putney Health System*, 568 US 216, 222 (2013).

Today, it seems that the agencies are willing to lessen the importance of market definition in merger cases. However, the courts do not appear to have any similar intentions.[101] In 2010, a District Court dismissed a case because 'plaintiffs ha[d] failed to sufficiently allege the existence of a cognizable product market',[102] ie, the relevant market(s) on which competition would be harmed by a merger between two pharmaceutical companies. That same year, another District Court granted summary judgment to two merging parties because the plaintiff, the City of New York, had failed to delineate the relevant antitrust market to the legal standard.[103] The courts' reluctance to analyse mergers without a proper relevant market having been defined can be understood when one considers the functions of market definition in merger control, as outlined below. In addition, a merger's alleged efficiencies also need to relate to the relevant market.[104]

E. The Functions of Market Definition under US Antitrust Law

Market definition serves two main functions under US antitrust law: first of all, it provides the basis for the subsequent analysis of market power. This is true for sections 1 and 2 of the Sherman Act as well as for merger control. Sometimes, this is seen as market definition's only function, particularly in merger and monopolisation cases.[105] The relevant market allows a court to make a connection between market power and market structure.[106] This link is frequently attacked as providing too vague a basis for the market power inquiry, a topic to which we will return below (see section IV.A). In mergers and attempted monopolisation, probabilistic market power needs to be shown, ie, market power that could be acquired through the merger or with the help of the anti-competitive conduct.[107] In merger control, market definition enables a court to assess what the likely effect of a merger will be on the relevant market(s) from a structural point of view.

As a second main function, market definition provides a background against which potentially anti-competitive market behaviour is assessed. Under sections 1

[101] See also C Beaton-Wells, 'Mergers without Markets? Unilateral Effects Analysis in the United States and its Prospects in Australia' (2006) 34 *Australian Business Law Review* 186, 198.

[102] *Golden Gate Pharmacy Services v Pfizer*, Case C-09-3854 2010 (ND Cal 16 April 2010).

[103] *City of New York v Group Health*, Case 06 Civ 13122 (RJS) (SDNY 11 May 2010); upheld in *City of New York v Group Health*, 649 F3d 151 (2d Cir 2011).

[104] *United States v Philadelphia National Bank*, 374 US 321, 371 (1963).

[105] LA Sullivan, *Handbook of the Law of Antitrust* (St Paul, West Academic Publishing, 1977) 41, quoting FM Scherer, *Industrial Market Structure and Economic Performance* (Chicago, Rand McNally, 1970); Areeda and Hovenkamp (n 25) §§ 500, 515; Maisel (n 28) 45, 48.

[106] Glick, Cameron and Mangum (n 10) 135.

[107] See already DA Crane, 'Market Power without Market Definition' (2014) 90 *Notre Dame Law Review* 31, 36.

and 2 of the Sherman Act, courts rely on the relevant market to analyse the effects of the alleged behaviour on the relevant market, ie, whether the behaviour is an anti-competitive conspiracy or contract under section 1 or an anti-competitive monopolisation or attempt at monopolisation under section 2 of the Sherman Act. Depending on the anti-competitive conduct under investigation, it might also be necessary to make out two separate relevant markets, for instance, in the case of tying.[108] Under section 1 of the Sherman Act, direct proof of harm without relying on market definition is a possibility, even though market definition continues to play an important role in the majority of cases. Under section 2 of the Sherman Act, market definition helps to characterise the market in question in order to determine whether, in the particular market environment, it constitutes anti-competitive behaviour monopolising or attempting to monopolise the relevant market. In the realm of merger control, a merger's competitive effects will be assessed against the background of the relevant market(s) with their particular characteristics.

There have been attempts to reduce the influence of market definition on the substantive analysis. The Supreme Court has found that one may substitute proof of actual anti-competitive effects for market delineation in cases under section 1 of the Sherman Act.[109] In the literature, it has been held that per se claims under section 1 of the Sherman Act do not require a relevant market to be defined.[110] Under section 2 of the Sherman Act, a Court of Appeals similarly found that 'a relevant market definition is not a necessary component of a monopolization claim' where there is direct evidence of monopoly power.[111] And under section 7 of the Clayton Act, competition economists favour an approach that directly estimates a merger's price effects rather than conducting a structural analysis based on market definition, even though the latter 'has effectively remained a legal requirement' and an important element of antitrust litigation.[112]

The view continues to be generally held that in any US antitrust case, 'the legally relevant market … is where the analysis must begin'.[113] Ultimately, the delineation of a relevant market remains a statutory requirement under US antitrust law.[114] Bearing in mind the two main functions that market definition fulfils under US antitrust law, it is likely that even if some of the attempts at supplanting market definition with other types of analysis are successful, there will remain functions that cannot properly be carried out by other analytical tools.

[108] See *Illinois Tool Works and Others v Independent Ink*, 547 US 28 (2006).

[109] *FTC v Indiana Federation of Dentists*, 476 US 447, 460 f (1986).

[110] Turner (n 5) 1145.

[111] *Pepsi v Coca-Cola*, 315 F3d 101, 107 (2d Cir 2002).

[112] M Remer and FR Warren-Boulton, '*United States v. H&R Block*: Market Definition in Court since the 2010 Merger Guidelines' (2014) 59 *Antitrust Bulletin* 599, 602 (direct quote), 618.

[113] Glick, Cameron and Mangum (n 10) 142.

[114] Areeda and Hovenkamp (n 25) § 531b.

F. Agency Guidance on Market Definition

Since 1968, the two principal antitrust enforcement agencies of the US, the Antitrust Division of the DoJ and the Federal Trade Commission (FTC), have issued a succession of guidelines in the area of mergers that touch upon the topic of market definition.[115] As soft law instruments, these guidelines are not legally binding upon courts, although the latter sometimes rely on the guidelines' teachings and adopt the analytical tools found therein.[116] Nevertheless, some discrepancy remains between established case law and agency guidance.[117] In this respect, the US guidance contrasts with the EU guidance, in that the European Commission explicitly sets out to reconcile the EU Courts' case law with its own, more economics-based stance - which can at times lead to its own inconsistencies.[118]

i. The Horizontal Merger Guidelines

Following the publication of its first Merger Guidelines in 1968,[119] the DoJ published a new set of Merger Guidelines in 1982 that some hailed as 'a revolution'.[120] Located in a footnote, these Guidelines contained the hypothetical monopolist test for the very first time. The DoJ defined the relevant antitrust market as follows:

> [A] group of products and an associated geographic area such that ([i]n the absence of new entry) a hypothetical, unregulated firm that made all the sales of those products in that area could increase its profits through a small but significant and non-transitory increase in price (above prevailing or likely future levels).[121]

This test can be understood as a way of framing cross-elasticity of demand, a factor that the Supreme Court had relied on for many years. Some saw it as '[t]he most important conceptual contribution of the 1982 Guidelines to merger analysis'.[122] Others, however, cautioned that the test was based on arbitrary values and provided guidance that could only pretend to be clear.[123] While the courts' initial reception

[115] The first guidelines were issued by the DoJ alone; since 1992, the agencies have issued them together.

[116] *FTC v PPG Industries*, 798 F2d 1500, 1503 fn 4 (DC Cir 1986); *United States v Engelhard*, 970 FSupp 1463, 1467 (MD Ga 1997); *FTC v Swedish Match*, 131 FSupp2d 151, 167 fn 12 (DDC 2000); *FTC v Heinz*, 246 F3d 708, 716 fn 9 (DC Cir 2001).

[117] Werden, 'History' (n 9) 171, 199.

[118] See section II.F.ii.

[119] US Department of Justice, 1968 Merger Guidelines (1968) § 3 (on market definition).

[120] C Shapiro, 'The 2010 Horizontal Merger Guidelines: From Hedgehog to Fox in Forty Years' (2010) 77 *Antitrust Law Journal* 49, 52.

[121] US Department of Justice, 1982 Merger Guidelines (1982) § II, fn 6.

[122] LJ White, 'Antitrust and Merger Policy: A Review and Critique' (1987) 1 *Journal of Economic Perspectives* 13, 14.

[123] P Areeda, 'Justice's Mergers Guidelines: The General Theory' (1983) 71 *California Law Review* 303, 308.

of the hypothetical monopolist test has been characterised as 'lukewarm',[124] the DoJ pointed out that the Guidelines were increasingly accepted in the courts.[125]

The DoJ explained that it would try to identify a group of products for which a hypothetical monopolist, ie, a company 'that was the only present and future seller of those products',[126] could profitably raise its price. Thereby, it would assess to which available substitutes buyers would switch within a year in reaction to a price increase of five per cent.[127] Supply substitutability would be taken into account if suppliers could reasonably shift their production so as to form part of the relevant market. Beyond a timeframe of six months, the Department would consider potential competition when evaluating entry conditions.[128]

The 1982 Guidelines also introduced the Herfindahl-Hirschman Index (HHI), which sums the squares of companies' respective market shares to depict the concentration prevailing on a market, giving respectively more weight to bigger market players.[129] This was also considered an important analytical improvement.[130]

Following a further revision in 1984,[131] 1992 saw the first joint publication of Horizontal Merger Guidelines by the FTC and the DoJ.[132] Under the new Guidelines, which were restricted to horizontal mergers, demand substitutability was to be the exclusive focus when delineating the relevant antitrust market.[133] Supply substitutability would be taken into account when analysing market participants and attributing market shares to them. In order to delineate the relevant antitrust market under the 1992/1997 Guidelines, the hypothetical monopolist test would be carried out again and again until 'the smallest group of products [is found] that satisfies this test'.[134]

The 2010 revision brought the Horizontal Merger Guidelines more into line with agency practice, which had come to place less and less emphasis on market definition.[135] Under the 2010 Guidelines, market definition has two main

[124] H Greene, 'Guideline Institutionalization: The Role of Merger Guidelines in Antitrust Discourse' (2006) 48 *William & Mary Law Review* 771, 798.

[125] SA Newborn and VL Snider, 'The Growing Judicial Acceptance of the Merger Guidelines' (1992) 60 *Antitrust Law Journal* 849, 851–54.

[126] US Department of Justice, 1982 Merger Guidelines (n 121) § II.A.

[127] ibid § II.A.

[128] ibid § II.B.

[129] ibid § III.A.

[130] Lopatka (n 92) 74.

[131] US Department of Justice, 1984 Merger Guidelines (1984) § 2.0.

[132] The 1997 revision of the 1992 Guidelines did not amend the section on market definition, which is why the following footnotes make reference to this updated version; see US Department of Justice and Federal Trade Commission, Horizontal Merger Guidelines (8 April 1997) note on p 2. The FTC had previously issued its own publication regarding mergers; see Federal Trade Commission, 'Statement Concerning Horizontal Mergers' (14 June 1982) § VI.

[133] US Department of Justice and Federal Trade Commission, 1997 Merger Guidelines (n 132) § 1.0.

[134] ibid § 1.11.

[135] US Department of Justice and Federal Trade Commission, Horizontal Merger Guidelines (19 August 2010) (US Horizontal Merger Guidelines 2010); DA Garza, 'Market Definition, the New Horizontal Merger Guidelines, and the Long March away from Structural Presumptions' [2010] *Antitrust Source* 1, 3 f.

purposes: that of delineating the relevant line of commerce and section of the country, and that of identifying market participants and allocating market shares to the latter.[136] Reflecting the Supreme Court dictum from 1966 cited above,[137] the agencies emphasise that '[r]elevant markets need not have precise metes and bounds'[138] and are therefore thought of as an approximation to the reality of the market. The agencies also point out that the analysis of a proposed merger 'need not start with market definition'.[139] Thereby, they give considerably less weight to relevant markets. The Guidelines' new approach to market definition remains contentious and contrasts with the one that prevails in courts.[140]

When defining a relevant market, the agencies will exclusively concentrate on demand substitutability. Supply substitutability will only be addressed during a further step, where suppliers are included amongst the competitors in the market based on the demand side.[141] This possibly contradicts the Supreme Court case law.[142] However, in the literature, it has been argued that it is a virtually equivalent approach.[143] In order to assess the concentration on a particular market and its supposed change through the proposed merger, the agencies will continue to make use of the HHI, which by definition requires previous market definition and the calculation of market shares, at least for the bigger market players.[144]

While courts regularly point out that they are not bound by the Horizontal Merger Guidelines, they have repeatedly adopted the hypothetical monopolist test.[145] In 1988, for instance, a Court of Appeals accepted the US government's market definition, which explicitly relied on the Guidelines' hypothetical monopolist test.[146] And in 2001, a District Court noted that while they were not legally bound by the Horizontal Merger Guidelines, 'courts have often adopted the [market definition] standards set forth'[147] therein. However, in areas apart from mergers, such as under section 2 of the Sherman Act, US courts generally do not rely on the Guidelines' market definition approach – although this could be done if the test were slightly adapted.[148]

[136] Shapiro, '2010 Horizontal Merger Guidelines' (n 120) 85 f.

[137] *United States v Pabst Brewing*, 384 US 546, 549 (1966).

[138] US Horizontal Merger Guidelines 2010 (n 135) § 4.

[139] ibid.

[140] See sections I.A and IV.

[141] US Horizontal Merger Guidelines 2010 (n 135) § 4.

[142] *United States v Columbia Steel Co*, 334 US 495 (1948).

[143] DW Carlton, 'Market Definition: Use and Abuse' (2007) 3 *Competition Policy International* 3, 13; Baker, 'Market Definition' (n 3) 134 f.

[144] US Horizontal Merger Guidelines 2010 (n 135) § 5.3.

[145] For a list of examples, see GJ Werden, 'The 1982 Merger Guidelines and the Ascent of the Hypothetical Monopolist Paradigm' (2003) 71 *Antitrust Law Journal* 253, 270–75. No Supreme Court judgment can be found amongst these.

[146] *United States v Archer-Daniels-Midland and Nabisco Brands*, 866 F2d 242 (8th Cir 1988).

[147] *California v Sutter Health System*, 130 FSupp2d 1109, 1120 (CD Cal 2001) (regarding geographic market delineation; containing further references).

[148] Glick, Cameron and Mangum (n 10) 122, 150.

ii. Further Agency Guidance

In their 2017 Antitrust Guidelines for the Licensing of Intellectual Property, the DoJ and the FTC highlight how important the characteristics of the relevant market(s) are for assessing licensing agreements.[149] This shows that market definition has a different function to fulfil in this scenario than simply establishing the competitive constraints. The Guidelines also contain market share thresholds in their safety zones.[150]

II. The Relevant Product Market under EU Competition Law

Market definition is a crucial aspect of EU competition law.[151] A recent study found that more than one in five appeals to Commission decisions in the area of competition law addresses the question of market definition. While parties raise market definition issues in over half of cases on abuse of dominance and in 42.5 per cent of merger cases, this only occurred in 15.8 per cent of the cases on anti-competitive agreements.[152] This sheds light on the perceived relative importance that parties attribute to market definition under the different areas of antitrust. In the following sections, the legal framework for market definition as developed by the CJEU is outlined against the background of the main EU competition provisions.

A. Substantive Tests in the Case Law

The TFEU does not provide any guidance regarding market definition. Article 101(1) and Article 102 TFEU merely refer to the internal market, a concept that is distinct from the relevant markets at issue in antitrust analysis.[153] On the other hand, the case law of the CJEU gives important guidance as to how the relevant market should be defined for the purposes of competition law.

[149] US Department of Justice and Federal Trade Commission, Antitrust Guidelines for the Licensing of Intellectual Property (14 January 2017) (US IP Guidelines 2017) § 4.1.1.

[150] ibid § 4.3. On these, see ch 9, section I.B.ii.

[151] European Commission, Notice on the definition of relevant market for the purposes of Community competition law [1997] OJ C372/5 (EU Market Definition Notice 1997) para 4.

[152] M Sousa Ferro, 'Judicial Review: Do European Courts Care about Market Definition?' (2015) 6 *Journal of European Competition Law & Practice* 400, 403. The same author will provide updated numbers in M Sousa Ferro, *Market Definition in EU Competition Law* (Cheltenham, Edward Elgar, 2019).

[153] The internal or single market designates the barrier-free marketplace among the Member States.

i. Early Case Law: Laying the Foundations of Market Definition

In the *Continental Can* judgment of 1973, the Court of Justice made a first attempt at sketching a legal test for delineating relevant antitrust markets. It held that

> the definition of the relevant market is of essential significance, for the possibilities of competition can only be judged in relation to those characteristics of the products in question by virtue of which those products are particularly apt to satisfy an inelastic need and are only to a limited extent interchangeable with other products.[154]

Not only did the Court underline the importance of market definition, but it also highlighted that product interchangeability and demand elasticity needed to be taken into account when defining markets. The Court emphasised that it was not only demand substitutability ('the use of a certain product') but also supply substitutability ('particular characteristics of production') that needed to be borne in mind when defining a relevant market, because market entry by likely competitors can act as a powerful competitive restraint.[155] Indeed, markets should only be regarded as distinct if it is shown that suppliers 'are not in a position to enter this market, by a simple adaptation, with sufficient strength to create a serious counterweight'.[156] The Court also required that potential competition should seriously be taken into account when delineating relevant antitrust markets.[157] However, it did not detail the timeframe that should be applied to potential competition. Based on this framework for delineating a relevant market, the Court held that rather than finding separate markets for light metal containers depending on the products to be packaged, the Commission should have perhaps found a larger market for light metal containers.[158] Market definition, the Court pointed out, was instrumental in grasping the market power of undertakings.[159] Without a careful market analysis, the Commission decision had to be annulled. It is interesting to contrast this early European decision with the US *Continental Can* merger case discussed above, in which the end use of cans was equally not considered an appropriate demarcation line for the relevant antitrust market.[160]

In *Michelin I* (1983), the Court considered that a market analysis that limits itself to the objective characteristics of the products concerned would not suffice. Beyond the characteristics of the product, market characteristics such as the conditions of competition prevailing on the relevant market and supply and demand structures were also relevant.[161]

[154] Case 6/72 *Europemballage and Continental Can v Commission* EU:C:1973:22, para 32.
[155] ibid para 33.
[156] ibid.
[157] ibid para 36.
[158] ibid paras 34–37.
[159] ibid para 35.
[160] See above, sections I.A.iv and I.D.
[161] Case 322/81 *Michelin v Commission (Michelin I)* EU:C:1983:313, para 37. See also Case C-333/94 P *Tetra Pak v Commission (Tetra Pak II)* EU:C:1996:436, para 13.

In *Consten & Grundig* (1966), the Court geographically distinguished the French market from the German market.[162] A few years later, in *United Brands* (1978), the Court explicitly highlighted that the relevant market had two dimensions: a product dimension and a geographical dimension.[163] When defining a relevant market, one needed to have 'regard to the particular features of the product in question and [establish] a clearly defined geographical area in which it is marketed and where the conditions of competition are sufficiently homogeneous'.[164]

Based on these principles of market definition, the Court has sometimes accepted very narrow product markets, eg, the market for spare parts for a particular brand of cash registers or the market for nails for the cartridge strips of a particular brand of nail guns.[165] Competition authorities in Europe have generally become known for relying on narrowly delineated product markets.[166] As market share thresholds continue to play an important role under EU competition law, the narrower the market definition is, the more powerful successful companies on that market will be found to be – with all the consequences of dominance that this entails.[167]

ii. The General Court on Market Definition: Consolidation of the Case Law

Since the establishment of the General Court as the first level of the EU judicature in 1989,[168] appeals to Commission decisions in the area of competition law need to be addressed to that court. The Court of Justice reviews the General Court's judgments on points of law; only where the evidence has been distorted may the Court revert to issues of fact.[169] This has made the General Court the prime addressee for parties challenging the Commission's market definition under Article 256 TFEU.

The General Court understands market definition as follows:

> The concept of the relevant market … implies that there can be effective competition between the products or services which form part of it and this presupposes that there is a sufficient degree of interchangeability between all the products or services forming part of the same market in so far as a specific use of such products or services

[162] Joined Cases 56 and 58/64 *Consten and Grundig v Commission* EU:C:1966:41, 343.

[163] Case 27/76 *United Brands v Commission* EU:C:1978:22, para 10.

[164] ibid para 11.

[165] Case 22/78 *Hugin v Commission* EU:C:1979:138; Case T-30/89 *Hilti v Commission* EU:T:1991:70, para 77, upheld on appeal in Case C-53/92 P *Hilti v Commission* EU:C:1994:77.

[166] G Monti, 'Article 82 EC and New Economy Markets' in C Graham and F Smith (eds), *Competition, Regulation, and the New Economy* (Oxford, Hart Publishing, 2004) 25.

[167] See European Commission, Guidance on the Commission's enforcement priorities in applying Article 82 of the Treaty to abusive exclusionary conduct by dominant undertakings [2009] OJ C45/7 (Guidance Paper) para 9.

[168] Initially, the General Court was called the Court of First Instance.

[169] Case C-439/11 P *Ziegler v Commission* EU:C:2013:513, paras 74 f.

> is concerned. The ... competitive conditions and the structure of supply and demand on the market must also be taken into consideration.[170]

The General Court thus clearly carries on the Court of Justice's traditional approach to market definition as outlined above. In addition, it regards demand-side substitutability as a more useful demarcation criterion than supply-side substitutability.[171]

In its *Telefónica* judgment (2012), the General Court highlighted some general principles of market definition. First of all, it subscribed to the Commission's view that substitutability must occur within the short term in order to be relevant for market definition.[172] This gives an indication of the timeframe for the relevant market, albeit not a very precise one. Second, it made clear that while under some circumstances substitutability 'on the margin' can be enough to include products within a product market, this is generally not the case.[173] This shows that it prefers to focus on the main throng of customers. Third, the General Court held that while 'asymmetric substitution' should be taken into account when delineating antitrust markets, 'a large discrepancy in the rates of switching between two products does not lend credence to the argument that they are interchangeable in the eyes of consumers'.[174]

On several occasions, the General Court did not object to market delineations that were based on the scope of the anti-competitive behaviour at issue.[175] This has been criticised as an error in law on the basis that 'an unlawful practice does not make a market smaller'[176] – or bigger, as it were.

iii. The Vague Concept of Sufficient Interchangeability

Already in *Hoffmann-La Roche* (1979), the Court of Justice underlined that in order to establish interchangeability between products, one needed to have regard to the degree with which products were interchangeable and to effective competition between those products.[177] Only products that exhibit 'a sufficient degree of interchangeability'[178] can form part of the same product market. What is sufficient,

[170] Case T-427/08 *CEAHR v Commission (Swiss Watchmakers)* EU:T:2010:517, para 67 (references to case law in original omitted).

[171] Case T-177/04 *easyJet Airline v Commission* EU:T:2006:187, para 99.

[172] Case T-336/07 *Telefónica España v Commission* EU:T:2012:172, para 123; Sousa Ferro, 'Judicial Review' (n 152) 408.

[173] Case T-336/07 *Telefónica España v Commission* EU:T:2012:172, para 128.

[174] ibid para 140 (reference to Case T-340/03 *France Télécom v Commission* EU:T:2007:22, paras 86–91, in original and omitted here). The Court held all of the appellant's pleas relating to market definition to be inadmissible; Case C-295/12 P *Telefónica España v Commission* EU:C:2014:2062, paras 81–90.

[175] Case T-9/89 *Hüls v Commission* EU:T:1992:31, paras 376–81; Case T-34/92 *Fiatagri and New Holland Ford v Commission* EU:T:1994:258, para 51; Case T-29/92 *VSPOB and Others v Commission* EU:T:1995:34, paras 76, 82.

[176] Sousa Ferro, 'Judicial Review' (n 152) 405 fn 44.

[177] Case 85/76 *Hoffmann-La Roche v Commission* EU:C:1979:36, para 28.

[178] Case T-504/93 *Tiercé Ladbroke v Commission* EU:T:1997:84, para 81; Case T-340/03 *France Télécom v Commission* EU:T:2007:22, para 80; Case T-427/08 *CEAHR v Commission (Swiss Watchmakers)* EU:T:2010:517, para 67.

of course, remains a matter of interpretation.[179] So far, neither court has provided a clear test for the degree of interchangeability that it considers appropriate.

The small but significant and non-transitory increase in price (SSNIP) test features prominently in the Commission's Market Definition Notice, where it is used in order to gauge the degree of interchangeability between products (see section II.F.ii below).[180] However, very few EU cases mention the SSNIP test. In *Telefónica*, the appellants alleged that the Commission had wrongly applied the SSNIP test in its temporal context, but the Court found this ground of appeal to be inadmissible.[181] And in *Swiss Watchmakers*, the General Court stated that the SSNIP test was but '[o]ne way of making this determination'[182] of whether or not products are regarded as substitutable by consumers. It called the SSNIP test 'a speculative experiment'.[183] The General Court made a similar remark in *Topps*, where it held that the Commission had several tools available for delineating the relevant market apart from relying on a SSNIP, for instance, market studies or a qualitative assessment of the points of view of consumers and competitors.[184] The widespread assumption that the SSNIP test is firmly established in EU case law is therefore somewhat misleading.[185]

iv. Some Specifics of Market Delineation under EU Competition Law

EU competition law does not rely on a sub-market concept comparable to the US *Brown Shoe* sub-market approach. Instead, when confronted with the possibility of a sub-market in *Compagnie Générale Maritime* (2002), the General Court emphasised that based on its own case law, the following was clear:

> [A] sub-market which has specific characteristics from the point of view of demand and supply, and which offers products which occupy an essential and non-interchangeable place in the general market of which it forms part, must be considered to be a distinct product market.[186]

This statement is in line with the interpretation given to *Brown Shoe* in more recent US judgments, as discussed above.[187]

[179] RJ van den Bergh, 'Modern Industrial Organization versus Old-Fashioned European Competition Law' (1996) 17 *European Competition Law Review* 75, 82; SD Anderman and HKS Schmidt, *EU Competition Law and Intellectual Property Rights: The Regulation of Innovation*, 2nd edn (Oxford, Oxford University Press, 2011) 40.

[180] EU Market Definition Notice 1997 (n 151) para 17.

[181] Case C-295/12 P *Telefónica España v Commission* EU:C:2014:2062, paras 81–90.

[182] Case T-427/08 *CEAHR v Commission (Swiss Watchmakers)* EU:T:2010:517, para 69.

[183] ibid. The SSNIP was also briefly mentioned in Case T-175/12 *Deutsche Börse v Commission* EU:T:2015:148, para 84.

[184] Case T-699/14 *Topps Europe v Commission* EU:T:2017:2, para 82.

[185] Sousa Ferro, 'Judicial Review' (n 152) 407.

[186] Case T-86/95 *Compagnie Générale Maritime v Commission* EU:T:2002:50, para 128 (references to other General Court case law in the original omitted).

[187] See section I.A.iv.

In one notable case, the General Court came close to adopting a cluster market concept. In *Lombard Club* (2006), it accepted the Commission's market definition consisting of a set of banking services, although these services were not interchangeable with one another.[188] This approach is reminiscent of the US Supreme Court's cluster market concept in *Philadelphia National Bank*.[189] While the banks in *Lombard Club* argued that the General Court's market definition was 'incorrect, inadequate and contradictory', the Court of Justice confirmed it in 2009 without going into any detail.[190] However, it appears that no such market definition has been adopted since.

Having outlined the substantive legal tests for market definition, the main EU competition law provisions are now scrutinised for the specific functions that they assign to the relevant market.

B. Anti-competitive Agreements and Market Definition

Article 101(1) TFEU prohibits anti-competitive agreements that have the object or effect of restricting competition, eg, by sharing markets. The analysis of a potentially anti-competitive agreement needs to look at whether and how competition is working on the relevant market as a whole, and how the agreement under scrutiny fares in this respect.[191] The Court of Justice has long held that agreements must be appraised in their economic and legal context because 'it would be pointless to consider an agreement … by reason of its effects if those effects were to be taken distinct from the market in which they are seen to operate'.[192] The relevant market therefore forms an integral part of the analysis of possible anti-competitive effects. At the same time, the economic context referred to by the Court can also go beyond the defined relevant market. In *Cartes bancaires* (2014), the Court underlined that confining the legal analysis to the economic or legal context as it relates to the relevant market alone would amount to an error in law, particularly if 'interactions between the two facets of a two-sided system' are not taken into account because the market sides in a particular case are not regarded as forming part of the same relevant market.[193] The Court of Justice therefore 'extend[ed] the contextual analysis to all the relevant sides of multi-sided markets'.[194]

[188] Joined Cases T-259 to T-264/02 and T-271/02 *Raiffeisen Zentralbank Österreich and Others v Commission (Lombard Club)* EU:T:2006:396, para 174.

[189] See above at section I.A.iv.

[190] Joined Cases C-125, C-133, C-135 and C-137/07 P *Raiffeisen Zentralbank Österreich and Others v Commission (Lombard Club)* EU:C:2009:576, para 59.

[191] Case C-234/89 *Stergios Delimitis v Henninger Bräu* EU:C:1991:91, paras 15 f; Case T-34/92 *Fiatagri and New Holland Ford v Commission* EU:T:1994:258, paras 49 f.

[192] Case 23/67 *Brasserie de Haecht v Wilkin* EU:C:1967:54, 415.

[193] Case C-67/13 P *Groupement des cartes bancaires v Commission* EU:C:2014:2204, paras 77–79 (direct quote at para 79); R Nazzini and A Nikpay, 'Object Restrictions and Two-Sided Markets in EU Competition Law after Cartes Bancaires' (2014) 10 *Competition Policy International* 157, 165.

[194] F Pradelles and A Scordamaglia-Tousis, 'The Two Sides of the Cartes Bancaires Ruling: Assessment of the Two-Sided Nature of Card Payment Systems under Article 101(1) TFEU and Full Judicial Scrutiny of Underlying Economic Analysis' (2014) 10 *Competition Policy International* 139, 142.

Apart from its importance for the analysis of anti-competitive effects, the relevant market also needs to be delineated when assessing whether a competition law infringement by object relates to the same relevant market, thus constituting horizontal behaviour.[195] In order to gauge an agreement's anti-competitive effects under Article 101(1) TFEU, one must assess whether the agreement in question, together with similar agreements present on the market, restricts competitors' access to the market.[196] This analysis must necessarily be preceded by market delineation.[197] In fact, market definition may be just as important under Article 101 TFEU as under Article 102 TFEU,[198] where the importance of market definition is perhaps more easily recognisable.

In *European Night Services* (1998), the General Court emphasised that an antitrust market had to be delineated in order to assess whether an agreement had an appreciable effect on trade between Member States.[199] However, only two years later, that same court chose a more cautious wording in *Volkswagen*, finding that under Article 101 TFEU, 'the reason for defining the relevant market, *if at all*, is to determine whether the [behaviour] at issue is liable to affect trade between Member States and has as its object or effect the prevention, restriction or distortion of competition'.[200] This shows that the General Court can envisage cases under Article 101 TFEU in which market definition is not necessary. However, it is difficult to imagine a full-fledged analysis under Article 101 TFEU that would ultimately not focus on whether or not the behaviour in question restricts competition.

The Commission has issued several binding block exemption regulations (BERs) for certain categories of agreements that are not subject to scrutiny under Article 101 TFEU, provided that certain market share thresholds are not surpassed. Market definition constitutes the very 'backbone'[201] of these safe harbours. In the case of research and development agreements, the Commission has set the market share threshold at 25 per cent for competitors.[202] For specialisation agreements, there is an aggregate market share threshold for all parties involved of

[195] Lemley and McKenna (n 58) 2077.

[196] Case C-234/89 *Stergios Delimitis v Henninger Bräu* EU:C:1991:91, para 15.

[197] ibid para 16.

[198] S Bishop and M Walker, *The Economics of EC Competition Law: Concepts, Application and Measurement*, 3rd edn (London, Sweet & Maxwell, 2010) para 5-063.

[199] Joined Cases T-374, T-375, T-384 and T-388/94 *European Night Services and Others v Commission* EU:T:1998:198, para 93.

[200] Case T-62/98 *Volkswagen v Commission* EU:T:2000:180, para 230 (emphasis added). See also Case T-29/92 *VSPOB and Others v Commission* EU:T:1995:34, para 74; Case C-111/04 P *Adriatica di Navigazione v Commission* EU:C:2006:105, para 31.

[201] B Lasserre, 'Market Definition: A Resilient Feature of Competition Enforcement?' in B Hawk (ed), *International Antitrust Law & Policy* (Huntington, Juris Publishing, 2013) 214.

[202] Commission Regulation (EU) 1217/2010 on the application of Article 101(3) of the Treaty on the Functioning of the European Union to certain categories of research and development agreements [2010] OJ L335/36 (R&D BER), arts 4(1), 4(2)(a) and (b).

20 per cent.[203] In the case of vertical agreements, the market share threshold is 30 per cent for the supplier and the buyer, respectively.[204] And for technology licensing, the Commission foresees an aggregate market share threshold of 20 per cent for competitors, and individual market share thresholds of 30 per cent for non-competitors.[205] These market share thresholds necessarily presuppose the calculation of market shares on a relevant market. As the Court has highlighted, 'it is not possible, by definition, to verify whether a market share threshold has been exceeded in the absence of any definition whatsoever of that market'.[206]

Market share thresholds also feature prominently in Commission guidance. The Notice on the effect on trade concept foresees an aggregate market share threshold of five per cent for all parties involved, combined with a turnover threshold of €40 million.[207] And the 2014 *De Minimis* Notice works with an aggregate market share threshold of 10 per cent for competitors and 15 per cent for non-competitors.[208] The relevant market is also an important element in the Commission's guidelines on applying individual exemptions under Article 101(3) TFEU, which foresee that consumer benefits need to relate to the same relevant market on which the anti-competitive agreement was concluded.[209] In *MasterCard*, the Court of Justice clarified that in the case of two-sided markets, one could not engage in a one-sided analysis of the benefits of an agreement; the benefits for the other market side also had to be taken into account.[210]

Where an agreement infringes Article 101(1) TFEU and cannot be block-exempted, it may be eligible for individual exemption under Article 101(3) TFEU. One of the conditions for applying this exemption is that the agreement may not put the parties in a position that would enable them to shun a considerable part of the relevant market from competition.[211] Such an assessment requires at least an approximate delineation of the market concerned.

[203] Commission Regulation (EU) 1218/2010 on the application of Article 101(3) of the Treaty on the Functioning of the European Union to certain categories of specialisation agreements [2010] OJ L335/43 (Specialisation BER), art 3.

[204] Commission Regulation (EU) 330/2010 on the application of Article 101(3) of the Treaty on the Functioning of the European Union to categories of vertical agreements and concerted practices [2010] OJ L102/1 (VABER), art 3(1).

[205] Commission Regulation (EU) 316/2014 on the application of Article 101(3) of the Treaty on the Functioning of the European Union to categories of technology transfer agreements [2014] OJ L93/17 (TTBER), art 3.

[206] Case C-439/11 P *Ziegler v Commission* EU:C:2013:513, para 63.

[207] European Commission, Guidelines on the effect on trade concept contained in Articles 81 and 82 of the Treaty [2004] OJ C101/81, para 52(a).

[208] European Commission, Notice on agreements of minor importance which do not appreciably restrict competition under Article 101(1) of the Treaty on the Functioning of the European Union [2014] OJ C291/1 (*De Minimis* Notice) para 8.

[209] European Commission, Guidelines on the application of Article 81(3) of the Treaty [2004] OJ C101/97, para 85. On this discussion, see also Pradelles and Scordamaglia-Tousis (n 194) 145 f.

[210] Case C-382/12 P *MasterCard v Commission* EU:C:2014:2201, paras 237, 241; Pradelles and Scordamaglia-Tousis (n 194) 147 f.

[211] Article 101(3)(b) TFEU.

Like in the case of per se restrictions under US antitrust law, one might be tempted to somewhat relax market definition standards when an agreement is found to restrict competition by object under Article 101(1) TFEU. But when the Commission submitted, in an appeal to its *Ziegler* decision, that it was under no obligation to delineate the relevant antitrust market where a restriction of competition by object was at issue,[212] the General Court was quick to rebuke this bold statement. It held that such an argument was not acceptable and that at the very least, the Commission had to delineate the relevant market if it was not otherwise possible to ascertain whether or not the agreement in question may affect trade between Member States.[213] Exceptionally, the Commission was in this case allowed to assume that certain turnover thresholds had been reached 'without expressly determining the market'.[214] The Court of Justice emphasised that the General Court had only allowed this because the Commission had described the relevant services market in sufficient detail. This should not be understood as a free pass to neglect market definition.[215]

C. The Relevant Market and Abuses of a Dominant Position

If it is found that market definition plays a key role under Article 101 TFEU, then this conclusion is even more forceful under Article 102 TFEU, which forbids the abuse of a dominant market position. According to the General Court, market definition is the first step in any competition law case on an infringement of Article 102 TFEU. The abuse of a dominant position presupposes the existence of such a dominant position, and finding a dominant position on a relevant market in turn presupposes market definition.[216] It is with reference to a relevant market that market power is analysed.[217] The Court of Justice holds dominance to be 'a position of economic strength enjoyed by an undertaking which enables it to prevent effective competition being maintained on the *relevant market*'.[218] Both the dominant position and the market behaviour of the company therefore need to be assessed with reference to the relevant market.

Case law has established the rebuttable presumption that companies with a market share above 50 per cent are dominant on the relevant market.[219] Below a

[212] Case T-199/08 *Ziegler v Commission* EU:T:2011:285, para 41.

[213] ibid paras 42, 44 f. As the Commission's Guidelines on the effect on trade concept are based on market share thresholds, it is difficult to see in which cases the Commission could effectively refrain from market definition without changing its Guidelines.

[214] Case T-199/08 *Ziegler v Commission* EU:T:2011:285, para 72.

[215] Case C-439/11 P *Ziegler v Commission* EU:C:2013:513, para 67.

[216] Case T-29/92 *VSPOB and Others v Commission* EU:T:1995:34, para 74; Case T-62/98 *Volkswagen v Commission* EU:T:2000:180, para 230; Case T-691/14 *Servier v Commission* EU:T:2018:922, para 1380.

[217] D Hildebrand, *The Role of Economic Analysis in EU Competition Law: The European School*, 4th edn (Alphen aan den Rijn, Kluwer Law International, 2016) 176.

[218] Case 27/76 *United Brands v Commission* EU:C:1978:22, para 65 (emphasis added).

[219] Case C-62/86 *AKZO Chemie v Commission* EU:C:1991:286, para 60.

market share of 40 per cent, the Commission will usually not consider a company to be dominant.[220] EU competition law puts a strong emphasis on market dominance. Article 102 TFEU cannot police monopolisation through anti-competitive means by a company that was not dominant at the time of its questionable behaviour.[221] This is a considerable difference from attempted monopolisation under US law.

A number of landmark cases on abuse of dominance also concerned market definition. In *United Brands*, the Court of Justice emphasised that the 'opportunities for competition … must be considered having regard to the particular features of the product in question and with reference to a clearly defined geographic area'.[222] The Court had to decide whether bananas formed part of a larger fresh fruit market 'because they are reasonably interchangeable by consumers with other kinds of fresh fruit'[223] or if they constituted their own product market. It reasoned that a number of locked-in consumers could not switch from bananas to other kinds of fresh fruit – eg, the very young, the old and the sick – leading to limited substitutability and the finding that bananas constituted their own relevant market.[224] Like the Supreme Court in *Grinnell* before it, the Court of Justice focused on captive consumers rather than on those who are willing and able to change to another product.[225] However, if the market behaviour of the latter group of consumers makes a price increase unprofitable, then the relevant market might have to be defined more broadly. *United Brands* can therefore be seen as an example of a market definition that results from relying on functional interchangeability rather than on economic evidence.[226] More recently, *Post Danmark II* (2015) concerned a possible abuse of a dominant position related to a rebate scheme in the distribution of direct advertising. The Court of Justice linked the relevant market to the substantive analysis, holding that the former's characteristics had to be carefully taken into account during the latter.[227]

In recent years, practical considerations might have played into the market definition exercise at the Commission, leading to a certain degree of pragmatism. In *Tomra* (2012), the Commission relied on a relevant market that encompassed high-end and low-end reverse vending machines. Despite its preference for a narrower market delineation, it settled on that broader one as this was more favourable to the defendant.[228] This might have been motivated by the fact that Tomra

[220] Guidance Paper (n 167) para 14.

[221] J Drexl, 'Anticompetitive Stumbling Stones on the Way to a Cleaner World: Protecting Competition in Innovation without a Market' (2012) 8 *Journal of Competition Law & Economics* 507, 529.

[222] Case 27/76 *United Brands v Commission* EU:C:1978:22, para 11.

[223] ibid para 12.

[224] ibid paras 31, 34 f.

[225] A Ezrachi, *EU Competition Law: An Analytical Guide to the Leading Cases*, 6th edn (Oxford, Hart Publishing, 2018) 38.

[226] PD Camesasca and RJ van den Bergh, 'Achilles Uncovered: Revisiting the European Commission's 1997 Market Definition Notice' (2002) 47 *Antitrust Bulletin* 143, 171.

[227] Case C-23/14 *Post Danmark v Konkurrencerådet (Post Danmark II)* EU:C:2015:651, para 30.

[228] *Prokent-Tomra* (Case COMP/E-1/38.113) Commission Decision of 29 March 2006 [2008] OJ C219/11, para 46; Case C-549/10 P *Tomra v Commission* EU:C:2012:221, para 8.

was likely to challenge a narrower market definition. Even on this wider market, Tomra had market shares at times exceeding 95 per cent.[229] In *Intel* (2014), the General Court noted that the Commission had relied on the x86 CPU market and had not concluded whether this market had to be segmented into three narrower markets. Intel's market shares on each of these markets was very high.[230]

D. European Merger Control and Market Definition

Market definition for the purposes of merger control asks what the market before the merger looks like and what it will look like once the merger has taken place. It necessarily has an ex ante component that neither Article 101 nor Article 102 TFEU requires. The EU Courts have observed that a 'proper definition of the relevant market is a necessary precondition for any assessment of the effect of a concentration on competition'.[231] While market definition is an intermediate step in antitrust assessments under Articles 101 and 102 TFEU, in merger control it directly concerns the competitive assessment as such because the substantive test includes the creation or strengthening of a dominant position.[232] In merger analysis, the merging companies' market shares before and after the merger, as well as the structure of the relevant market(s), are of utmost relevance, although non-coordinated effects outside of dominance will also be considered if they constitute a significant impediment of effective competition.[233] The relevant market is also used as a decisive parameter in order to ascertain whether the merger in question is horizontal or not; depending on its nature, it will be subject to a different kind of substantive analysis.[234]

In *Continental Can* (1973), the Court of Justice concluded that contradictions in the Commission's market delineation had to lead to the annulment of a Commission decision.[235] Market definition, in the Court's view, was 'of essential significance' in order to establish the consequences of the merger in question.[236] In more recent merger cases, market definition has continued to play an important role. In *Microsoft/Skype* (2011), the Commission assessed the relevant markets in

[229] Case C-549/10 P *Tomra v Commission* EU:C:2012:221, para 10.

[230] Case T-286/09 *Intel v Commission* EU:T:2014:547, para 23.

[231] Joined Cases C-68/94 and C-30/95 *France v Commission (Kali & Salz)* EU:C:1998:148, para 143 (direct quote); Case T-342/99 *Airtours v Commission* EU:T:2002:146, para 19; Case T-151/05 *NVV and Others v Commission* EU:T:2009:144, para 51.

[232] Council Regulation (EC) 139/2004 on the control of concentrations between undertakings [2004] OJ L24/1 (EU Merger Regulation), art 2(3); Bishop and Walker (n 198) para 7-013.

[233] EU Merger Regulation (n 232) Recital 25.

[234] European Commission, Guidelines on the assessment of horizontal mergers under the Council Regulation on the control of concentrations between undertakings [2004] OJ C31/5, para 5; European Commission, Guidelines on the assessment of non-horizontal mergers under the Council Regulation on the control of concentrations between undertakings [2008] OJ C265/6, para 2.

[235] Case 6/72 *Europemballage and Continental Can v Commission* EU:C:1973:22, paras 32–37.

[236] ibid para 32.

great detail. It found that there were two relevant services markets at issue, namely consumer communication services and enterprise communication services. As in the *Intel* case discussed above, it did not decide whether these broader markets should be further segmented into narrower markets, for instance, by functionality, platform, operating system or size of customers; instead, it concluded that even under the narrowest possible market, the acquisition would not lead to anti-competitive effects.[237]

The proposed merger between the Deutsche Börse and the New York Stock Exchange was one of the few cases in which the Commission outright prohibited a merger as incompatible with the internal market.[238] In that case, the Commission divided the stock exchanges' business activities into market groups, and carried out detailed market definitions and competitive assessments within each of these.[239] The case was therefore organised around the multiple relevant markets that the Commission distinguished. On appeal, the General Court upheld the prohibition.[240] In *Niki Luftfahrt* (2015), the Commission again 'identified a number of [relevant product markets], namely passenger air transport services, air cargo transport services, the sale of airline seats to tour operators, maintenance, repair and overhaul services, in-flight catering and ground handling services'.[241] On each of these markets, the Commission carried out a competitive assessment that was organised around the multiple relevant markets that the Commission distinguished.[242] Similarly, the European Commission considered several dozen product markets in the merger case of *Bayer/Monsanto* (2018), which effectively organised the case.[243]

Under the Commission's Horizontal Merger Guidelines, the purpose of market definition is to 'identify in a systematic way the immediate competitive constraints facing the merged entity'.[244] The HHI, which gives economists a first indication of the market's pre- and post-merger concentration, also relies on market shares as a proxy and needs to be preceded by market definition.[245]

[237] *Microsoft/Skype* (Case COMP/M.6281) Commission Decision of 7 October 2011, paras 17, 43, 63. Upheld on appeal in Case T-79/12 *Cisco Systems & Messagenet v Commission* EU:T:2013:635.

[238] *Deutsche Börse/NYSE Euronext* (Case COMP/M.6166) Commission Decision of 1 February 2012 [2014] OJ C254/8. Overall, the Commission has only prohibited 30 mergers since 1990; see European Commission, 'Merger Statistics: 21 September 1990 – 31 August 2019' (31 August 2019), ec.europa.eu/competition/mergers/statistics.pdf.

[239] *Deutsche Börse/NYSE Euronext* (Case COMP/M.6166) Commission Decision of 1 February 2012 [2014] OJ C254/8, paras 22, 28–1132.

[240] Case T-175/12 *Deutsche Börse v Commission* EU:T:2015:148.

[241] Case T-162/10 *Niki Luftfahrt v Commission* EU:T:2015:283, para 13.

[242] *Lufthansa/Austrian Airlines* (Case COMP/M.5440) Commission Decision of 28 August 2009 [2010] OJ C16/11.

[243] *Bayer/Monsanto* (Case M.8084) Commission Decision of 21 March 2018 [2018] OJ C456/10, paras 304 ff, 710 ff, 861 ff, 1346 ff, 2045 ff, 2243 ff, 2289 ff, 2331 ff, 2379 ff, 2555 ff.

[244] Guidelines on the assessment of horizontal mergers under the Council Regulation on the control of concentrations between undertakings (n 234) para 10.

[245] ibid para 16.

E. Functions of Market Definition under EU Competition Law

Market definition represents the first step in virtually any competition law case in the EU. The legal tests of Articles 101 and 102 TFEU and of the Merger Regulation presuppose the delineation of a relevant antitrust market, fulfilling three main functions. First of all, market definition is used as the basis for determining market power, particularly under Article 102 TFEU. European merger control often uses market definition as the basis for assessing the companies' market power pre- and post-merger, and to establish whether a merger creates or strengthens a single or collective dominant position.

Second, and closely related to the first function, market share thresholds are relied upon in many different contexts under EU competition law, and in all three areas of competition law – either as rebuttable presumptions devised by the Court, in binding block exemption regulations issued by the Commission or by way of the Commission's guidance. These thresholds require previous market delineation. They work as a filter in order to separate cases that are probably unproblematic from others that are possibly problematic, before applying a full substantive analysis to the possibly problematic cases. Under Article 102 TFEU, EU competition law relies on market share thresholds as a rebuttable presumption to indicate a dominant position. In merger control, market shares allow the calculation of HHI levels.

Finally, the relevant market provides the necessary economic context for the behaviour under investigation. This is true irrespective of the type of competition law infringement. Under Article 101 TFEU, it provides the context for the agreement under investigation, situating the alleged anti-competitive effects within the real world and allowing for their legal assessment within the given context. Even where a market is not explicitly delineated, 'many of the facts relevant to such a definition play a role in the Commission's competitive analysis' in an Article 101 TFEU case.[246] Market definition also enables an assessment of whether an agreement affects trade between Member States and allows for an analysis of whether or not market behaviour has the object or effect of restricting competition.[247] Under Article 101(3) TFEU, market definition is also necessary for analysing a possible individual exemption. Under Article 102 TFEU, the relevant market gives context to a dominant company's commercial behaviour. Some abuses, such as tying, require the finding of two separate markets. In merger control, market definition is relied upon to highlight competitive concerns in a specific market below the threshold of market dominance. Market definition may also provide some natural guidance to the analysis of a merger's competitive effects, thus acting as an organising principle for the framing of the case.

[246] TE Kauper, 'The Problem of Market Definition under EC Competition Law' (1997) 20 *Fordham International Law Journal* 1682, 1692.

[247] Case C-439/11 P *Ziegler v Commission* EU:C:2013:513, para 71.

The EU Courts consistently rely on market definition for its various functions. Contrary to the US, there have been no serious attempts to abolish market definition under EU competition law, despite a growing awareness of the pitfalls of relying on static market definitions.[248]

F. Commission Guidance on Market Definition

i. General Remarks

The European Commission's soft law instruments in the area of competition law specify how the Commission interprets the legal provisions and the case law. As long as they do not contradict the TFEU, the CJEU accepts that 'the Commission may adopt a policy as to how it will exercise its discretion in the form of measures such as guidelines'.[249] In addition, the Commission's understanding remains subject to any interpretation by the Court of Justice.[250]

While Commission guidance is not legally binding, it is binding on the Commission itself. In *General Electric* (2005), the General Court agreed with the applicant that 'the Commission may not depart from rules which it has imposed on itself'.[251] Where the Commission's Market Definition Notice precisely sets out the method that the Commission will rely upon in order to delineate a relevant market, the Commission must follow its own guidance.[252] This ensures legal certainty for companies subjected to the competition laws.

The General Court has repeatedly held that 'in so far as the definition of the product market involves complex economic assessments on the part of the Commission, it is subject to only limited review by the [Union] judicature'.[253] Such marginal review is limited to manifest errors of assessment, although the latter criterion remains unclear.[254] The Commission enjoys considerable autonomy and weight in market definition, and the Court's reticence to fully review

[248] Camesasca and van den Bergh (n 226) 154.

[249] Case C-439/11 P *Ziegler v Commission* EU:C:2013:513, para 59 (containing further references).

[250] EU Market Definition Notice 1997 (n 151) para 6; Guidelines on the effect on trade concept contained in Articles 81 and 82 of the Treaty (n 207) para 5; *De Minimis* Notice (n 208) para 7.

[251] Case T-210/01 *General Electric v Commission* EU:T:2005:456, para 516 (citing a number of cases to this effect).

[252] ibid para 516.

[253] Case T-301/04 *Clearstream v Commission* EU:T:2009:317, para 47; Case T-201/04 *Microsoft v Commission* EU:T:2007:289, para 482. See also Case T-342/99 *Airtours v Commission* EU:T:2002:146, para 26; Case T-57/01 *Solvay v Commission* EU:T:2009:519, para 250; Case T-151/05 *NVV and Others v Commission* EU:T:2009:144, para 53.

[254] See H Schweitzer, 'The European Competition Law Enforcement System and the Evolution of Judicial Review' in C-D Ehlermann and M Marquis (eds), *European Competition Law Annual 2009: The Evaluation of Evidence and its Judicial Review in Competition Cases* (Oxford, Hart Publishing, 2011) 99; A Kalintiri, 'What's in a Name? The Marginal Standard of Review of "Complex Economic Assessments" in EU Competition Enforcement' (2016) 53 *CML Rev* 1283, 1285.

its complex economic assessments has been criticised as judicial deference.[255] In *AstraZeneca* (2010), the General Court appeared to address this criticism when holding that despite the Commission's margin of assessment, it could very well review that authority's interpretation of economic data.[256] In *Tetra Laval* (2005), the Court of Justice also emphasised that the EU Courts could and would review the Commission's interpretation of economic data.[257] This particularly applies to the Commission's 'legal classification of economic data'.[258] In *UPS/TNT* (2019), the Court held that the Commission needed to fully disclose its economic analyses to the parties so as to preserve their rights of defence.[259] While an econometric price concentration model was at issue in the case at hand, this ruling also applies to economic modelling for the purposes of market definition.

ii. The Market Definition Notice

In its Market Definition Notice of 1997, the Commission applies a more economics-based approach to market definition.[260] Market definition is understood as 'a tool to identify and define the boundaries of competition between firms … [t]he main purpose of [which] is to identify in a systematic way the competitive constraints that the undertakings involved face'.[261] The Commission also underlines the importance of providing transparency and legal certainty with respect to market definition.[262] The Notice largely complies with established case law, but gives less weight to supply-side substitution.[263]

The Market Definition Notice applies to Articles 101 and 102 TFEU as well as to merger control, thereby aligning market definition in all three areas.[264] While the methodology for market definition is the same for all antitrust provisions, it 'might lead to different results depending on the nature of the competition issue being examined'.[265] Such differences may be due to the fact that merger control relies on a prospective analysis, while Articles 101 and 102 TFEU require an ex post analysis.[266]

The Commission starts from the premise that a relevant market can be assessed by linking the product and the geographical market dimensions.[267]

[255] Sousa Ferro, 'Judicial Review' (n 152) 402.
[256] Case T-321/05 *AstraZeneca v Commission* EU:T:2010:266, para 33. See also M Jaeger, 'The Standard of Review in Competition Cases Involving Complex Economic Assessments: Towards the Marginalisation of the Marginal Review?' (2011) 2 *Journal of European Competition Law & Practice* 295, 296, 305.
[257] Case C-12/03 P *Tetra Laval v Commission* EU:C:2005:87, para 39.
[258] Case T-210/01 *General Electric v Commission* EU:T:2005:456, para 63.
[259] Case C-265/17 P *Commission v UPS* EU:C:2019:23.
[260] On this intention of the Notice, see also Nevo (n 19) 3.
[261] EU Market Definition Notice 1997 (n 151) para 2 (footnote in original omitted).
[262] ibid para 5.
[263] Sousa Ferro, 'Judicial Review' (n 152) 401.
[264] EU Market Definition Notice 1997 (n 151) para 1.
[265] ibid para 12.
[266] ibid.
[267] ibid para 9.

The relevant product market encompasses 'all those products … which are regarded as interchangeable or substitutable by the consumer, by reason of the products' characteristics, their prices and their intended use'.[268] The interchangeability criterion focuses on consumer views, thus relying on actual consumers and their (possibly irrational) choices. This is in line with a behavioural approach to antitrust law.[269] However, as it focuses on short-term demand substitutability, the Notice misses out on identifying innovation as an important competitive parameter in dynamic industries.[270]

In earlier documents, the Commission had supplied a list of factors that should be taken into account when defining a relevant product market, such as the similar appearance of the products, their end use, their price, switching costs, consumer preferences and any existing product classifications.[271] The Notice no longer includes these factors.

In order to ascertain product interchangeability, the Notice relies on a Europeanised SSNIP test.[272] This has been praised as a 'substantial [methodological] improvement'.[273] The SSNIP test supposes a small but significant non-transitory increase in price, in the range of five to 10 per cent, in order to ascertain whether customers would switch to a different product/a supplier located elsewhere in the event of such a price increase.[274] In order to carry out the SSNIP test, a significant amount of econometric data and analysis is required. This is both time-consuming and expensive, and the Commission will often refrain from such an extensive analysis. Instead, it has in the past relied on a number of other factors that indicate interchangeability, such as actual substitution, the views expressed by customers and competitors, the product's intended use, consumer preferences and barriers to switching demand.[275] The Commission warns of applying a SSNIP to a company that may be dominant, as the current price might already be above a competitive level due to the company's market dominance (the cellophane fallacy; see section I.C above).[276]

[268] ibid para 7.

[269] See N Petit and N Neyrinck, 'Behavioral Economics and Abuse of Dominance: A Fresh Look at the Article 102 TFEU Case-Law' [2010] *Österreichische Zeitschrift für Kartellrecht* 203, 204; A Heinemann, 'Behavioural Antitrust: A "More Realistic Approach" to Competition Law' in K Mathis (ed), *European Perspectives on Behavioural Law and Economics* (Berlin, Springer, 2015) 218 f.

[270] C Ahlborn, DS Evans and AJ Padilla, 'Competition Policy in the New Economy: Is European Competition Law up to the Challenge?' (2001) 22 *European Competition Law Review* 156, 161.

[271] Commission Regulation (EC) 3385/94 on the form, content and other details of applications and notifications provided for in Council Regulation No 17 [1994] OJ L377/28 (Form A/B), Operational Part – secs 6 and 11.

[272] On the slight theoretical differences between the US SSNIP test ('would profitably increase') and the EU SSNIP test ('could profitably increase'), see Niels, Jenkins and Kavanagh (n 26) §§ 2.72–2.74.

[273] L Wu and S Baker, 'Applying the Market Definition Guidelines of the European Commission' (1998) 19 *European Competition Law Review* 273, 273.

[274] EU Market Definition Notice 1997 (n 151) para 17.

[275] Anderman and Schmidt (n 179) 39.

[276] EU Market Definition Notice 1997 (n 151) para 19.

As for competitive restraints, the Commission will mainly take into account demand substitutability, eg, determined based on the SSNIP test. However, it may also take supply-side substitutability into account, even though it has been more reluctant in this respect.[277] Potential competition is recognised as a competitive restraint, but is not immediately taken into consideration when defining a relevant market. Instead, it might be referred to once the relevant market has been established.[278]

A practical aspect that the Commission frequently relies upon in merger control is found in paragraph 27 of the Notice: where a narrow relevant market does not raise any competitive concerns, the Commission will not try to 'get it right' by broadening the relevant market in accordance with the Notice, but will instead leave the exact market definition open.[279]

The Notice can be criticised for the way in which it combines the market definition approach developed in the Court's case law with modern economic thinking and quantitative techniques.[280] From a practical point of view, the Commission does not always follow the more economics-based approach to market definition that it has advocated itself,[281] instead relying on the more qualitative analysis of product interchangeability and characteristics that can be derived from the Court's case law.[282] In its *Virgin/British Airways* decision, the Commission stated that the SSNIP test was primarily an instrument to 'explain the concept of a relevant market'.[283]

The Commission's Market Definition Notice turned 20 in 2017 – a considerable age for competition law guidance. Bearing in mind the growing importance of high-innovation markets and the technological advances witnessed over the last two decades, dynamic markets might require not only a re-evaluation of the market definition method, but also a realistic reappraisal of what market definition can achieve.

iii. Further Commission Guidance

Several of the Commission's guidelines rely on market share thresholds and thereby presuppose market definition.[284] Once an infringement of the competition provisions has been established, the Commission will also calculate the basic amount of

[277] European Commission, 'XXIVth Report on Competition Policy 1994' (Luxembourg, 1995) para 280; Anderman and Schmidt (n 179) 39.

[278] EU Market Definition Notice 1997 (n 151) paras 13–24, especially para 24.

[279] ibid para 27.

[280] Camesasca and van den Bergh (n 226) 156 f; Nevo (n 19) 78.

[281] Nevo (n 19) 187.

[282] C Veljanovski, 'EC Merger Policy after *GE/Honeywell* and *Airtours*' (2004) 49 *Antitrust Bulletin* 153, 167.

[283] *Virgin/British Airways* (Case IV/D-2/34.780) Commission Decision 2000/74/EC [2000] OJ L30/1, para 70.

[284] On these, see sections II.B and II.D above.

the fine based on the value of product sales to which the infringement relates.[285] This presupposes that the relevant market has been determined in its product and geographical dimension.

III. The Functions of the Relevant Product Market in the EU and the US: A First Comparative Look

A. Market Definition between Convergence and Divergence

Courts, competition authorities and academics alike repeatedly find market definition to play a central role in US and EU antitrust law. In the US, but perhaps even more so in the EU, 'legal analysis generally requires market definition as part of an [antitrust] enforcement action'.[286] Although the precise functions vary across the jurisdictions and their respective antitrust provisions, market definition is 'a central component of antitrust cases'[287] and is often regarded as a 'legal necessity'.[288]

Market definition is not an aim in itself,[289] but an analytical tool within the legal framework. It should not serve any particular goal on the competition policy agenda, but should remain a neutral analytical step.[290] EU and US competition laws rely on near-identical functions of market definition. In the US, market definition mainly serves as a basis for the assessment of market power and as the background for assessing the (anti-)competitiveness of an agreement, conduct or concentration. In the EU, market definition is also central in establishing market power, but contrary to the US, market share thresholds have manifold roles to play under EU competition law. This second function can be integrated into the first in that the market share thresholds are attempting to quantify (the lack of) market power. In addition, EU competition law strongly relies on market definition to provide the economic background to the substantive analysis of antitrust investigations.

Although the last few decades have seen a gradual convergence of the methods of antitrust market definition in the EU and the US, and despite their similar functions, one can perceive important differences in the way in which EU and

[285] European Commission, Guidelines on the method of setting fines imposed pursuant to Article 23(2)(a) of Regulation No 1/2003 [2006] OJ C210/2, para 13.

[286] Coate and Simons (n 32) 682 (referring to the US).

[287] DS Evans and R Schmalensee, 'Markets with Two-Sided Platforms' in ABA Section of Antitrust Law (ed), *Issues in Competition Law and Policy* (Chicago, ABA Publishing, 2008) 688.

[288] RC Lind, P Muysert and M Walker, 'Innovation and Competition Policy: Part I – Conceptual Issues' (OFT377, Economic Discussion Paper 3, March 2002) § 4.46.

[289] See P Areeda, 'Market Definition and Horizontal Restraints' (1983) 52 *Antitrust Law Journal* 553, 553; L Peeperkorn and V Verouden, 'Market Definition' in J Faull and A Nikpay (eds), *The EU Law of Competition*, 3rd edn (Oxford, Oxford University Press, 2014) § 1.188.

[290] Similarly, see Podszun (n 1) 130.

US courts and competition authorities rely on market definition. Today, there is a broad consensus that antitrust needs to delineate relevant markets, that this delineation should be based on product interchangeability, and that demand and supply-side substitutability need to be taken into account. However, a more detailed comparative analysis unearths considerable divergence, particularly with regard to the way in which product interchangeability is to be established, the role of supply-side substitution and of potential competition. In addition, this formal convergence has not smoothly translated into a convergence in actual antitrust practice.[291] This divergence could have its roots in the different philosophical underpinnings of these two competition law systems.[292] Market definition was developed and shaped under different antitrust rules in the two jurisdictions, with the EU primarily focusing on abuse of dominance and on anti-competitive agreements, while merger control played a more prominent role in the US. This might also account for some of the nuanced differences in the roles that are ascribed to market definition in each jurisdiction, with the US being more focused on market power assessment and the EU tirelessly highlighting market definition as the necessary context to understand a competition law case.

B. The Relevant Market as a Tool for Assessing Market Power

Market dominance under EU competition law and monopoly power under US antitrust law are two distinct legal concepts.[293] While this difference is not further explored here, it should be stressed that market power is a term borrowed from the realm of economics, while market dominance and monopoly power are legal concepts from competition law that are based on the notion of a market.[294] In the following, a position of economic strength on a relevant market will be referred to as market power, a term which is meant to encompass both the EU and the US legal (rather than economic) concepts.

In its original conception, the analysis of the relevant market within the framework of antitrust law can be conceived as a legal test to establish the degree of market power for legal purposes. While there are many different methods for ascertaining market power, the more technical ones rely on a substantial amount of data. In any case, there is 'a strong judicial preference for power measures based

[291] This is particularly true in innovation-intense market environments; see Nevo (n 19) 248 and the analysis in Part II.

[292] These differences are sharpest in relation to innovative markets; see DJ Gifford and RT Kudrle, 'Antitrust Approaches to Dynamically Competitive Industries in the United States and the European Union' (2011) 7 *Journal of Competition Law & Economics* 695, 715.

[293] On the EU concept, see L Ortiz Blanco, *Market Power in EU Antitrust Law* (Oxford, Hart Publishing, 2012).

[294] J Drexl, 'The Relationship Between the Legal Exclusivity and Economic Market Power: Links and Limits' in I Govaere and H Ullrich (eds), *Intellectual Property, Market Power and the Public Interest* (Brussels, Peter Lang, 2008) 15.

on market definition'.[295] This is true not only of the EU and the US, but also of other jurisdictions.[296] The notion of market power only makes sense when it relates to a relevant market.[297] However, methods that economists have devised in order to directly estimate market power have been said to neglect innovation.[298]

The market share thresholds that EU competition law likes to rely on are connected to a structural understanding of competition law analysis, for they quantify static thresholds which are seen as limits for market power. Market share thresholds are thus stubborn remnants of the structure-conduct-performance paradigm advocated by the Harvard School,[299] and their usefulness for assessing market power continues to be debated amid general agreement that they are often misleading, particularly in innovative markets.

C. The Relevant Market's Functions Outside of Market Power Assessments

As far as its function of informing assessments of market power and of providing the basis for market share thresholds is concerned, market definition is geared towards a structure-oriented antitrust analysis. This raises the question whether market definition may also be useful for an antitrust analysis that is less geared towards market structure. This can be answered in the affirmative. One of market definition's essential functions is that of providing context; in other words, it 'adds clarity and power to the narrative'[300] of many antitrust cases. It is the reference to a specific relevant market that allows courts to apply an abstract antitrust theory of harm to the case at hand.[301] In order to argue an antitrust case in a coherent fashion, market definition and the subsequent substantive analysis need to be in line with each other.[302]

Another function of the relevant market is that it characterises the market so as to lead to a comprehensive understanding of complex market behaviour and the

[295] HJ Hovenkamp et al, *IP and Antitrust: An Analysis of Antitrust Principles Applied to Intellectual Property Law*, 3rd edn, vol I (New York, Wolters Kluwer, 2018) § 4.01[B] fn 7.

[296] Eg, see Beaton-Wells, 'Mergers without Markets?' (n 101) 186.

[297] MM Dabbah, *International and Comparative Competition Law* (Cambridge, Cambridge University Press, 2010) 70 f.

[298] HJ Hovenkamp, 'Response: Markets in IP and Antitrust' (2012) 100 *Georgetown Law Journal* 2133, 2135.

[299] On the Harvard School and its main tenets, namely the structure-conduct-performance (SCP) paradigm, its focus on structural problems and overall welfare, and its more interventionist approach, see the references in n 55 above.

[300] GJ Werden, 'Why (Ever) Define Markets? An Answer to Professor Kaplow' (2013) 78 *Antitrust Law Journal* 729, 740.

[301] JB Baker and TF Bresnahan, 'Economic Evidence in Antitrust: Defining Markets and Measuring Market Power' in P Buccirossi (ed), *Handbook of Antitrust Economics* (Cambridge, MA, MIT Press, 2008) 7.

[302] See HJ Kahwaty and CB Tyler, 'Market Definition – Achieving an Integrated Analysis' (2014) 59 *Antitrust Bulletin* 667.

workings of competition on a given market. By allowing insights into the market as such, market characterisation contributes to a better understanding of existing theories of harm and to the development of novel theories of harm. Market characterisation may also help in understanding a certain behaviour's anti- or pro-competitive effects. For instance, the welfare effects of parity clauses in online markets will depend, amongst other things, on the relevant market's characteristics.[303] More broadly, when a case wants to take efficiencies into account, these must relate to the relevant market. In addition, a comprehensive understanding of how competition works on the relevant market is essential in order to design remedies that successfully address the anti-competitive harm.

Antitrust liability cannot arise in the void – 'to monopolize or to reduce competition, there has to be a defined market'.[304] The antitrust market provides the background and sometimes even the basis for antitrust legal analysis; 'identifying markets is part of a decision making structure'.[305] It is a common misconception that market definition serves the primary goal of unveiling market power through the attribution of market shares;[306] its functions go much beyond this one task – a task which, in itself, is losing significance thanks to econometric models that directly deduce market power. Market definition needs to adapt to changing economic realities.[307] As market definition for antitrust law purposes is in a state of flux, it is in constant need of redefining itself.

Market definition attempts to translate complex economic market realities into (relatively) simple and straightforward legal terms.[308] However, this does not mean that market definition cannot meaningfully inform antitrust analysis; on the contrary, a certain amount of abstraction is needed in order to be able to apply antitrust laws to a given case. Only a workable market definition can serve its functions under the antitrust rules, meaning that overly complex market delineations might be ill-suited to fulfil their role of providing context to antitrust cases, as the way in which such context is presented needs to be both as detailed as necessary and as simple as possible.

D. Further Reflections on the Functions of the Relevant Market

Market definition is considered a basic analytical tool in antitrust laws across the globe, as is shown by instruments drawn up within both the International

[303] A Ezrachi, 'The Competitive Effects of Parity Clauses on Online Commerce' (2015) 11 *European Competition Journal* 488, 489.

[304] K Bernard, 'Innovation Market Theory and Practice: An Analysis and Proposal for Reform' (2011) 7 *Competition Policy International* 159, 165.

[305] Camesasca and van den Bergh (n 226) 144.

[306] ibid 146.

[307] Podszun (n 1) 130.

[308] See also ibid.

Competition Network[309] and the OECD.[310] The relevant market provides both 'a focus' and 'an initial screen for the competitive assessment'.[311] In order to overcome some of the current limitations of the relevant market as a legal concept, it needs to be freed from its preoccupation with market power and price effects, instead adopting a more autonomous and dynamic nature. This approach would allow the concept of the relevant market to be more flexible and would not predetermine the outcome of the market analysis.[312] Chapter 10 is a first attempt at reconceptualising market definition in such a way, with dynamic and innovative markets in mind. The guidance options that are discussed want to harness market delineation's ability to characterise innovative markets so as to best portray the market dynamics at work, acknowledging competitive relationships and partial substitutability where the need arises. At the same time, these suggestions recognise the inherent limits that the relevant market concept is confronted with: its sensitivity to changing circumstances, its dependence on (sometimes irrational) customer response and its uneasy relationship with market power.

Market definition cannot be carried out in the abstract, but needs to relate to the facts of a specific case.[313] However, it is controversial whether market definition should differ depending on the competitive harm that is being investigated.[314] From a legal perspective, such an approach is problematic insofar as the legal provisions do not in themselves provide a credible reason for differing market definitions – where they do differ, a policy motive may be suspected.[315] It defies the purpose of market definition if it is carried out with a particular theory of harm or a particular outcome in mind. Yet, the type of anti-competitive behaviour at issue in a case may very well be related to the characteristics of the relevant market. This justifies differences in the relevant markets that are arrived at rather than differences in methodology.

[309] International Competition Network, 'Recommended Practices for Merger Analysis' (2010) 5; International Competition Network, 'Unilateral Conduct Workbook: Chapter 3: Assessment of Dominance' (May 2011) § 23; International Competition Network, 'ICN Merger Guidelines Workbook' (April 2006) § A.16.

[310] See OECD (ed), *Policy Roundtable: Market Definition* (2012) DAF/COMP(2012)19.

[311] Bishop and Walker (n 198) paras 4-002 f.

[312] See also R Podszun and B Franz, 'Was ist ein Markt? – Unentgeltliche Leistungsbeziehungen im Kartellrecht' (2015) 3 *Neue Zeitschrift für Kartellrecht* 121, 126.

[313] Case C-439/11 P *Ziegler v Commission* EU:C:2013:513, para 72; A Katz, 'Making Sense of Nonsense: Intellectual Property, Antitrust, and Market Power' (2007) 49 *Arizona Law Review* 837, 880.

[314] See *Soda-ash – Solvay* (Case IV/33.133-C) Commission Decision 91/299/EEC [1991] OJ L152/21, para 42; Baker, 'Market Definition' (n 3) 129; Bishop and Walker (n 198) para 4-038; MH Morse, 'Product Market Definition in the Pharmaceutical Industry' (2003) 71 *Antitrust Law Journal* 633, 634. Arguing for a 'purposive market definition' that is congruent with the behaviour being investigated, see RL Smith, 'Defining and Proving Markets and Market Power' in J Duns, A Duke and BJ Sweeney (eds), *Comparative Competition Law* (Cheltenham, Edward Elgar, 2015) 28.

[315] Similarly, see Ortiz Blanco (n 293) 4, 13 f.

IV. No More Antitrust Market Definition?

A. The Debate in the Literature: Kaplow, Markovits and Beyond

A number of US scholars and practitioners have questioned whether antitrust law should continue to rely on relevant markets. As early as 1978, Richard Markovits questioned the usefulness of the market-oriented approach for the antitrust analysis of horizontal mergers, and particularly the market definition/market share/market concentration arithmetic.[316] In a follow-up paper, he criticised market definition's 'all-or-nothing' mentality, which tries to impose strict boundaries around relevant markets when in fact there can be none.[317] In 2014, he acknowledged that '[t]he concrete antitrust-law analyses of every country are virtually always market-oriented' as the legal analysis heavily relies on factors to do with the particular market, which in turn presupposes 'that markets can be defined non-arbitrarily'.[318] The latter, he argues, is a misguided assumption.[319]

Louis Kaplow developed another critique of market definition in a series of articles published since 1982.[320] His argument has evolved from criticising market 'redefinition' (as occurs under the SSNIP test) and the market definition/market share paradigm to a general critique of antitrust market definition as such. He focused on the role of market definition in assessing market power, while disregarding other functions of market definition.[321] In a 2010 article, he argued that competition law's inference of market power from market shares is fundamentally flawed, including when market share thresholds are relied upon.[322] He held that 'there is no way to define relevant markets in the first instance that does not presume the conclusion' as to the prevailing market power, and it would thus be best to forego market definition altogether.[323] He acknowledged that using a

[316] RS Markovits, 'Predicting the Competitive Impact of Horizontal Mergers in a Monopolistically Competitive World: A Non-Market-Oriented Proposal and Critique of the Market Definition-Market Share-Market Concentration Approach' (1978) 56 *Texas Law Review* 587, 590.

[317] RS Markovits, 'International Competition, Market Definition, and the Appropriate Way to Analyze the Legality of Horizontal Mergers under the Clayton Act: A Positive Analysis and Critique of Both the Traditional Market-Oriented Approach and the Justice Department's Horizontal Merger Guidelines' (1988) 64 *Chicago-Kent Law Review* 745, 788–90.

[318] RS Markovits, *Economics and the Interpretation and Application of US and EU Antitrust Law*, vol I (Berlin, Springer, 2014) 165.

[319] ibid.

[320] L Kaplow, 'The Accuracy of Traditional Market Power Analysis and a Direct Adjustment Alternative' (1982) 95 *Harvard Law Review* 1817.

[321] PC Carstensen, 'Introduction' (2012) 57 *Antitrust Bulletin* 655, 656.

[322] L Kaplow, 'Why (Ever) Define Markets?' (2010) 124 *Harvard Law Review* 438, 459 ff, 498.

[323] ibid 515.

'market metaphor' can be useful from time to time, but believed that this only applies outside of market power assessments.[324]

In the eyes of Mark Lemley and Mark McKenna, 'market definition draws an arbitrary line when what we need is a continuum' along which partial substitutability amongst products and geographical areas can be shown.[325] And David Evans believes that while market definition will remain an important part of antitrust analysis 'in the sense of understanding the environment in which a firm operates', economics provides no basis for the drawing of steadfast market boundaries or the strict reliance on market shares.[326]

Scepticism towards market definition can also be found at the competition authorities. In its Market Definition Notice of 1997, the European Commission held that no precise market definition needs to be adopted in cases in which neither of the possible market definitions gives rise to any antitrust concern.[327] And the 2010 US Horizontal Merger Guidelines do not require exact market boundaries,[328] thus catering to the blurry lines of actual markets.

There have also been a number of important rebukes to these attacks on market definition. Some of the counter-arguments can be understood as a direct criticism of non-market-oriented analysis, others as stemming from legal pragmatism or from a teleological and historically informed reading of the legal provisions. Some of these will be highlighted in the following.

Ian Ayres argued that while Markovits' critique was compelling, it lost momentum when faced with the Horizontal Merger Guidelines, which he regarded as a step forward over the previous substitutability test.[329] From a realistic perspective, Herbert Hovenkamp has cautioned that despite the value in many of Markovits' criticisms of market definition, 'one should not expect courts to drop market-definition and market-share measures in antitrust cases anytime soon'.[330] Similarly, two German economists warned that 'the notion of the relevant market may be intrinsically embedded in competition law analysis, in which case abandoning it would be extremely difficult' and could impede legal certainty.[331] Likewise, Daniel Crane found that if antitrust market definition were to fall, 'so does the entire structure of analysis built on top of it – which is to say, a whole lot

[324] L Kaplow, 'Market Definition and the Merger Guidelines' (2011) 39 *Review of Industrial Organization* 107, 110 fn 9.

[325] Lemley and McKenna (n 58) 2098. On the arbitrariness of drawing lines on the continuum of market definition, see already Maisel (n 28) 49.

[326] DS Evans, 'Lightening up on Market Definition' in E Elhauge (ed), *Research Handbook on the Economics of Antitrust Law* (Cheltenham, Edward Elgar, 2012) 89.

[327] EU Market Definition Notice 1997 (n 151) para 27.

[328] US Horizontal Merger Guidelines 2010 (n 135) § 4.

[329] I Ayres, 'A Private Revolution: Markovits and Markets' (1988) 64 *Chicago-Kent Law Review* 861, 864.

[330] HJ Hovenkamp, 'Reimagining Antitrust: The Revisionist Work of Richard S Markovits' (2016) 94 *Texas Law Review* 1221, 1237.

[331] Schwalbe and Maier-Rigaud (n 16) 84.

of antitrust law'.[332] Essentially, therefore, the relevant market is seen as the basis for so many aspects of antitrust analysis that it is difficult to imagine the latter without the former. Abandoning market definition might require a complete reconceptualisation of antitrust analysis, also at the level of the legal provisions. In other words, market definition is currently too big to fail.

Malcolm Coate and Joseph Simons have suggested that market definition can serve as a useful 'organizing structure for the economic analysis' that the assessment of competitive effects requires.[333] As was seen, this is how the European Commission has organised some of its recent merger cases. They also argue that Kaplow and others represent post-Chicago attempts at eroding the significance of market definition by relying on unilateral effects models that, in themselves, have not been proved to be reliable.[334]

Gregory Werden has shown that the significance of market definition goes beyond market power, as it includes the ability of the relevant market to convey the essence of the anti-competitive nature of the behaviour in question.[335] A further trio of commentators regard Kaplow's critique as being 'based on a misperception of the purpose and practice of market definition, an overstatement of the flaws of market definition, and an overstatement of the workability of the direct measurement of market power'.[336] Market definition, they assert, provides an important background in order to analyse whether a given behaviour is indeed anti-competitive.[337]

B. Market Definition as a Legal Concept

The present study regards the relevant market as highly relevant to competition law. The criticisms of antitrust market definition outlined above approach the issue of market definition from a purely economics perspective and lose sight of the fact that the relevant market is a legal concept that antitrust law relies upon for a variety of purposes.[338] As a legal concept, it has a lot to contribute to antitrust legal analysis. In the following, a European counter-point of view on the need for market definition in antitrust law is formulated. As a caveat, it should be underlined that this is merely *one* possible European standpoint; it is not purported that this point of view represents all European competition law scholars. In addition,

332 Crane, 'Market Power' (n 107) 33.

333 Coate and Simons (n 32) 669.

334 ibid 668 f.

335 Werden, 'An Answer' (n 300) 741.

336 D Cameron, MA Glick and D Mangum, 'Good Riddance to Market Definition?' (2012) 57 *Antitrust Bulletin* 719, 719.

337 ibid 741.

338 See VHSE Robertson, 'The Relevant Market in Competition Law: A Legal Concept' (2019) 7 *Journal of Antitrust Enforcement* 158.

some US scholars will also subscribe to this point of view. This European critique will be applied as a counter-argument to Kaplow's and Markovits' contributions, showing that antitrust law and antitrust economics may well need to be more carefully disentangled than is generally thought.

i. Economics and Antitrust Law

Economists and lawyers agree that antitrust decision-making is enriched by market definition because it allows enforcers to understand the market beyond what econometric models can show.[339] While the purposes of market definition have changed over time, this analytical step has acquired new roles, such as the 'identification of competitive forces and the description of the competitive landscape'.[340] Linking back to the legal provisions, Rupprecht Podszun asserts that 'from a pure legalistic European perspective, it is impossible to avoid market definition'.[341] Despite some commentators' assertions to the contrary, the same holds true for US antitrust law.[342]

The critiques of market definition outlined above approach their object of contention through the lens of economics. In doing so, they miss a key point: antitrust market definition is an integral part of the *legal* analysis. While economics may contribute to an enhanced understanding of markets, this does not necessarily mean that there is no friction between economics and antitrust law, and how the two disciplines conceptualise competition and its regulation. Indeed, in many instances there will be quite important discrepancies. While the more economic approach to competition law largely dominated EU competition law in the 2000s, there now is a growing perception that antitrust law might need to re-evaluate its relationship with economics. This is particularly so in light of the legal uncertainty which the more economic approach has infused EU competition law with.[343] While legal scholarship has long accepted that market power should not necessarily be deduced from high market shares, there remain antitrust issues that can only properly be understood within the context of a relevant market. Werden has referred to this as 'narrative',[344] but arguably this also involves developing a coherent approach to the theory of harm.

[339] WH Boshoff, 'Why Define Markets in Competition Cases?' (Stellenbosch Economic Working Papers 10/2013) 18; D Zimmer, 'The Emancipation of Antitrust from Market-Share-Based Approaches' (2016) 61 *Antitrust Bulletin* 133, 137.

[340] Zimmer (n 339) 137, 147, 150 (direct quote).

[341] Podszun (n 1) 128. See also S Marco Colino, *Competition Law of the EU and UK*, 8th edn (Oxford, Oxford University Press, 2019) 21.

[342] For the analysis of US law to this effect, see section I.

[343] On the legal uncertainty that the different approaches by the CJEU and the European Commission have adduced, see AC Witt, *The More Economic Approach to EU Antitrust Law* (Oxford, Hart Publishing, 2016) 309.

[344] Werden, 'An Answer' (n 300) 740.

With reference to economic theories, Kaplow asserts that 'the concept of a relevant market does not exist in industrial organization economics'.[345] This finding is very much in line with the argument that the relevant antitrust market is a *legal* concept that has some of its roots in economics, but cannot be expressed in purely economic terms.[346] The question, then, is who is in a better position to carry out the legal tests of market definition. While some judges feel confident in delineating markets,[347] others will ask economic experts to assist them, particularly where empirical data is required to carry out the market definition. For instance, in Austrian competition law, market definition is seen as a matter of law insofar as it is concerned with an assessment of the method used for the market definition.[348] Otherwise it is considered a question of fact. As one author has criticised, US antitrust law 'has had to flounder around in the … mists of legal-economic confusion' ever since the US Supreme Court intertwined economics-based and legal tests of market definition.[349] What competition law needs to establish is not the market as it makes sense to economists, but the antitrust market as it is relevant for antitrust law. That is the essence of the construct of the relevant antitrust market.

From a comparatist's perspective, it is notable that many US antitrust scholars base their views on economics rather than on a legal theory of antitrust. One reason for this strong emphasis on economics in US antitrust scholarship might be that US antitrust law, which is predominantly privately enforced, might prove more fruitful a ground for the development of alternatives to market definition in order to support antitrust claims.[350] Another reason might be the fact that most legal academics who have joined the antitrust market definition debate from the US side hold a PhD or an undergraduate degree in economics, while in European legal academia they generally hold a PhD in law.[351] The reason for this divide can be seen in the different systems of higher legal education prevailing in these two jurisdictions.[352] This important difference may account for differing views on economics-related antitrust law issues across the Atlantic. The academic training

[345] L Kaplow, 'Market Definition Alchemy' (2012) 57 *Antitrust Bulletin* 915, 918. Nevertheless, econometrics and other economic methods have been used in order to delineate markets; see, eg DT Scheffman and PT Spiller, 'Econometric Market Delineation' (1996) 17 *Managerial and Decision Economics* 165.

[346] See Robertson, 'The Relevant Market' (n 338) 175.

[347] Here it is relied upon a comment by a German judge (2015) who spoke on the condition of anonymity.

[348] Austrian Supreme Court, *RIS-Justiz RS0124421* AT:OGH0002:2008:RS0124421.

[349] Carstensen (n 321) 659.

[350] Schwalbe and Maier-Rigaud (n 16) 84; WE Kovacic, 'The Influence of Economics on Antitrust Law' (1992) 30 *Economic Inquiry* 294, 295.

[351] For this, the CVs of Louis Kaplow (Professor of Law and Economics; PhD in economics from Harvard University), Richard Markovits (Professor of Law; PhD in economics from the London School of Economics) and Gregory Werden (PhD in economics from the University of Wisconsin) were sifted.

[352] Law is an undergraduate degree in Europe, while it is a graduate degree in the US, where it is generally taught as a professional doctorate (Juris Doctor or JD) upon completion of an undergraduate degree.

of these scholars quite naturally has an impact on their analysis of legal issues, as does the law and economics movement which has been very influential in the US and less so in the EU – possibly for the very same reasons. This opens up entire avenues of comparative research possibilities that can merely be hinted at here.

ii. A European Point of View on Market Definition

From a European perspective, those who argue in favour of abandoning market definition may misjudge its importance for competition law.[353] One can perceive a certain consensus amongst the European antitrust community that market definition is so deeply embedded in EU competition law – as well as in the national competition laws of the EU Member States – that it constitutes an indispensable element of competition law doctrine. Without any doubt, many US scholars share this view regarding their own antitrust laws. In addition, in Europe it is held that economics 'is deeply rooted in the foundations of the Lisbon Treaty', but the more economic approach 'was never supposed to change the rule of law'; it is the legal system which represents 'the anchor of the European approach'.[354] In Europe, the application of economic theory to antitrust law is seen much more sceptically than in the US.[355] On the other hand, the power of legal analysis in competition law cases is often underestimated.[356] Particularly in areas in which there is no consensus among economists about which economic theory best describes economic reality, one cannot expect competition law to evaluate the credibility of different strands of economics.

The legal framework for market definition may rely on insights from economics, but it nevertheless pertains to the legal sphere.[357] This legal concept of the market depends on legal semantics much more than on economics.[358] Although some argue that antitrust markets today cannot be delineated without the support of an economic expert,[359] it remains a legal issue that judges and competition lawyers need to fully grasp in greater depth. While it certainly represents a major challenge to translate economic market definition concepts into law,[360] competition law may have an autonomous understanding of what an antitrust relevant market entails and how it should be delineated. Therefore, it is essential that legal

[353] Bishop and Walker (n 198) para 4-025.

[354] Hildebrand (n 217) 33 (direct quotes from the main text and from fn 94 on that page).

[355] P Ibáñez Colomo, 'Beyond the "More Economics-Based Approach": A Legal Perspective on Article 102 TFEU Case Law' (2016) 53 *CML Rev* 709, 711.

[356] ibid 738.

[357] M Sousa Ferro, '"Ceci n'est pas un marché": Gratuity and Competition Law' [2015(1)] *Concurrences* 1, para 83; Robertson, 'The Relevant Market' (n 338) 175.

[358] A Früh, *Immaterialgüterrechte und der relevante Markt: Eine wettbewerbsrechtliche und schutzrechtliche Würdigung technologischer Innovation* (Cologne, Carl Heymanns, 2012) 37.

[359] Hovenkamp et al (n 295) § 4.03[A][1].

[360] WH Boshoff, 'Market Definition as a Problem of Statistical Inference' (2014) 10 *Journal of Competition Law & Economics* 861, 862.

academics and practitioners deal with antitrust market definition in detail, solving the many legal issues involved for the sake of the rule of law and legal certainty.[361] At a bare minimum, competition lawyers must have enough insight into market definition so as to determine whether the economic experts remain 'within the legal bounds already established by judicial or administrative action'.[362]

Some important insights can be gained from the criticism levied at market definition. For one thing, market share thresholds are an arbitrarily drawn line, as so much depends on the specific circumstances of the case. Despite this, market share thresholds can continue to establish safe harbours in the form of rebuttable presumptions,[363] thereby providing some general guidance for the practical application of competition law. Similarly, the European *AKZO* presumption, according to which companies with a market share of at least 50 per cent are presumed to be dominant, is a rebuttable presumption that does not necessarily need to be changed. The rebuttable nature of these presumptions means that where the market shares at issue clearly do not reflect the (absence of) market power which they are intended to reflect, the presumption can be rebutted. A broader lesson to take away is that, particularly within the framework of innovative markets that this study is concerned with, it will be worthwhile to reflect on the aims and purposes of market definition, asking what the analytical tool of market definition is used for and how accurate and reliable it must be in order to fulfil its functions. Another path is to consider ways in which antitrust can become less dependent on market definition as an analytical tool.[364] This latter avenue is usually economics-based and should be assessed with care in order to consolidate it with legal antitrust doctrines. A proper legal theory, at least from a European point of view, should be informed by economics but nevertheless remain within the legal framework. An approach that is exclusively grounded in economics, no matter how persuasive, can ultimately not produce the kind of legal rules and concepts that a legal framework necessarily needs to consist of in order to satisfy constitutional principles and in order to promote the goals that the competition laws should aim at from a regulatory point of view.

Market definition constitutes an important analytical step that provides the playing field within which the competitive effects of market behaviour are assessed.[365] In order to allow the referees to do their work, it needs to be as predictable – ie, legally certain – as possible. It would be short-sighted to assert that market definition's only purpose is to determine market power. Its market characterisation function should not be underestimated.

[361] See Sousa Ferro, 'Gratuity and Competition Law' (n 357) para 83.
[362] Kauper (n 246) 1682.
[363] This is also suggested by Zimmer (n 339) 150.
[364] Markovits, *US and EU Antitrust Law*, vol I (n 318); RS Markovits, *Economics and the Interpretation and Application of US and EU Antitrust Law*, vol II (Berlin, Springer, 2014).
[365] Turner (n 5) 1145; Ezrachi, *EU Competition Law* (n 225) 35.

Another lesson that perhaps needs to be learned is that antitrust analysis should not exclusively rely on market definition so as to ensure that antitrust assessments do not collapse like a sandcastle if the market definition has gone wrong. In other words, the foundation that the relevant market often represents in antitrust law cases needs to be as strong as possible, *and* competition lawyers should rely on additional pillars that do not stand and fall with a particular market definition, but that fit well into a market's (dynamic) characteristics. In a way, this approach reconciles purely economics-based approaches with purely market-oriented approaches. It is a challenging task, but one with important rewards.

V. Conclusion

The relevant product market represents a legal concept that shares its roots with economics, but that equally takes on a specific legal meaning when incorporated into competition law. Under the EU and US antitrust provisions, the relevant market fulfils two main functions: it serves as the basis to assess market power and it allows the characterisation of a market. Under the market power function, the competition rules rely on market definition in order to inform a structural analysis. For instance, courts rely on market shares in order to decide on issues such as market dominance or monopolisation, or they use them as the basis to calculate concentration ratios within the framework of merger control. In this area, forceful scholarly objections to market definition have been levied. Under the market characterisation function, the relevant market allows enforcers to understand the market environment and its (actual and potential) participants, as well as to grasp how the theory of harm relates to the specific market in question. Furthermore, it provides some much-needed background to the competition case and allows for insights into the effects of market behaviour.

Depending on the antitrust violation or the merger in question, a competition law case will have a certain focal point that also extends to market delineation. While some antitrust enquiries are ex post analyses, others concern future developments; this outlook can also impact the market delineation. In general, however, there is no legal basis for relying on a different methodology when delineating relevant markets for different antitrust violations, as this can increase market definition's arbitrariness and lead to legal uncertainty. This is most visible in cases in which one and the same market behaviour raises questions under several antitrust provisions.

As the relevant market is a legal concept that is required for a legally sound antitrust analysis, competition law will continue to rely on market definition and, in particular, on its market characterisation function, both in the EU and US. In addition, the relevant market provides the necessary grounding in reality that furthers our understanding of the competition issues at stake.

3

Innovation and Competition Law

In both the EU and the US, the analytical framework for delineating antitrust markets was developed in static market environments with largely homogeneous goods, a constant number of producers, modest rates of innovation and stable demand.[1] This raises the question of whether market definition can cope with the specific characteristics of dynamic market contexts. Market definition allows but a snapshot of the current market situation and is susceptible to considerable change within a short period of time.[2] Some have therefore called traditional antitrust market definition 'inadequate … in knowledge-intensive high technology industries',[3] while others have urged that it must be adapted where innovation is at issue.[4]

In the following, we will establish what particular understanding of innovation lies at the heart of the present study. Then, we will engage in a discussion of the characteristics of dynamic markets that make market definition such a 'daunting task'[5] in the face of high-innovation industries. Dynamic markets display a multitude of features that, in their combination, constitute the real challenge of market definition in these contexts.[6] The discussion of these features is followed by a reflection on the incorporation of innovation considerations into the antitrust legal framework.

Before embarking on this enquiry, a caveat is in order: the aim of the present study is *not* the promotion of innovation as such. While the promotion of

[1] JG Sidak and DJ Teece, 'Dynamic Competition in Antitrust Law' (2009) 5 *Journal of Competition Law & Economics* 581, 602 f; RA Posner, *Economic Analysis of Law*, 9th edn (New York, Wolters Kluwer, 2014) 388.

[2] A Fatur, *EU Competition Law and the Information and Communication Technology Network Industries: Economic versus Legal Concepts in Pursuit of (Consumer) Welfare* (Oxford, Hart Publishing, 2012) 116.

[3] C Pleatsikas and DJ Teece, 'The Analysis of Market Definition and Market Power in the Context of Rapid Innovation' (2001) 19 *International Journal of Industrial Organization* 665, 666.

[4] See, eg, OECD Secretariat, 'Executive Summary' in OECD (ed), *Policy Roundtable: Merger Review in Emerging High Innovation Markets* (2002) DAFFE/COMP(2002)20, 8; Antitrust Modernization Commission, 'Report and Recommendations' (April 2007) 31.

[5] DA Balto and R Pitofsky, 'Antitrust and High-Tech Industries: The New Challenge' (1998) 43 *Antitrust Bulletin* 583, 584.

[6] See also DS Evans, 'The Middle Way on Applying Antitrust to Information Technology' [2009] *Antitrust Chronicle* 1, 3; H Nevo, *Definition of the Relevant Market: (Lack of) Harmony between Industrial Economics and Competition Law* (Cambridge, Intersentia, 2015) 86.

innovation may constitute an important goal of competition policies,[7] this study asks how antitrust markets are, can and should be defined in the presence of innovation, under EU and US antitrust law. A competition law framework that actively wants to promote innovation may use such an innovation-conscious market definition framework as a starting point, but would then need to go much further.

I. The Notion of Innovation

The nature of innovation is difficult to grasp. Innovation has been described as 'the creation of something qualitatively new, via processes of learning and knowledge building'.[8] According to the *Oxford English Dictionary*, innovation is 'the introduction of novelties', which not only encompasses standalone innovations but also innovations that build on already-existing ones.[9] In a commercial sense, innovation stands for the marketing of a new product. Joseph Schumpeter once asserted that innovation and invention were not necessarily interconnected, as innovation could exist without a proper invention, and not every invention leads to an innovation.[10] In a similar vein, innovation can be seen as 'the bringing of an invention into widespread, practical use'.[11] In this sense, an invention is often just the first step towards innovation.

The definition of innovation relied upon in the present context is based on the OECD's Oslo Manual, which regards innovation as 'the implementation of a new or significantly improved product (good or service), or process, a new marketing method, or a new organisational method in business practices, workplace organisation or external relations'.[12] This broad definition captures a wide range of innovations from many different areas, from modest improvements to downright technical revolutions. The OECD distinguishes four main types of innovation, namely product innovations, process innovations, marketing innovations and organisational innovations.[13] By product innovation, the Manual understands new or significantly improved products. Novelty or improvement may relate to the product's characteristics or its intended use. This includes features that relate to the product's 'technical specifications, components and materials, incorporated

[7] See M Glader, 'Innovation Economics: The Antitrust Guidelines on Horizontal Co-operation' (2001) 24 *World Competition* 513, 522 f.

[8] K Smith, 'Measuring Innovation' in J Fagerberg, DC Mowery and RR Nelson (eds), *The Oxford Handbook of Innovation* (Oxford, Oxford University Press, 2006) 149.

[9] 'innovation, n.' in *Oxford English Dictionary Online* (Oxford, Oxford University Press), www.oed.com/view/Entry/96311.

[10] JA Schumpeter, *Business Cycles: A Theoretical, Historical, and Statistical Analysis of the Capitalist Process*, vol I (New York, McGraw-Hill, 1939) 84.

[11] JA Allen, *Scientific Innovation and Industrial Prosperity* (Amsterdam, Elsevier, 1967) 8.

[12] OECD and Eurostat (eds), 'Oslo Manual: Guidelines for Collecting and Interpreting Innovation Data' (2005) para 146.

[13] ibid para 155.

software, user friendliness or other functional characteristics'.[14] Process innovations, then, relate to new or significantly improved methods of production or of delivery. This improvement can relate to anything ranging from techniques to equipment or software.[15] In the digital environment, business model innovations have come to play a particularly important role.[16]

Depending on their impact, innovations can be characterised as disruptive or sustaining. Disruptive innovations have an unsettling impact on current technologies, products or processes, while sustaining innovations build on the status quo.[17] Disruptive innovations outperform and often replace existing innovations, for they are 'typically cheaper, simpler, smaller, and, frequently, more convenient to use' than already-existing products.[18] An example of this can be seen in the music business, where the digital economy has produced disruptive innovations that have had a profound influence on this industry.[19] One can classify both disruptive and sustaining innovations as subsequent, as they build on prior innovations. While disruptive innovations are substitutive, ie, they replace the prior innovation, sustaining innovations are derivative because they incorporate and improve prior innovations.[20]

One can also distinguish between radical and incremental innovation, based on the changes that the innovation brings about. In incremental or evolutionary innovation, changes over the previous state of the art are only gradually and slowly introduced, and the changes thus arguably take place within the same antitrust market.[21] At the same time, such incremental innovations may slowly change the antitrust market itself. Radical innovation, on the other hand, overthrows the current system,[22] thereby leading to the creation of new demand and possibly a new market in the antitrust sense. Further categorisations are possible,[23] and an innovation may fall into several of these categories. In the present context, these categorisations primarily serve to illustrate the wide range of innovations that are possible.

[14] ibid para 156.

[15] ibid para 163.

[16] H Schweitzer, 'Preserving Digital Innovation through Competition Policy' (DG Competition Conference on 'Shaping Competition Policy in the Era of Digitisation', Brussels, 17 January 2019).

[17] JL Bower and CM Christensen, 'Disruptive Technologies: Catching the Wave' (1995) 73 *Harvard Business Review* 43, 45.

[18] CM Christensen, *The Innovator's Dilemma: The Revolutionary Book that Will Change the Way You Do Business* (New York, Collins, 2003) xviii.

[19] OB Arewa, 'YouTube, UGC, and Digital Music: Competing Business and Cultural Models in the Internet Age' (2010) 104 *Northwestern University Law Review* 431, 462.

[20] G Ghidini, *Innovation, Competition and Consumer Welfare in Intellectual Property Law* (Cheltenham, Edward Elgar, 2010) 69.

[21] J Drexl, 'Anticompetitive Stumbling Stones on the Way to a Cleaner World: Protecting Competition in Innovation without a Market' (2012) 8 *Journal of Competition Law & Economics* 507, 513.

[22] See MA Carrier, *Innovation for the 21st Century: Harnessing the Power of Intellectual Property and Antitrust Law* (Oxford, Oxford University Press, 2009) 27.

[23] eg, the distinction between user and manufacturer innovation or between discrete and complex innovation; E von Hippel, *Democratizing Innovation* (Cambridge, MA, MIT Press, 2005) 8; Carrier, *Innovation* (n 22) 26 ff.

II. Distinctive Features of Innovative Markets as Challenges for Antitrust Market Definition

A number of industries display high rates of innovation: digital market environments with their multi-sided platforms, new business models and app-based economy; R&D-driven pharmaceuticals; high-technology industries with their standard-essential patents, software innovations and hardware advancements. In addition, digitisation is impacting every sector, from farming to care for the elderly. Chapter 4 adds many more examples to these. While innovative markets are highly diverse, they also possess a number of characteristics that are markedly different from the more static and homogeneous goods and services markets that traditional market definition had in mind:[24] They are fast-moving with competition working in a more Schumpeterian fashion, intellectual property rights are often key, product differentiation plays an important role, price is no longer the most important competitive parameter, network effects are at play and multi-sided platforms are an issue to be considered. These features represent a stylised dynamic market environment and are not present in every single innovative market, but they are frequently present in various combinations. These characteristics are further analysed below, highlighting the particular challenges they pose for the basic pillars of antitrust market definition as outlined in Chapter 2. Part II then uncovers ways in which EU and US competition laws take these characteristics of dynamic markets into account when delineating antitrust markets.

A. Features of the Innovative Market

Markets that exhibit high innovation are frequently fast-moving and have short product cycles.[25] An established market order can quickly be succeeded by a new market order, and market boundaries may need to be questioned on an ongoing basis,[26] with new aspects entering the analytical framework and others becoming superfluous. The notion of innovation contains an element of constant flux, which has unsettling consequences for market definition. The more iterative the process of innovation is,[27] the less certain can antitrust market definition be. This necessarily influences the extent to which antitrust legal analysis can build upon static

[24] See G Hewitt, 'Background Note' in OECD (ed), *Policy Roundtable: Merger Review in Emerging High Innovation Markets* (2002) DAFFE/COMP(2002)20, 20 f (containing further references); RC Lind, P Muysert and M Walker, 'Innovation and Competition Policy: Part I – Conceptual Issues' (OFT377, Economic Discussion Paper 3, March 2002) § 3.20; Antitrust Modernization Commission (n 4) 32 f.

[25] Hewitt (n 24) 20.

[26] E Mackaay, *Law and Economics for Civil Law Systems* (Cheltenham, Edward Elgar, 2013) 144.

[27] S Ranchordas, 'Innovation Experimentalism in the Age of the Sharing Economy' (2015) 19 *Lewis & Clark Law Review* 871, 882.

market definition. Of course, it is not only market delineations which shift, but old market power may also be replaced by new market power at short notice.[28] Economists have urged that competition policy need not be overly concerned with such market power in dynamic markets, as it will sooner or later be eroded by new innovations.[29]

While timing is generally of the essence in business, this is particularly true where technological innovation is concerned.[30] Incremental innovation occurring in a market will sometimes be caught off-guard by radical innovation that can have a disruptive impact on the market and can lead to paradigm shifts in the industrial and technological sphere.[31] Also, an innovator successful in one round of innovation might not fare so well in the next.[32] Hand in hand with this more dynamic conception of competition, once applicable market definitions will be overthrown and will need to be replaced by new ones. This requires changes in the analytical framework for market definition,[33] and a clear approach to dealing with this ever-present flexibility.

A common feature in innovative markets is the inherent uncertainty that comes with an innovative environment. This uncertainty stems from the fact that input decisions need to be taken at a time when the outcome deriving from that input cannot yet be adequately foreseen.[34] This insight is also closely connected to the bounded rationality that behavioural economics has at its heart, which refers to a state in which 'agents have an imperfect understanding of the environment they live in, and what the future will deliver'.[35] Behavioural economics thus proposes a counterpoint to the perfectly rational and informed *homo economicus*. There are numerous reasons why firms may not act as the profit-maximising players they are often set out to be.[36] This, in turn, undermines many antitrust theories of harm that assume firms' rationality. At the same time,

[28] R Pitofsky, 'New Definitions of Relevant Market and the Assault on Antitrust' (1990) 90 *Columbia Law Review* 1805, 1812.

[29] C Caffarra and B Bishop, 'Dynamic Competition and Aftermarkets' (1998) 19 *European Competition Law Review* 265, 266.

[30] ibid 265.

[31] DJ Teece and M Coleman, 'The Meaning of Monopoly: Antitrust Analysis in High-Technology Industries' (1998) 43 *Antitrust Bulletin* 801, 804, 808.

[32] R Hartman et al, 'Assessing Market Power in Regimes of Rapid Technological Change' (1993) 2 *Industrial and Corporate Change* 317, 317.

[33] U Schwalbe and F Maier-Rigaud, 'Background Note' in OECD (ed), *Policy Roundtable: Market Definition* (2012) DAF/COMP(2012)19, 57 (speaking of 'conceptual difficulties' in this respect).

[34] KJ Arrow, 'Economic Welfare and the Allocation of Resources to Invention' in National Bureau of Economic Research (ed), *The Rate and Direction of Inventive Activity* (Princeton, Princeton University Press, 1962) 610.

[35] DJ Teece, 'Favoring Dynamic over Static Competition: Implications for Antitrust Analysis and Policy' in GA Manne and JA Wright (eds), *Competition Policy and Patent Law under Uncertainty: Regulating Innovation* (Cambridge, Cambridge University Press, 2011) 214.

[36] See M Armstrong and S Huck, 'Behavioral Economics and Antitrust' in RD Blair and DD Sokol (eds), *The Oxford Handbook of International Antitrust Economics*, vol 1 (Oxford, Oxford University Press, 2015).

this understanding befits innovative environments that are characterised by inherent uncertainty. While some market conditions are said to be more conducive to innovation than others,[37] innovation as such remains highly uncertain. Once an innovation has occurred, one cannot be certain of its success either: one patented invention might turn into a blockbuster, while another will go unnoticed. And technical complexity in itself can lead to considerable uncertainty.[38] Overall, the instability of innovative markets is sometimes seen as the major challenge for market definition.[39]

Innovation is confronted with many factors of uncertainty, including acceptance uncertainty, commercial uncertainty, legal uncertainty and techn(olog)ical uncertainty.[40] Companies can typically not forecast the expected returns on their R&D investment.[41] With the ultimate outcome of many R&D efforts – and their subsequent success on the market – highly uncertain, this makes it very difficult to reliably delineate future markets or to single out potential competitors. Uncertainty also has a bearing on how competition policy is formulated. For instance, in industrial organisation, economists adhering to the Austrian School believe that 'uncertainty makes the perfectly competitive model irrelevant as a welfare ideal'.[42] Uncertainty, of course, can also surround competition law enforcement authorities when innovative industries are concerned.[43]

B. Features of the Innovative Product

While innovative markets are fast-moving, time-sensitive and uncertain, products in these markets are often R&D-intensive, IP-protected and differentiated. R&D is at the heart of many innovations, as companies hope to bring forth innovation through their R&D endeavours – and to receive IP protection in due course.[44]

Economists view intellectual property rights 'as a policy tool to ensure adequate private returns to innovation and creative activities'.[45] IPRs allow innovators to

[37] On the conditions that are advantageous for innovation, see Teece, 'Favoring Dynamic over Static Competition' (n 35) 215, 219.

[38] Lind, Muysert and Walker (n 24) § 3.58.

[39] Europe Economics, 'The Development of Analytical Tools for Assessing Market Dynamics in the Knowledge Based Economy' (12 September 2003) 39.

[40] H Jalonen, 'The Uncertainty of Innovation: A Systematic Review of the Literature' (2011) 4 *Journal of Management Research* 1, table 1 (at 11).

[41] Lind, Muysert and Walker (n 24) § 3.22.

[42] J Ellig, 'Industrial Organization' in PJ Boettke (ed), *The Elgar Companion to Austrian Economics* (Cheltenham, Edward Elgar, 1994) 244.

[43] Ranchordas (n 27) 885.

[44] RA Posner, 'Antitrust in the New Economy' (2001) 68 *Antitrust Law Journal* 925, 926; DS Evans and R Schmalensee, 'Some Economic Aspects of Antitrust Analysis in Dynamically Competitive Industries' in AB Jaffe, J Lerner and S Stern (eds), *Innovation Policy and the Economy*, vol 2 (Cambridge, MA, MIT Press, 2002) 3.

[45] C Greenhalgh and M Rogers, 'The Value of Intellectual Property Rights to Firms and Society' (2007) 23 *Oxford Review of Economic Policy* 541, 541.

appropriate the gains associated with their work for a limited amount of time.[46] They are often regarded as one of the major drivers of innovation.[47] Sometimes, they are even equated with innovation itself.[48] However, this view is too short-sighted; not all innovations are protected by IPRs and not all IPRs protect highly innovative inventions. Where IPRs enable an innovator to appropriate the profits that stem from its innovation, this can seem like anti-competitive foreclosure and may lead to tensions between antitrust and IP law that need to be resolved.

IPRs can add to the complexity of market definition. A single IPR – through the innovation underlying it – can, depending on the specific circumstances, delineate the relevant product market, confer market power on its owner or overturn old market definitions. IPRs also lead to product differentiation,[49] an aspect that static market definition struggles to accommodate. This requires IPRs to be carefully factored into market definition. In addition, technical standards can play an important role in innovation-intensive markets.[50] Taking the example of standard-essential patents (SEPs), one might have to distinguish three separate product markets: the technology or licensing market on which the SEP is competing with other technologies, the standards market in which the SEP is competing with other available standards, and the actual product market in which products incorporating the SEP are competing with other available products.[51] This complicates market definition, particularly as the relevant markets are very much interconnected.

While static markets are known for their homogeneous products (eg, class I raspberries or a particular grade of steel), innovative market environments are frequently characterised by product differentiation. Product differentiation alludes to the heterogeneity of products, ie, the perceived differences in products despite their belonging to the same product category. As stated above, IPRs adduce product differentiation as they make exact copying illegal.[52] Other features of innovative markets also lead to considerable product differentiation. New technological capabilities enable increasing levels of product differentiation – ranging from personalised children's books to personalised software. The more differentiated

[46] C Shapiro, 'Competition and Innovation: Did Arrow Hit the Bull's Eye?' in J Lerner and S Stern (eds), *The Rate and Direction of Inventive Activity Revisited* (Chicago, University of Chicago Press 2012) 388.

[47] T Nicholas, 'What Drives Innovation?' (2011) 77 *Antitrust Law Journal* 787, 788.

[48] For a narrow view of the scope of innovation, limiting it to innovations protected by patents and copyright, see Carrier, *Innovation* (n 22) 20.

[49] MA Lemley and MP McKenna, 'Is Pepsi Really a Substitute for Coke? Market Definition in Antitrust and IP' (2012) 100 *Georgetown Law Journal* 2055, 2101; HJ Hovenkamp et al, *IP and Antitrust: An Analysis of Antitrust Principles Applied to Intellectual Property Law*, 3rd edn, vol I (New York, Wolters Kluwer, 2018) § 4.02[B] (IP 'guarantees at least limited product differentiation').

[50] Hewitt (n 24) 20.

[51] See further D Telyas, *The Interface between Competition Law, Patents and Technical Standards* (Alphen aan den Rijn, Kluwer Law International, 2014) 53 f.

[52] See also HJ Hovenkamp, 'Response: Markets in IP and Antitrust' (2012) 100 *Georgetown Law Journal* 2133, 2145.

a product range is, the more difficult it becomes to gauge the question of interchangeability from the point of view of the demand side, which lies at the heart of antitrust market definition.

While product differentiation might limit product interchangeability for the purposes of antitrust market definition, this is not always the case. If there is an alternative that customers would readily turn to, then even differentiated products based on entirely different technologies belong to the same relevant product market.[53] What matters is the degree to which products exercise a competitive constraint on each other. However, it is far from settled which degree of interchangeability amongst differentiated products is sufficient in order to place them in the same relevant market.

C. The Dynamics of Competition in Innovative Markets

The nature of competition is a noteworthy feature of innovative markets, in particular as it relates to network effects, multi-sided markets, performance as an important parameter of competition and competition for the market.

Networks connect users with users, and buyers with sellers. While old world examples of such networks include the telephone and fax machines,[54] virtual networks have gained prominence over the last two decades, such as internet search engines, social networking platforms and operating systems.[55] These virtual networks are typically characterised by high innovation. Apart from the examples just cited, a large number of important industries are regarded as network industries, for instance, air transport, email, the internet more generally, computer hardware, music and video players, banking services, legal services and many more. These share a set of common characteristics that relate to complementarity, compatibility and standards, consumption externalities, switching costs and lock-in effects, and significant economies of scale in production.[56]

Networks such as the ones characterised above frequently exhibit positive externalities called network effects, ie, a link amongst users where the value of that link is affected by links amongst other users.[57] The benefit to the network's users is not only determined by what services the network directly provides, but also by the

[53] Hartman et al (n 32) 334.

[54] SJ Liebowitz and SE Margolis, 'Network Effects and the *Microsoft* Case' in J Ellig (ed), *Dynamic Competition and Public Policy: Technology, Innovation, and Antitrust Issues* (Cambridge, Cambridge University Press, 2001) 160.

[55] DF Spulber and CS Yoo, 'Antitrust, the Internet, and the Economics of Networks' in RD Blair and DD Sokol (eds), *The Oxford Handbook of International Antitrust Economics*, vol 1 (Oxford, Oxford University Press, 2015) 381.

[56] O Shy, *The Economics of Network Industries* (Cambridge, Cambridge University Press, 2001) 1.

[57] GL Priest, 'Networks and Antitrust Analysis' in ABA Section of Antitrust Law (ed), *Issues in Competition Law and Policy* (Chicago, ABA Publishing, 2008) 641.

number of users that the network joins together.[58] There is a positive correlation between the network's value and the number of individuals using the network.[59] Social networking platforms are usually more valuable to a user the more users the platform can attract (direct network effect), and increasing numbers of users will, in turn, also increase the platform's attractiveness for advertisers (indirect network effect). Direct network effects therefore relate to the network's increasing value due to more users joining it. Indirect network effects occur where the network's value increases through complementary products, and this increase in value is equally driven by an increase in users.[60] This is referred to as a positive feedback loop.

By their nature, network effects can lead to high switching costs:[61] A user who leaves a network not only loses the network's services, but also his or her connection to a multitude of other users. And an adopter of a technology will have incurred considerable sunk costs that typically cannot be recuperated when switching to an alternative technology. High switching costs may, in turn, prompt a lock-in.

When a commercial environment exhibits strong network effects, this needs to be reflected in the antitrust market definition and, subsequently, in the assessment of market power. The links that network effects create cannot simply be ignored when delineating an antitrust market. This would mischaracterise the market under investigation and lead to erroneous findings as to product substitutability. However, one important challenge in network-driven markets is that network effects cannot be measured accurately.[62]

A special case of a market with strong network effects, both direct and indirect, are two- or multi-sided markets. Examples of multi-sided markets can be found in both the online and the offline realms: credit cards, hotel booking platforms, internet search engines, newspapers partially or exclusively sustained through advertising, operating software platforms, social networking platforms, video gaming platforms etc.[63] As the example of newspapers shows, these types of markets are not new. However, the digital environment has led to a renaissance of multi-sided markets. These consist of two or more product markets that are connected in such a way that it would not make sense to consider one market without simultaneously taking the other into account.[64] A particular feature of many digital platforms is that one market side – typically the user side – receives

[58] Spulber and Yoo (n 55) 385.

[59] Balto and Pitofsky (n 5) 587.

[60] Lind, Muysert and Walker (n 24) § 3.24.

[61] Teece and Coleman (n 31) 828; Nevo (n 6) 86.

[62] Priest (n 57) 645.

[63] DS Evans and R Schmalensee, 'The Antitrust Analysis of Multisided Platform Businesses' in RD Blair and DD Sokol (eds), *The Oxford Handbook of International Antitrust Economics*, vol 1 (Oxford, Oxford University Press, 2015) 404 f.

[64] DS Evans, 'Two-Sided Markets' in ABA Section of Antitrust Law (ed), *Market Definition in Antitrust: Theory and Case Studies* (Chicago, ABA Publishing, 2012) 437 ff.

certain services (eg, hotel booking, online search, social networking) ostensibly for free, while the other market side (eg, advertisers, hotels, online retailers) pays for the entire service. This also raises interesting questions as to whether a 'free' market side can constitute an antitrust market. It should be noted that users usually pay in kind – with their personal data and their attention to advertising. Either market side of a multi-sided market might be competing with different players. For instance, a social platform will compete for advertising revenue on the one side and for social media users on the other. Based on the network effects at play, the tipping effect, cross-subsidisation and practices aimed at expanding the network, these markets will often be quite concentrated.[65] Against this background, competition policy needs to assess whether or not such situations require antitrust intervention[66] and, if so, at what point in time intervention is opportune.

Traditionally, price and quality rank highest as competitive parameters and are therefore the focus when defining relevant markets.[67] In innovation-driven industries, price is often not the most important competitive parameter, while innovation and performance can prove essential.[68] The development of new or improved products or processes, particularly through new technologies, becomes a much more important competitive factor.[69] In such markets, consumers will be best off if companies continue their innovative efforts.[70]

In innovative markets, companies frequently 'compete *for* a future market and not *in* an existing market'.[71] This means that competition does not concentrate on the products – goods or services – that are currently forming part of the market, but relates to future or potential products. Such 'winner takes all'[72] scenarios occur because of the tipping effect associated with networks, meaning that demand concentrates on the successful competitor,[73] who in turn expects high profits.[74] These markets are called Schumpeterian,[75] as Schumpeter argued that dynamic

[65] Priest (n 57) 665.

[66] See A Italianer, 'Innovation and Competition' in B Hawk (ed), *International Antitrust Law & Policy* (Huntington, Juris Publishing, 2013) 319.

[67] GJ Stigler and RA Sherwin, 'The Extent of the Market' (1985) 28 *Journal of Law and Economics* 555, 555; Lemley and McKenna (n 49) 2058; BR Kern, 'Innovation Markets, Future Markets, or Potential Competition: How Should Competition Authorities Account for Innovation Competition in Merger Reviews?' (2014) 37 *World Competition* 173, 174.

[68] TM Jorde and DJ Teece, 'Competing through Innovation: Implications for Market Definition' (1988) 64 *Chicago-Kent Law Review* 741, 742; PA Geroski, 'Competition in Markets and Competition for Markets' (2003) 3 *Journal of Industry, Competition and Trade* 151, 159.

[69] Hartman et al (n 32) 319.

[70] P Crocioni, 'Leveraging of Market Power in Emerging Markets: A Review of Cases, Literature, and a Suggested Framework' (2007) 4 *Journal of Competition Law & Economics* 449, 451.

[71] Drexl, 'Anticompetitive Stumbling Stones' (n 21) 510 (emphasis in original).

[72] Schwalbe and Maier-Rigaud (n 33) 58.

[73] Lind, Muysert and Walker (n 24) § 3.30.

[74] Teece and Coleman (n 31) 811.

[75] For instance, see R Schmalensee, 'Antitrust Issues in Schumpeterian Industries' (2000) 90 *AEA Papers and Proceedings* 192, 192.

competition 'acts not only when in being but also when it is merely an ever-present threat'.[76] In these types of markets, competition law needs to provide a market definition that can serve as a useful basis for this future-oriented competition law analysis.

D. Big Data, Artificial Intelligence, the Internet of Things: What Drives Innovative Markets Today

From a geographical point of view, digitisation has led to a consolidation of a range of markets.[77] From a product point of view, the digital realm and the many applications it has enabled – including big data, artificial intelligence and the internet of things (IoT) – have led to unique challenges for grasping these markets for antitrust purposes.

In today's data-driven economy, access to real-time information on the market – be it information on buyers, competitors, products or users – is vital in order to succeed. Data has become an important aspect of current innovation, far beyond the digital platforms that are often at the centre of the antitrust debate. It has been called the 'new currency' with which consumers pay on the internet. Data is generally not understood to constitute an antitrust market in a traditional sense, unless specific data sets are sold.[78] However, data plays an important role as an irreplaceable input in many digital markets (and beyond). Combined with the network effects present in these markets, data can represent a powerful competitive advantage for those in its possession. Amongst many other things, personal data allows companies to target specific individuals, be it through personalised search results, particular offers, discriminatory pricing etc. Of course, data is only useful if it can be analysed in a way that serves a company's goals, meaning that big data and big analytics go hand in hand.[79]

Self-learning algorithms (artificial intelligence or AI) are another aspect that is gaining importance in many of today's innovative markets and may become an ever more powerful tool for competition.[80] AI shares an intricate connection with big data and big analytics, as it relies on this input in order to learn about patterns, correlations and interactions that can then represent a competitive advantage. AI is already used for a plethora of applications. For instance, Google uses AI to

[76] JA Schumpeter, *Capitalism, Socialism and Democracy*, 5th edn (London, Allen & Unwin, 1976) 85.

[77] ABA Section of Antitrust Law, *Market Definition in Antitrust: Theory and Case Studies* (Chicago, ABA Publishing, 2012) 5.

[78] I Graef, 'Market Definition and Market Power in Data: The Case of Online Platforms' (2015) 38 *World Competition* 473, 489 f; AP Grunes and ME Stucke, 'No Mistake about it: The Important Role of Antitrust in the Era of Big Data' [2015] *Antitrust Source* 1, 3.

[79] ME Stucke and AP Grunes, *Big Data and Competition Policy* (Oxford, Oxford University Press, 2016).

[80] A Ezrachi and ME Stucke, *Virtual Competition: The Promise and Perils of the Algorithm-Driven Economy* (Cambridge, MA, Harvard University Press, 2016).

improve its search results, its advertising, speech translation, video and photo recognition.[81] The British retailer Marks & Spencer has recently teamed up with Microsoft in order to improve its operations through AI.[82]

Another important feature of today's innovative markets, particularly in the digital economy, is that systems are becoming ever more prevalent.[83] This means that several complementary markets are linked through interfaces, enabling them to function together. An example of such a system are repair parts and the services relating to them, or ATMs and ATM cards. From an economic perspective, each of these products may have little to offer on a standalone basis – imagine what use an ATM card is without an ATM within reach. But as a system, these products will carry value.[84] Network effects can powerfully alter the balance amongst such systems. This naturally raises the question of how antitrust markets need to be defined in these systems. Some of these issues may be solved by resorting to market definition in aftermarkets.

A particular kind of system is the IoT, which connects everyday objects such as air conditioning, electrical appliances, fridges or radiators with the internet, thereby making them an important part of the data-driven economy.[85] This allows these everyday objects to interact with each other, based on the far-reaching digital possibilities available today, such as sensors, software and wireless connectivity. For instance, our radiators may tell a control system when they are leaking – with a plumber sent on his or her way before the homeowner has even acknowledged the problem. In such markets, interoperability will play a central role. These interoperable devices might provide scope for considerable product differentiation and network effects, thus also impacting market definition.

III. Perspectives on Incorporating Dynamic Competition into Antitrust

With the characteristics of innovation and dynamic markets in mind, it remains to be seen why antitrust law in general, and market definition in particular, should be concerned with dynamic competition.

[81] P Larrey, *Connected World: From Automated Work to Virtual Wars the Future, by Those Who are Shaping it* (London, Portfolio Penguin, 2017) 38.

[82] C Rigby, 'Marks & Spencer Partners with Microsoft on Artificial Intelligence' *Internet Retailing* (22 June 2018), https://internetretailing.net/themes/themes/marks--spencer-partners-with-microsoft-on-artificial-intelligence-17931.

[83] J Reitzes and D Moss, 'Airline Alliances and Systems Competition' (2008) 45 *Houston Law Review* 293, 296.

[84] ML Katz and C Shapiro, 'Systems Competition and Network Effects' (1994) 8 *Journal of Economic Perspectives* 93, 93.

[85] GG Wrobel, 'Connecting Antitrust Standards to the Internet of Things' (2014) 29 *Antitrust* 62, 62; 'internet of things, n.' in *Oxford English Dictionary Online* (Oxford, Oxford University Press), www.oed.com/view/Entry/248411#eid332666668.

A. Schumpeter and Arrow on Dynamic Markets

In order to illustrate the relationship between competition and innovation, the positions of two economists, Joseph Schumpeter and Kenneth Arrow, are often juxtaposed. Quite frequently, it is then argued that a middle ground must be found between the two.[86] As this topic has been covered many times before,[87] it shall only very briefly be outlined here, from a competition lawyer's perspective. Although framing the relationship between competition and innovation as the 'Schumpeter v Arrow' debate falls short of grasping the complexity of this question, it allows us to showcase opposing views in economics as regards market structure and its influence on innovation.

The Austrian School of economics, including scholars such as Friedrich Hayek, Carl Menger, Ludwig von Mises and Schumpeter[88] emphasises the importance of the dynamic nature of the market rather than the static outcome to be achieved.[89] Austrians regard innovation as an important driver for social welfare and think about ways in which to attain innovation in the market.[90] They see competition as a process rather than a state of affairs, thus highlighting its dynamic component.[91] Hayek famously conceptualised competition as a process of discovery by which producers offer innovations and discover which products consumers will buy and at which price.[92] In Hayek's words, 'competition is by its nature a dynamic process whose essential characteristics are assumed away by the assumptions underlying static analysis'.[93] The Austrian School only calls for antitrust intervention where there is no substantial innovation, despite considerable concentration in the market. Concentration is seen as the motor for innovative effort.[94] It has been argued that 'the evolution of industrial economics from static to dynamic analysis entailed a rediscovery of much of the central Austrian argument'.[95] The Austrian

[86] Teece, 'Favoring Dynamic over Static Competition' (n 35) 206; DA Crane, 'Rationales for Antitrust: Economics and Other Bases' in RD Blair and DD Sokol (eds), *The Oxford Handbook of International Antitrust Economics*, vol 1 (Oxford, Oxford University Press, 2015) 11.

[87] See, eg, JB Baker, 'Beyond Schumpeter vs Arrow: How Antitrust Fosters Innovation' (2007) 74 *Antitrust Law Journal* 575; MA Carrier, 'Two Puzzles Resolved: Of the Schumpeter–Arrow Stalemate and Pharmaceutical Innovation Markets' (2008) 93 *Iowa Law Review* 393.

[88] PJ Boettke, 'Austrian School of Economics' in DR Henderson (ed), *The Concise Encyclopedia of Economics*, 2nd edn (Carmel, Liberty Fund, 2008).

[89] See also Glader, 'Innovation Economics' (n 7) 524.

[90] Similarly, see RJ van den Bergh and A Giannaccari, 'L'approcio più economico nel diritto comunitario della concorrenza: Il più è troppo o non (ancora) abbastanza?' (2014) XVI *Mercato concorrenza regole* 393, 413 ff.

[91] DB Audretsch, WJ Baumol and AE Burke, 'Competition Policy in Dynamic Markets' (2001) 19 *International Journal of Industrial Organization* 613, 618.

[92] Mackaay (n 26) 143; D Hildebrand, *The Role of Economic Analysis in EU Competition Law: The European School*, 4th edn (Alphen aan den Rijn, Kluwer Law International, 2016) 140.

[93] FA Hayek, 'The Meaning of Competition' in FA Hayek, *Individualism and Economic Order* (New York, Routledge, 1949) 94.

[94] Audretsch, Baumol and Burke (n 91) 619.

[95] ibid 621.

understanding of competition as a process rather than a static market structure fits well within the current emphasis on dynamic competition. The Austrian School believes that competition in the market is not merely desirable because it leads to increased efficiency, but because the competitive process in itself is a 'legitimate regulatory value'.[96]

As Schumpeter asserted:

> [The] kind of competition which counts [is] the competition from the new commodity, the new technology, the new source of supply, the new type of organization … competition which commands a decisive cost or quality advantage and which strikes not at the margins of the profits and the outputs of the existing firms but at their foundations and their very lives.[97]

Schumpeter termed this form of competition creative destruction.[98] In his view, the new and innovative is in a much better position to positively affect the overall competitive process than everyday competition (or the lack thereof). Only innovation can bring about social welfare.

A second proposition gathered from Schumpeter's writings is that monopoly should be favoured as a market structure because it is more conducive to innovation, and thus ultimately to social welfare.[99] This belief is also reflected in the US Supreme Court's judgment in *Trinko* (2004), where it held that 'monopoly power … induces risk taking that produces innovation and economic growth'.[100] Following Schumpeter's understanding, accepting big companies that have 'a considerable degree of market power [is] the price society must pay for rapid technological advance'.[101]

Schumpeter's assertions are 'difficult to substantiate or disprove'.[102] Some, while agreeing with Schumpeter's assertion that dynamic efficiency is of greater significance for social welfare than static allocative efficiency, are wary of his insistence on concentration as conducive to greater innovation.[103]

Often depicted as an antithesis to Schumpeter's propositions is Arrow, who believed that competition rather than concentration would foster innovation. He argued that a monopolist did not have the same incentive to invest in innovation as a company finding itself in a competitive market environment, as the monopolist's

[96] O Andriychuk, 'The Dialectics of Competition Law: Sketching the Ordo-Austrian Approach to Antitrust' (2012) 35 *World Competition* 355, 356 (direct quote), 373.

[97] Schumpeter, *Capitalism, Socialism and Democracy* (n 76) 84.

[98] ibid 83.

[99] For instance, see ibid 106.

[100] *Verizon Communications v Trinko*, 540 US 398, 407 (2004).

[101] RR Nelson and SG Winter, 'The Schumpeterian Tradeoff Revisited' (1982) 16 *American Economic Review* 114, 114.

[102] JR Eiszner, 'Innovation Markets and Automatic Transmissions: A Shift in the Wrong Direction?' (1998) 43 *Antitrust Bulletin* 297, 318.

[103] WF Baxter, 'The Definition and Measurement of Market Power in Industries Characterized by Rapidly Developing and Changing Technologies' (1984) 53 *Antitrust Law Journal* 717, 726.

pre-innovation monopoly profits acted as a disincentive towards new innovation: by innovating, the monopolist only stands to lose some of its own monopoly profits.[104] However, Arrow's model assumes that a monopoly is shielded by barriers to entry and that in a monopolistic situation, 'only the monopoly itself can invent'.[105] Where there is scope to challenge a monopoly through new market entry, this is more akin to a competitive rather than a monopolistic market situation.[106] Where radical innovation can challenge an established monopoly, perhaps even displacing an old market order, this may alleviate concerns of market concentration in relation to radical innovation.

Several views have been put forward that try to reconcile the positions advocated by Arrow and Schumpeter from a competition (law) perspective.[107] Economics, it would seem, has not yet arrived at a 'final microeconomic theory that explains … the relationship between competition and innovation, and, in the end, whether and how long monopolies should last in order to promote competition and innovation'.[108] Different industries require different incentives to promote innovation – with some performing better with strong IPRs and others faring better in the face of strong competition.[109] This makes it difficult – if not virtually impossible – to craft competition law rules with the primary goal of promoting innovation for long-term consumer welfare. Nevertheless, there is 'general agreement that antitrust enforcement should account for innovation'.[110] As was highlighted above, the present study does not purport to promote innovation. Its aim is to depict it in relevant antitrust markets as accurately as possible so as to allow for a pertinent antitrust analysis. An antitrust policy that aims to promote innovation in the marketplace will need to use different tools – particularly in relation to the theory of harm – in order to pursue such a goal.

B. Competition Policy in Dynamic Markets

Dynamic competition, driven by innovation, product differentiation and fast-moving markets, offers a more accurate portrayal of real-world economies than static competition.[111] While static competition focuses on market outcomes,

[104] Arrow (n 34) 619–22.

[105] ibid 619.

[106] ibid.

[107] Shapiro, 'Competition and Innovation' (n 46) 362, 363–65, 401; Teece, 'Favoring Dynamic over Static Competition' (n 35) 210 f; Carrier, 'Schumpeter-Arrow' (n 87) 404–10.

[108] M Maggiolino, 'The Economics of Antitrust and Intellectual Property Rights' in SD Anderman and A Ezrachi (eds), *Intellectual Property and Competition Law: New Frontiers* (Oxford, Oxford University Press, 2011) 87 (direct quote).

[109] MA Lemley, 'Industry-Specific Antitrust Policy for Innovation' [2011] *Columbia Business Law Review* 637, 640.

[110] JD Wright, 'Antitrust, Multi-dimensional Competition, and Innovation: Do We Have an Antitrust-Relevant Theory of Competition Now?' in GA Manne and JD Wright (eds), *Competition Policy and Patent Law Under Uncertainty: Regulating Innovation* (Cambridge, Cambridge University Press, 2011) 235.

[111] Teece, 'Favoring Dynamic over Static Competition' (n 35) 210 f.

and particularly allocative and productive efficiency gains such as lower prices, dynamic competition focuses on the process of competition and favourably views dynamic efficiency gains in the form of innovation.[112] There clearly is a trade-off between the various types of efficiency, and 'antitrust may have to tolerate some degree of *static inefficiency* in order to promote *dynamic efficiency*'.[113] For a discipline that has largely relied on static efficiency and static analysis in the past, this is a tall order. In addition, competitive harm that arises from restrictions on innovation is less palpable than harm that arises when price competition is impaired.[114] For these reasons, '[i]nnovation has never fit very comfortably into antitrust enforcement policy'.[115]

Market definition, particularly through its function of market characterisation, can be of great value to antitrust analysis as it allows for insights into how innovation-driven markets operate. Once it is well understood that competition in such markets works differently from markets in which prices and output are key, it also becomes clear that it would be misguided to simply apply economic wisdom from static markets to dynamic markets.[116] For the time being, however, antitrust law continues to rely on largely static parameters, including homogeneous products, a given technology level and price competition as its main focus.[117] The mostly static training of economists has been made responsible for this 'static-ization of antitrust'.[118]

Commentators have discerned a '[s]cholarly enthusiasm for increased consideration in antitrust analysis of dynamic competition and of innovation' over the last quarter of a century.[119] Different policy suggestions have emerged on how to incorporate insights as to the relationship between competition and innovation into antitrust law and policy. Some have suggested that one should stick to the simpler, more familiar – but also more unrealistic – static model of competition in antitrust where possible, and only revert to the dynamic model of competition where this is necessary.[120] Others have instead argued that a continued focus on static competition in the short run rather than on dynamic competition in the long run is myopic.[121] Amongst antitrust authorities, the view appears to

[112] ibid 217; Crane, 'Rationales for Antitrust' (n 86) 10.

[113] Spulber and Yoo (n 55) 392 (emphasis in original).

[114] Eiszner (n 102) 298; HJ Hovenkamp, 'Schumpeterian Competition and Antitrust' (2008) 4 *Competition Policy International* 273, 279.

[115] C Bohannan and HJ Hovenkamp, *Creation without Restraint: Promoting Liberty and Rivalry in Innovation* (Oxford, Oxford University Press, 2012) 238.

[116] Eiszner (n 102) 300; D Encaoua and A Hollander, 'Competition Policy and Innovation' (2002) 18 *Oxford Review of Economic Policy* 63, 64.

[117] Glader, 'Innovation Economics' (n 7) 522 f.

[118] DS Evans and KN Hylton, 'The Lawful Acquisition and Exercise of Monopoly Power and its Implications for the Objectives of Antitrust' (2008) 4 *Competition Policy International* 203, 238.

[119] DH Ginsburg and JD Wright, 'Dynamic Analysis and the Limits of Antitrust Institutions' (2012) 78 *Antitrust Law Journal* 1, 12.

[120] Maggiolino (n 108) 87 fn 60.

[121] TF Cotter, 'Innovation and Antitrust Policy' in RD Blair and DD Sokol (eds), *The Oxford Handbook of International Antitrust Economics*, vol 2 (Oxford, Oxford University Press, 2015) 132.

prevail that their analytical tools are flexible enough in order to adapt to different market environments, in particular innovation-driven industries.[122]

The usefulness of market delineation has been questioned in the context of rapid technological change.[123] Such an argument, of course, stands in stark contrast to the antitrust laws as they currently stand, which rely heavily on legal tests, including market definition.[124] In addition, the market characterisation function of the relevant market can be particularly useful in innovative settings.

While economics cannot yet tell us the optimal level of innovation for promoting consumer and overall welfare, and while the research on the relationship between innovation and competition is far from conclusive, this does not pre-empt an innovation-sensitive delineation of the relevant antitrust market. Although the development of an innovation-sensitive theory of harm on which competition law can rely may be the logical and necessary next step, that step needs to build on a strong foundation: robust and consistent market definition.

C. Market Definition's Influence on Innovation: Error Costs

As a legal concept,[125] the relevant market does not promote innovation as such. Nevertheless, it can have an influence on innovation through its impact on antitrust analysis. A brief look at error cost analysis shows ways in which our approach to market delineation might impact competition and innovation.

Antitrust scholars have frequently relied on error cost analysis, a type of decision theory that has found its way into law and economics, in order to assess whether a certain antitrust rule minimises social cost.[126] They rely on several factors in order to determine this: the costs of false positives (type I errors), the costs of false negatives (type II errors) and broadly understood transaction costs.[127] False positives are situations in which an antitrust violation is found, although the underlying market behaviour does not restrict competition ('convicting the innocent'), while

[122] United States in OECD (ed), *Policy Roundtable: Market Definition* (2012) DAF/COMP(2012)19, 331; DL Rubinfeld and J Hoven, 'Innovation and Antitrust Enforcement' in J Ellig (ed), *Dynamic Competition and Public Policy: Technology, Innovation, and Antitrust Issues* (Cambridge, Cambridge University Press, 2001) 90; European Commission, Guidelines on the application of Article 101 of the Treaty on the Functioning of the European Union to technology transfer agreements [2014] OJ C89/3, para 9; European Commission, 'Online Platforms and the Digital Single Market – Opportunities and Challenges for Europe' (25 May 2016) COM(2016) 288 final, 13.

[123] Teece, 'Favoring Dynamic over Static Competition' (n 35) 220; J Crémer, Y-A de Montjoye and H Schweitzer, 'Competition Policy for the Digital Era' (Report for DG Competition, 3 April 2019) 42–50.

[124] For a discussion of this, see ch 2, section IV.

[125] See ch 2, section IV.B.

[126] CF Beckner III and SC Salop, 'Decision Theory and Antitrust Rules' (1999) 67 *Antitrust Law Journal* 41, 45.

[127] FH Easterbrook, 'The Limits of Antitrust' (1984) 63 *Texas Law Review* 1; FS McChesney, 'Easterbrook on Errors' (2010) 6 *Journal of Competition Law & Economics* 11, 14–17. While US antitrust law followed Easterbrook's aversion to type I errors, the EU did not; Easterbrook, 'Limits of Antitrust' (at 3, 15); A Devlin and M Jacobs, 'Antitrust Error' (2010) 52 *William & Mary Law Review* 75, 96, 97.

false negatives are situations in which no antitrust violation is found, despite anti-competitive market behaviour being present ('acquitting the guilty').[128] In the US, today's competition rules are moulded in the Chicago School tradition,[129] while some push for even less antitrust intervention. This push relies on assumptions that 'systematically overstate the incidence and significance of false positives, understate the incidence and significance of false negatives, and understate the net benefits of various rules by overstating their cost'.[130] Type I errors in the context of innovation have been characterised as particularly harmful as they could have negative effects on innovation.[131] While an exclusive focus on static competition would certainly harm innovation, so would an approach that is too scared to intervene because of the risk of harming innovation, and therefore tolerates anti-competitive and anti-innovative behaviour to the detriment of consumers.[132] At the same time, economists have highlighted that they 'cannot accurately measure the effects of over-enforcement or under-enforcement', meaning that although economic analysis can help in appreciating possible effects of certain company behaviour, it is for competition policy to carry out the balancing act.[133]

While error cost analysis will not be relied upon as such in the following chapters, it is useful to engage in some preliminary thoughts about the potential error costs of market definition as the first stage of antitrust analysis. Antitrust markets that are defined too narrowly because a good substitute is excluded from the relevant market bear a risk of over-enforcement (false positive or type I error), while antitrust markets that are defined too broadly because a bad substitute is included in the relevant market bear a risk of under-enforcement (false negative or type II error).[134] Against this background, market definitions do not appear objective as the inclusion or exclusion of a possible substitute will always reflect the antitrust enforcer's preference for or against type I or type II errors.[135] Where markets are characterised by high innovation, forecasts and predictions might easily turn out

[128] For the direct quotes, see K Heyer, 'A World of Uncertainty: Economics and the Globalization of Antitrust' (2005) 72 *Antitrust Law Journal* 375, 380. On the factors of error cost analysis, see JB Baker, 'Taking the Error out of "Error Cost" Analysis: What's Wrong with Antitrust's Right' (2015) 80 *Antitrust Law Journal* 1, 5.

[129] For instance, see the laissez-faire approach in KN Hylton, '*Brown Shoe* versus the Horizontal Merger Guidelines' (2011) 39 *Review of Industrial Organization* 95, 105.

[130] Baker, 'Error Cost' (n 128) 37. Both the modern Harvard and the Chicago Schools disfavour a risk of over-enforcement in the realm of unilateral conduct; WE Kovacic, 'The Intellectual DNA of Modern US Competition Law for Dominant Firm Conduct: The Chicago/Harvard Double Helix' [2007] *Columbia Business Law Review* 1, 70.

[131] HA Shelanski, 'Information, Innovation, and Competition Policy for the Internet' (2013) 161 *University of Pennsylvania Law Review* 1663, 1666.

[132] Ginsburg and Wright (n 119) 2.

[133] G Niels, H Jenkins and J Kavanagh, *Economics for Competition Lawyers*, 2nd edn (Oxford, Oxford University Press, 2016) § 1.53.

[134] PD Camesasca and RJ van den Bergh, 'Achilles Uncovered: Revisiting the European Commission's 1997 Market Definition Notice' (2002) 47 *Antitrust Bulletin* 143, 144; WH Boshoff, 'Market Definition as a Problem of Statistical Inference' (2014) 10 *Journal of Competition Law & Economics* 861, 871.

[135] Boshoff, 'Market Definition' (n 134) 882.

to be wrong, risking a type I error.[136] This argument can directly be transferred to the realm of market definition in highly innovative environments. For instance, broader market definitions in innovative environments could solve the concern that antitrust is over-enforced in these markets.[137]

Where innovative markets are concerned, the risk of false positives entails the risk of negatively impacting innovation to the detriment of customers and consumers at large – ultimately restricting competition. Innovators faced with a significant risk of antitrust liability for their innovative enterprise might decide not to engage in that innovation in the first place. On the other hand, false negatives in the area of innovative markets may mean that anti-competitive conduct which has a negative impact on innovation and technological progress may not be challenged, again negatively impacting innovation to the detriment of customers. Where antitrust rules are inherently uncertain, this may in itself lead to situations in which the antitrust rules cannot achieve their goals.[138] One of these goals may be the promotion of innovation.

It is sometimes believed that the costs associated with false positives outweigh those of false negatives, particularly in innovative market environments where disruptive innovation constantly challenges current market situations.[139] However, bearing in mind the inherent uncertainty in innovation, an approach that favours false negatives over false positives may let anti-competitive, innovation-hindering market behaviour continue unscathed for many years. One of the most apparent differences between EU competition and US antitrust law is, in fact, that the former may be biased towards short-term competition benefits and over-enforcement, while the latter may focus more on long-term competition benefits and favour under-enforcement.[140] The Court of Justice pointed out in *TeliaSonera* that it prefers a reactionary approach in innovative markets so as to ensure that no anti-competitive effects materialise.[141] This points towards a willingness to make type I errors. European economists have also warned that considering innovation as important does not mean that competition law should not intervene – quite the contrary.[142] Translated into the realm of antitrust market definition, this

[136] A Ezrachi and M Maggiolino, 'European Competition Law, Compulsory Licensing, and Innovation' (2012) 8 *Journal of Competition Law & Economics* 595, 599 (stated within the context of balancing IP and antitrust law).

[137] AC Hruska, 'A Broad Market Approach to Antitrust Product Market Definition in Innovative Industries' (1992) 102 *Yale Law Journal* 305, 305. While this may lead to a bias in favour of under-enforcement, such may be justified in the presence of innovation (ibid 310). Arguing instead for rigorous enforcement in digital markets, see Crémer, de Montjoye and Schweitzer (n 123) 51.

[138] H First and SW Waller, 'Antitrust's Democracy Deficit' (2013) 81 *Fordham Law Review* 2543, 2571.

[139] DA Crane, 'The Economics of Antitrust Enforcement' in KN Hylton (ed), *Antitrust Law and Economics* (Cheltenham, Edward Elgar, 2010) 4.

[140] A Devlin and M Jacobs, 'Antitrust Divergence and the Limits of Economics' (2010) 104 *Northwestern University Law Review* 253, 258, 266, 270.

[141] Case C-52/09 *Konkurrensverket v TeliaSonera Sverige* EU:C:2011:83, para 108. Similarly, see Crémer, de Montjoye and Schweitzer (n 123) 51.

[142] T Curzon Price and M Walker, 'Incentives to Innovate v Short-Term Price Effects in Antitrust Analysis' (2016) 7 *Journal of European Competition Law & Practice* 475, 477.

might explain why EU competition law finds narrow markets – and, subsequently, dominance – more often than US antitrust law. However, it cannot be said that this result is inherent in the different ways that EU and US law approach market definition, which were discussed above; rather, it is a result of the different aims informing the respective approaches.

Frank Easterbrook suggested that antitrust courts 'should use the economists' way out. They should adopt some simple presumptions that structure antitrust inquiry'.[143] This would give greater certainty to companies. While this was written with substantive antitrust rules in mind, simple and reliable rules on antitrust market definition would also increase legal certainty in this domain. The DoJ has remarked that the 'market-definition requirement brings discipline and structure to the monopoly power inquiry [in section 2 cases], thereby reducing the risks and costs of error'.[144] Error costs in market definition can only be assessed indirectly, namely through the market definition's influence on the final outcome. Nevertheless, it can be useful to consider the implications of a narrow/broad market definition at the policy stage.

IV. Conclusion

The conventional way of conceptualising the relevant market is based on static, homogeneous goods markets. As such, it contrasts with the Schumpeterian vision of dynamic, heterogeneous product markets. Dynamic competition is intrinsically 'future-oriented' as it expects innovation to deliver new products and processes.[145] Innovative market environments are fast-moving and today's market definitions may no longer hold tomorrow. There is considerable uncertainty inherent in innovation, leading to difficulties when delineating markets and discerning market participants. Research and development is central to innovative market environments, and its outcome is frequently IP-protected. IPRs lead to product differentiation, a feature which market definition must be prepared to accommodate. In addition, performance rather than price is an important competitive parameter in such markets, and the question arises as to how price-based analytical tools from economics can apply in them. Network effects, interoperability and the role of big data, as well as advances in artificial intelligence and the IoT, are further issues that need to be addressed when delineating dynamic markets. These features cannot easily be accommodated by traditional antitrust market definition, particularly as their interplay needs to be carefully considered.

[143] Easterbrook, 'Limits of Antitrust' (n 127) 14. However, he also argued that market definition is 'a subject that has bedevilled the law of mergers' and that any inquiry into the correct market was 'a fool's errand'; ibid 22.

[144] US Department of Justice, 'Competition and Monopoly: Single-Firm Conduct under Section 2 of the Sherman Act' (September 2008) 25.

[145] Sidak and Teece, 'Dynamic Competition' (n 1) 600.

While industrial organisation theory has not yet provided a final view on the market structure that is most conducive to innovation, it is not necessary to settle this debate for the present purposes. The accurate, realistic depiction of innovative markets through market delineation – and an understanding of how competition works in such dynamic markets – is the first step in antitrust analysis and is at the centre of the present study. A coherent, innovation-based theory of harm may be constructed upon this basis.

PART II

The Innovation Factor in Market Delineation under EU Competition and US Antitrust Law

4

Innovative Product Markets

Innovation has become a competition law catchword and it continues to be debated whether innovation considerations can – or even should – form part of the competition law assessment.[1] By relying on a comparative analysis, Part II asks how different aspects of innovative markets – be it innovative products, IPRs or platforms – are currently dealt with in antitrust analysis in the EU and the US. Its findings may be relevant far beyond these two jurisdictions.

The product is one of the central dimensions when delineating a relevant antitrust market.[2] It is only with reference to a particular product that demand and supply substitutability can be appreciated. Both the initial product that is taken as a starting point and the substitutes that are considered for delineating the market that it forms part of are essential to the relevant market inquiry. Section I highlights some of the conceptual issues that arise when trying to grasp the nature of a product in new or emerging innovative markets, often focusing on markets in which digital technology is key. However, similar issues also arise in more traditional markets when they are confronted with disruptive or even just sustaining innovation. Subsequently, section II undertakes a comparative analysis of several innovation-related aspects of product differentiation, while section III deals with issues of timing. Section IV looks at future markets.

I. Innovative Products as a Challenge for Market Definition

Innovation leads to new or significantly improved processes, technologies or products.[3] Innovative goods and services have an important impact on our daily lives, yet we often do not know where innovation is heading, which innovations will be successful and which ones will come to nothing. This uncertainty explains some of the excitement surrounding innovation. At the same time, it injects considerable uncertainty into antitrust law. In the following, we explore to what

[1] P Ibáñez Colomo, 'Restrictions on Innovation in EU Competition Law' (2016) 41 *EL Rev* 201, 201.
[2] Products encompass anything that is traded, such as goods, services, technology and data.
[3] See section I in ch 3.

extent and how various aspects of innovation (may) influence antitrust market definition, and at what point it is appropriate to consider them.

A. Policy Approaches to Innovative Product Markets

Opinions are divided on the matter of principle of how broadly innovative markets should be delineated. A more laissez-faire approach analyses innovative products under broad market definitions, thereby perhaps more accurately reflecting the economic reality that dynamic competition can move fast and in unexpected directions.[4] A company wanting to avoid antitrust scrutiny will certainly advocate such an approach.

A competition authority wanting to get involved in a given case will advocate a more interventionist attitude. One of the arguments put forward in such circumstances is that there is no good reason to spare big companies from antitrust intervention in technology-driven industries; on the contrary, entry barriers for new innovators – both structural and behavioural – should be kept to a minimum.[5] Where a competition authority enjoys a wide margin of discretion as to which relevant product it starts its antitrust analysis with, particularly where complex products are involved, this may lead to the finding of very narrow product markets. Such a market definition outcome influences not only whether a company is found to exercise market power, but can also result in consequences on the substantive antitrust analysis relating to the anti-competitive harm.[6]

When pondering the policy question of how broadly or narrowly product markets in innovative environments should be defined to provide a reliable basis for antitrust analysis, there is thus an additional layer that may bias antitrust enforcement towards narrower or broader markets. The characterisation of competition facing online platforms, for instance, depends heavily on who carries it out: independent academics, competition authorities with stakes in an ongoing investigation, or researchers sponsored by one of the leading tech companies.[7] While it is extremely difficult to get a hold of this intellectual capture,[8] its possibility should be borne in mind when assessing arguments made by all sides. Market definition is a fertile breeding ground for lobbyists – if they are persuasive when it comes to the relevant antitrust market, the competition authority may no longer

[4] RC Lind et al, 'Report on Multiparty Licensing' (22 April 2003) 46 f.

[5] FM Scherer, 'Technological Innovation and Monopolization' in ABA Section of Antitrust Law (ed), *Issues in Competition Law and Policy* (Chicago, ABA Publishing, 2008) 1068.

[6] See SD Anderman and J Kallaugher, *Technology Transfer and the New EU Competition Rules: Intellectual Property Licensing after Modernisation* (Oxford, Oxford University Press, 2006) § 10.06.

[7] On this issue, see A Ezrachi and ME Stucke, *Virtual Competition: The Promise and Perils of the Algorithm-Driven Economy* (Cambridge, MA, Harvard University Press, 2016) 245–47.

[8] But see P Ibáñez Colomo and G de Stefano, 'Protecting the Integrity and Reputation of Legal Research: JECLAP's New Rules on Disclosure' (2017) 8 *Journal of European Competition Law & Practice* 623.

have a case. Of course, the extent to which lobbying or intellectual capture can be successful also depends on how open antitrust enforcement is to receiving input from lobbyists, from intellectually captured researchers or from politicians who might have different interests altogether.

B. Market Definition for Digital Innovation: Examples of Innovative Markets

Over the last two decades, innovation in everything digital has advanced in great leaps and bounds. The following characterises some of these digital innovations and highlights the market definition challenges they bring with them.

i. Interoperability, Network Effects and Data in Digital Platforms

Network effects, data harvesting and data analytics are highly relevant to today's digital market environments. These factors are also responsible for the emergence of only a handful of super-platforms that greatly impact how competition takes place in these dynamic markets: GAFAM – Google, Amazon, Facebook, Apple and Microsoft. Their relationships with each other, with application developers and with users will increasingly become central to understanding the digital ecosystem.[9] Surrounding this data-driven market setting are issues such as interdependence, asymmetry, network effects and consumer lock-in,[10] all of which need to be taken into account at the stage of market definition. In many cases, this will complicate the delineation of *the* relevant product market, thus forcing us to de-emphasise market definition's function of allowing for insights into market power, and instead emphasising its usefulness for characterising a market environment so as to inform our understanding of how these markets operate and how an antitrust theory of harm may apply to them.

In the increasingly digitised economy, interoperability is key. Many of today's smart devices need to interact with other devices in order to unfold their innovative potential. While the European *Microsoft* case, which was first launched in 1998,[11] already broached the issue of interoperability amongst computer networks, interoperability was also at the focus of the European *Google Android* case (2018), which looked at the licensing terms of Google's Android mobile operating system.[12]

[9] Ezrachi and Stucke (n 7) 149, 157.

[10] ibid 222.

[11] *Microsoft* (Case COMP/C-3/37.792) Commission Decision 2007/53/EC [2007] OJ L32/23, para 3 (Sun Microsystems' complaint of 1998); Case T-201/04 *Microsoft v Commission* EU:T:2007:289.

[12] *Google Android* (Case AT.40099) Commission Decision of 18 July 2018; European Commission, 'Antitrust: Commission Fines Google €4.34 Billion for Illegal Practices Regarding Android Mobile Devices to Strengthen Dominance of Google's Search Engine' (IP/18/4581, 18 July 2018).

In mobile operating systems, issues have arisen in relation to Apple's App Store, with allegations that Apple is using its control of its operating system in order to extract higher profits from app developers distributing their apps via its App Store.[13] An application store running on a particular mobile operating system may be used as a bottleneck for accessing users on that particular platform. For market definition, the question becomes to what extent different mobile operating systems and their corresponding app stores constitute separate relevant markets in and of themselves.[14] For instance, a user wishing to have a mobile fitness application such as Runtastic on her smartphone will need to go through her smartphone's app store in order to download and install it – which in turn requires that Runtastic as an app developer can write its app for that particular operating system (which requires interface information) and that it can subsequently distribute its app through that particular app store. For an iPhone or iPad, this is the iOS App Store, for a smartphone or tablet running on Android, this primarily is the Google Play Store, and for a smartphone or tablet running on Windows, this is the Microsoft Store. In terms of market coverage, recent data shows that 85.9 per cent of all smartphones sold worldwide in 2017 ran on Android, 14 per cent ran on iOS and only 0.1 per cent ran on another operating system.[15] Can these findings inform our market definition, leading to increasingly narrow product markets?

A feature which makes many (super-)platforms so hard to grasp for market definition is the fact that they do not in themselves make any innovative contribution in a substantive sense. Instead, it is the users who generate content on Facebook, Instagram, LinkedIn, Snapchat, Twitter and YouTube, use the search function on Baidu, Bing and Google, use the messaging function on Facebook, Skype, WhatsApp and their smartphones, and review products and sellers on Amazon after shopping or on TripAdvisor after travelling. The user contributions in turn provide these platforms with enormous amounts of data that they can subsequently monetise.[16] This can be likened to an art gallery which exhibits several artists' works (except in the digital sphere the users' works are usually much less sophisticated): the gallery provides the space in which the artists exhibit their works and can get in touch with people interested in their work, just like these digital platforms provide the space in which the users can post content and interact with others. The difference, of course, is that unlike art galleries, digital

[13] See especially *Apple v Pepper*, 587 US ___ (2019), 139 SCt 1514 (2019). While the case before the US Supreme Court focused on the standing of the class, the substantive issue in this case revolves around Apple's alleged monopolisation of the aftermarket for iPhone apps. See also an ongoing case brought by app developers: *Cameron v Apple*, Case 4:19-cv-03074 (ND Cal, case filed 4 June 2019) (pending).

[14] In addition, the distinction between mobile and desktop operating systems is blurring.

[15] 'Gartner Says Worldwide Sales of Smartphones Recorded First Ever Decline during the Fourth Quarter of 2017' (22 February 2018), www.gartner.com/newsroom/id/3609817.

[16] See Ezrachi and Stucke (n 7) 234 f; OB Arewa, 'YouTube, UGC, and Digital Music: Competing Business and Cultural Models in the Internet Age' (2010) 104 *Northwestern University Law Review* 431, 431.

platforms are a mass phenomenon, gathering data on users and exploiting it for advertisement and analytics purposes, amongst other things. The digital platforms are also much more short-lived. From a market definition perspective, it is intriguing to ask which initial candidate product market can capture these digital platforms. Can we say there is a market for the digital content users post? A market for the data users are generating – even though these platforms usually only use their data in-house rather than selling it on?[17] Or is there a market for the infrastructure that those platforms provide users with?

Big data plays a central role in digital platforms. This becomes especially evident when taking third-party tracking into account, the practice through which a digital company harvests large amounts of personalised data about its users from multiple online sources, combining it to provide stunningly detailed individual user profiles that can be used for a multitude of purposes.[18] At a later stage, we will consider whether such data can constitute its own product market (see section III.D.ii in Chapter 8).

A possibility when it comes to digital platforms on which users generate content is to replace the focus on the product that users engage with (eg, messaging, online search, social networking) with that aspect of the business model with which those online platforms generate income: user attention. The argument goes that irrespective of how online platforms attract users' eyeballs, they are competing with other platforms that are also trying to attract user attention.[19] The actual products that online platforms provide (usually a service) may become largely irrelevant for antitrust purposes. Similarly, Google's main product may not be its search engine, YouTube, its maps or even Android phones – its main product may consist in online search advertising.[20] This analysis shows that providers of very different services can nevertheless compete with each other – at least on one of the market sides (ie, the revenue-generating market side, such as online advertising). However, it does not account for freemium models (discussed below), where a service can be enjoyed for free if ads are endured, or is fee-based in return for enhanced functionalities and/or no more ads. Therefore, while certainly an interesting and important aspect of market definition, solely focusing on attention rivalry would be too limited an approach. However, framing this issue around online advertising brings together the strands through which online platforms

[17] On the absence of a market where data is used for in-house purposes only, see I Graef, 'Market Definition and Market Power in Data: The Case of Online Platforms' (2015) 38 *World Competition* 473. On data brokers, see Federal Trade Commission, 'Data Brokers: A Call for Transparency and Accountability' (May 2014) 3.

[18] See A Ezrachi and VHSE Robertson, 'Competition, Market Power and Third-Party Tracking' (2019) 42 *World Competition* 5.

[19] DS Evans, 'Attention Rivalry among Online Platforms' (2013) 9 *Journal of Competition Law & Economics* 313, 314, 330. This research was funded by Google; ibid 313 fn *.

[20] BJ Smith, 'Vertical vs. Core Search: Defining Google's Market in a Monopolization Case' (2012) 9 *NYU Journal of Law & Business* 331; N Newman, 'Search, Antitrust, and the Economics of the Control of User Data' (2014) 31 *Yale Journal on Regulation* 401, 409, 412.

generate income in competition with each other, while providing various types of (often free) services to the other market side.

ii. E-Commerce and Digital Platforms

The digital market environment has led to a paradigm shift: not only have innovative digital products seen the light of day, but conventional goods and services are now also being offered in many innovative ways.[21] We can distinguish at least three types of digital platforms that are engaged in the sale of goods or services: e-commerce platforms for goods, sharing economy platforms for goods and services, and digital service providers. Each of these poses challenges to our traditional conceptualisation of substitutability.

E-commerce platforms have changed the way in which business is conducted on the internet. There are different kinds of these platforms: websites such as the geographically widespread Craigslist or the Oxford-centred Daily Info and apps such as Shpock simply digitise traditional classifieds, ie, they leave delivery and payment to be arranged by the parties to the transaction, without support from the platform. Transactional websites such as eBay go a considerable step further and provide a platform for buyers and sellers to directly interact until the deal is sealed, ie, including payment and delivery arrangements. Amazon goes beyond eBay in that it both provides a transactional platform through its Marketplace and acts as an online retailer itself, thus effectively competing with some of its own customers, the Marketplace sellers.[22] What is the relevant product market on which these e-commerce platforms are active?[23] Is it the platform as such, the goods that are being traded or both? Is there a significant difference between transactional and non-transactional platforms, or do consumers regard them as interchangeable?

An aspect related to e-commerce platforms are price comparison websites.[24] These are often not the neutral, consumer-friendly enterprises they make themselves out to be – they need to generate revenue just like any other business. As such, they charge fees or commissions to the sellers listed on their website, which in turn may introduce a bias into the information consumers are shown.[25] Where demand substitutability is analysed in such market environments, the focus of the enquiry

[21] VD Roman, 'Digital Markets and Pricing Algorithms: A Dynamic Approach towards Horizontal Competition' (2018) 39 *European Competition Law Review* 37, 44.

[22] This competitive relationship is currently under antitrust scrutiny in the EU, Germany and Austria.

[23] See also J Furman et al, 'Unlocking Digital Competition' (Report of the Digital Competition Expert Panel, March 2019) para 1.57.

[24] On these, see European Commission, 'Comparison Tools: Report from the Multi-stakeholder Dialogue – Providing Consumers with Transparent and Reliable Information' (March 2013); European Commission, 'Study on the Coverage, Functioning and Consumer Use of Comparison Tools and Third-Party Verification Schemes for Such Tools: Final Report Prepared by ECME Consortium (in Partnership with Deloitte)' (EAHC/FWC/2013 85 07, 2013).

[25] Ezrachi and Stucke (n 7) 131, 136.

must rest on actual consumer behaviour. If a user obtains price information on a particular good from one source (such as a particular price comparison website, A) and could theoretically also get price information from other sources (such as other price comparison websites B, C and D, the retailer's own website, another retailer's website or the retailer's brick-and-mortar shop), but doesn't, then this consumer behaviour indicates that a narrower market needs to be delineated that more accurately reflects the anti-competitive effect that price comparison website A's market behaviour may have.[26]

A relatively new type of online-based business model has recently conquered the business world: the sharing economy, a world in which online platforms connect users with providers of mobility services (eg, Lyft and Uber), accommodation (eg, Airbnb), car sharing or rental (eg, easyCar club, Liftshare and Zipcar) and other services such as dog walking, freelance work or private lending.[27] Originally, many of these new markets were peer-to-peer (P2P) or consumer-to-consumer markets, with the good or service being shared against payment.[28] However, the sharing economy has experienced rapid professionalisation. Despite the altruistic connotation of the term 'sharing', the sharing economy refers to a system which is largely market-based and relies on high-impact capital and crowd-based, decentralised networks. In this environment, the lines between the personal and the professional are blurring and employment structures do not fall into clear categories.[29] There are certainly areas of overlap between the more traditional e-commerce platforms mentioned above and some sharing economy platforms. For instance, a used book can be sold through a conventional e-commerce platform or a neighbourhood sharing platform. What characterises these markets is that the sharing platform itself does not provide any goods or services, but merely acts as an intermediary. As such, it is quite distinct from the digital service providers discussed below. It is also interesting to note that sharing platforms with similar purposes may be built on quite different business models.[30] The question, then, is how to define the relevant market within the sharing economy. For instance, does Uber compete with regular taxis (or minicabs in the UK)? Does Airbnb compete with regular hotels or with hotel booking platforms such as Booking.com? Do Uber or Airbnb – as platforms – compete at all, or do we need to look at the market at the level of each individual driver or of each individual host? If an anti-competitive cartel is

[26] See also MR Patterson, *Antitrust Law in the New Economy: Google, Yelp, LIBOR, and the Control of Information* (Cambridge, MA, Harvard University Press, 2017) 66.

[27] On the latter three, see, for instance, Rover, Upwork Global and LendingClub.

[28] JJ Horton and RJ Zeckhauser, 'Owning, Using and Renting: Some Simple Economics of the "Sharing Economy"' (NBER Working Paper, vol 22029, 2016) 1.

[29] See A Sundararajan, *The Sharing Economy: The End of Employment and the Rise of Crowd-Based Capitalism* (Cambridge, MA, MIT Press, 2016) 27 f; A Ross, *The Industries of the Future* (New York, Simon & Schuster, 2017) 92; J Prassl, *Humans as a Service: The Promise and Perils of Work in the Gig Economy* (Oxford, Oxford University Press, 2018).

[30] F Russo and ML Stasi, 'Defining the Relevant Market in the Sharing Economy' (2016) 5 *Internet Policy Review* 1, 6.

alleged as stemming from the Uber app and pricing algorithm, on which market is that cartel active?[31] In a recent case, a New York district judge urged Uber's legal counsel to 'walk from their offices to the courthouse to put their theory [that the relevant market comprised not only mobile app-generated ride-share services but also taxis, car services, public transport, personal vehicle use and walking] to the test'.[32]

Next to sharing economy platforms and e-commerce platforms, both of which (frequently) act as mere intermediaries, another type of online platform directly provides services, for instance, entertainment such as music streaming (eg, Apple Music and Spotify) or movie and video streaming (eg, Amazon Prime, Netflix and YouTube). The various platforms are not limited to providing users with a search function for films/shows/music; they also offer categorisation, recommendations based on the user profile the platform has built, music playlists based on certain themes etc. For these digital service providers, an intriguing question is to what extent the entertainment package they provide can be seen as the relevant product market, or whether certain musical or cinematographic works constitute their own markets.

iii. Digital Assistants and the Internet of Things: Integration of Markets versus Narrow Market Delineation

Digital assistants are voice-controlled little helpers that carry out tasks such as reading you the news, telling you the weather forecast, buying items you request, booking you a hotel room, playing your favourite music playlist, setting your alarm or sending your friend a message. Digital assistant software can be plugged into smartphones, tablets and desktops (with speaker and microphone) or into purpose-built hardware such as speakers with internet connectivity. Examples are Amazon's Echo devices incorporating the Alexa digital assistant software, Apple's Siri software incorporated in its iPhone, iPad and HomePod, Facebook's M, Microsoft's Cortana and the Google Assistant, which runs on Android smartphones and forms part of Google Home. However, the technology incorporated into virtual digital assistants is no longer merely carrying out user commands; it is increasingly taking charge.[33] In terms of the relevant product market, digital assistants may entirely shut out competing offers when answering the user's search queries. When you ask your digital assistant to buy a certain good, order take-out or book a hotel room in city X, the market definition question lurking behind these simple requests is which data the digital assistant relies on in order to fulfil

[31] For a preliminary reference to the European Court of Justice on the question whether proportionality precludes Uber's services being regarded as taxi services, which was judged to be inadmissible, see Case C-526/15 *Uber Belgium v Taxi Radio Bruxellois* EU:C:2016:830.

[32] *Spencer Meyer v Travis Kalanick (Uber)*, 174 FSupp3d 817, 828 (SDNY 2016).

[33] Ezrachi and Stucke (n 7) 192.

them, and to what extent it is biased towards 'partners' of the digital assistant's company or towards that company's own products. One merely need to think of Amazon's own range of batteries, which Alexa might readily ship when requested to order AAA replacement batteries. Siri will happily recommend a book and suggest downloading it from the iBook store. And the Google Assistant might be biased towards its mother company Alphabet's vast range of products. Another option, of course, is for third parties (eg, pizza delivery companies or publishing houses) to buy their way into digital assistants' favours, eg, by becoming 'partners'. All these scenarios are possibilities.[34] If digital assistants were to engage in any such practices, the question for antitrust market definition would become to what extent this drastically narrows down any relevant product market to the choices that the digital assistant allows consumers to make, perhaps even breaking relevant markets down to an individual consumer's level based on the highly customised, data-driven answers a consumer would get. Overall, such practices would render any digital assistant's actions highly susceptible to antitrust enforcement under the provisions on anti-competitive agreements and/or single firm conduct. It is yet to be determined how competition authorities would analyse all the data necessary in order to conclude that digital assistants are, in fact, engaging in practices that narrow consumer choice and thus the relevant market. Ultimately, this calls for innovative antitrust enforcement, possibly itself relying on artificial intelligence.[35]

As could be seen from the example of the digital assistant, it is conceivable that algorithms contained in these assistants (or elsewhere) will affect substitutability as we know it and thereby have a direct impact on how antitrust markets need to be delineated. While the discussion on digital assistants suggested that this may result in a narrowing of relevant product markets, the opposite may also be the case. Where users rely on multi-task algorithms, which make multiple interdependent choices for consumers based on personal preferences, this might ultimately expand market boundaries. The idea is that such multi-task algorithms can 'completely change the overall bundle that the consumer purchases [and thereby] expand the boundaries of substitutability and market definitions'.[36] While the user may not have arrived at this exact product bundle on his or her own, it may perfectly satisfy his or her various needs. In such a case, product market definitions are pushed to include larger sets of substitutes in something that resembles a moveable system. How this can be reflected with our current tools of market definition is an

[34] Some first-hand experience gives a rather gloomy picture, however. When this author asked her iPhone's Siri, on 10 June 2017, to buy a copy of the book *Virtual Competition* by Ariel Ezrachi and Maurice Stucke, Siri said 'OK, here's iBooks…' and opened that application on the iPhone. When asked to get a hard copy or a paper copy of that same book instead, Siri stated: 'I don't know what that means. If you like, I can search the web for "I want a paper copy of this book."' The digital has simply shut out the world of hard-copy books.

[35] See, eg, F Yun Chee, 'EU Considers Using Algorithms to Detect Anti-Competitive Acts' *Reuters* (4 May 2018), www.reuters.com/article/us-eu-antitrust-algorithm/eu-considers-using-algorithms-to-detect-anti-competitive-acts-idUSKBN1I5198.

[36] MS Gal and N Elkin-Koren, 'Algorithmic Consumers' (2017) 70 *Harvard Journal of Law & Technology* 309, 328.

open question. At the same time, of course, substitutability is limited to the options that the algorithm is willing to take into account, thus leading to the same possible restricting factor as discussed in the case of digital assistants.

The budding arena of the internet of things (IoT) consists of hardware components as well as extensive data analytics and services based on the latter, and is a complex candidate for market delineation.[37] It is closely related to digital assistants – in fact, many digital assistants already have the necessary functionalities to be used to take control of your smart home. Here, we encounter similar market definition issues to the ones discussed above. Smart home systems powered by the IoT rely on the interoperability between the app through which the smart home is controlled and the hub that is connected to a multitude of electronic devices or 'things' – ranging from alarm systems to blinds, CCTV, coffee machines, door locks, fridge and freezer, garage gates, light switches, motion sensors, thermostats, water boilers, windows etc. Interoperability and standardisation are central to the IoT infrastructure and might also come to bear on the market delineation.[38] Beyond these questions, however, the IoT poses an altogether different kind of challenge to antitrust market definition. When an IoT application detects a faulty part (eg, a leaking radiator), consumables running low (eg, a printer running out of toner or a vacuum cleaner soon requiring a fresh filter) or a service being required (eg, an upcoming annual inspection for a gas boiler), the application may either suggest to order the necessary replacement parts and maintenance services or do so automatically. In both cases, the question for antitrust market definition becomes whether these maintenance services, consumables and replacement parts for an IoT device still form part of the wider markets for maintenance services, consumables and replacement parts, or whether these markets need to be narrowly defined around the choices (if any) the IoT app makes available to the consumer. This can lead to extremely narrow antitrust markets, with the strict antitrust standards that apply to these.

A feature of the growing interconnectedness of people, devices and applications described here is that we witness an increasingly 'seamless integration of markets' based on technology.[39] This integration of markets may have important repercussions on how competition works in these markets and how competition law needs to be applied to these markets, including how markets need to be defined. While the technology-driven integration of markets may expand substitutability patterns, technology also has the potential to artificially restrict substitutability options. The unprecedented possibilities that algorithms contained in digital assistants – together with IoT applications connected to these digital assistants – have to literally disconnect their users from third-party competition seems daunting.

[37] GG Wrobel, 'Connecting Antitrust Standards to the Internet of Things' (2014) 29 *Antitrust* 62, 62.

[38] ibid 63, 65; European Commission, 'Advancing the Internet of Things in Europe' (19 April 2016) SWD(2016) 110 final, 16, 19.

[39] TJ McInturff, 'LOMA Interviews Dr Canton in Emerging Technology: Shaping the Future' (on file with the author).

This can be countered by consumers who are prepared to make use of more competitive offerings from third parties through other channels. However, if users were happy to shop around and spend time comparing offers, then many of them would not use their digital assistant in the first place.

From a market definition perspective focused on demand and supply substitutability, this leaves two options: to either delineate very narrow markets at the product level, centred around the choices the digital assistant allows the user to make, or to look at this issue from the point of view of the digital platform that the digital assistant is running on and ask whether competition amongst those platforms may be such as to remedy any anti-competitive conduct at the downstream level. The latter, one could argue, is an approach that competition law is familiar with from its analysis of aftermarkets (see Chapter 7). The IoT scenario goes even further than digital assistants, as the IoT may act without the user's direct request, simply upon detecting the need for a replacement part, a consumable or a service. In such circumstances, it is equally conceivable that IoT-related markets need to be delineated based on an analogy with the market definition standards for innovative aftermarkets. Thereby, one can look at the IoT hub as the primary market, with the connected devices as the aftermarkets and the consumables related to those aftermarkets as sub-aftermarkets. The question then essentially becomes whether competition in the connected devices, or indeed in the IoT hub itself, can remedy market power – or perhaps even a monopoly position – in the narrowly defined sub-aftermarkets.

C. How Innovation Impacts Markets: A Tour d'Horizon

Although digital markets have attracted a lot of attention from antitrust scholarship in recent years due to their impressive rise and important repercussions for users' daily lives, innovation may extend to any market, through innovative products, better processes, new technologies or improved applications. A great number of areas that have not yet been touched upon are currently considered to be highly innovative and possibly paradigm-shifting in the not-so-distant future, with other innovations looming on the horizon.[40] The digitisation of the economy sometimes has far-reaching consequences for the antitrust markets that we rely upon in competition law analysis. Some of these exciting areas were pointed out above, while others are discussed in the following, together with their possible impact on antitrust market delineation. This short tour d'horizon covers areas as diverse as

[40] See E Hiltunen and K Hiltunen, *Technolife 2035: How Will Technology Change Our Future?* (Newcastle upon Tyne, Cambridge Scholars Publishing, 2015); K Kelly, *The Inevitable: Understanding the 12 Technological Forces that Will Shape Our Future* (New York, Viking, 2016); P Larrey, *Connected World – From Automated Work to Virtual Wars: The Future, by Those Who are Shaping it* (London, Portfolio Penguin, 2017); Ross (n 29).

robotics, the professions, self-driving cars, payment systems, the life sciences and data analytics. One could add an endless number of areas to this list.

Robotics is an area that has seen great advances over the last decades. It is imaginable that robots or 'intelligent machines'[41] could be used in very diverse settings, upsetting the established market order of things. To give but one example, robots are being developed that can take on the role of caregivers for the elderly,[42] thus replacing an increasingly scarce human resource for which demand is steadily growing, namely nurses. Another type of robot developed for nursing homes are therapeutic robots, one of which mimics animal therapy for patients with memory issues.[43] Such a technological innovation might then directly compete with a human (or animal) service. Although the self-learning algorithms that power these robots can already carry out astonishing tasks, humans continue to 'outperform' robots with regard to creativity and cultural sensitivity.[44] For the time being, there thus remain some areas in which caregiver robots are not interchangeable with human nurses, and therapeutic robots are not interchangeable with human-provided therapies involving live animals. Acceptance by the elderly is another factor that has an impact on the extent to which caregiver robots or therapeutic robots are seen as (partial) substitutes for human nurses and human therapists. Overall, however, the prospect of properly developed caregiver and therapeutic robots is an exciting possibility for geriatric care and an interesting one from the perspective of substitutability, which, as was discussed in Chapter 2, lies at the very heart of antitrust market delineation. Many more settings are thinkable, such as humanoid robots that assist in the household, performing varied tasks ranging from carrying items and cleaning to conversation.[45] While we have already gotten used to robots mowing our lawns or vacuum cleaning our homes, many more developments are expected in this area. Is your robotic mower competing with your gardener and is your robot vacuum cleaner in competition with your housekeeper? Just like industrialisation entailed far-reaching change in previous centuries, many services markets are set to change significantly based on these technological advances.[46] For market definition purposes, it will be important to focus on demand and supply substitutability rather than on the means through which certain services are delivered. This may indeed lead to broader market definitions if (partial) substitutability between robots and human service providers is deemed sufficient.

The technology-induced changes that are conceivable as regards geriatric care also seem tangible in many other professions, even though technology might affect

[41] Hiltunen and Hiltunen (n 40) 123.

[42] Ross (n 29) 16.

[43] See Hiltunen and Hiltunen (n 40) 1250. The PARO therapeutic robot looks like a baby harp seal. Toyota is also developing a range of social, human support and rehabilitation robots.

[44] See Ross (n 29) 182.

[45] Honda is developing ASIMO, a humanoid with possible helper applications.

[46] Hiltunen and Hiltunen (n 40) 131.

every single one of them differently: Whether in architecture, education, health, journalism, law, management consulting or tax consulting, technology has already considerably transformed the professions in each one of these – and many more – fields, with more change yet to come.[47] Innovation in the professions might inevitably lead to a new perception of antitrust markets in particular services, with technology filling in where it was previously thought inconceivable. The same applies to areas such as farming, where sophisticated farm technology is introducing smart farming based on systems equipped with sensors and algorithms.[48] Only recently, the acquisition of Monsanto by Bayer led to parallel merger investigations in the EU and the US in which both merging parties' activities in digital agriculture services were amongst the product markets concerned.[49]

Self-driving cars are another area of innovation for which hopes are high, building upon a range of different technologies such as radar, sensors, GPS, infrared, 3D imaging and many more.[50] As was discussed above,[51] in innovative markets performance is often a more relevant competitive parameter than price. Self-driving cars are a particularly good example of an innovative market where performance relating to safety is crucial,[52] and arguably much more important a criterion than price. In the near future, self-driving cars may prove highly disruptive to the established car industry, leading to new and highly innovative product markets that could gradually replace old ones.[53] At the same time as self-driving cars are being developed, e-mobility is already replacing fuel-powered mobility in both cars and mopeds (while replacing muscle power in bicycles), a transformative technology that also brings with it interesting market definition challenges. Substitutability in this respect may still only be partial due to the limited range that e-cars currently have as compared to those powered by fossil fuels, but this will equally evolve over time.

3D printing, as a new production process, allows for a broad range of applications and could entirely disrupt the supply chain in many markets.[54] The particular challenge for market delineation that stems from 3D printing will be this technology's impact on unrelated markets that could lead to new competition dynamics.

How we make payments or money transfers is also evolving based on technological innovation.[55] How do Apple Pay, TransferWise and Square compete

[47] See RE Susskind and D Susskind, *The Future of the Professions: How Technology Will Transform the Work of Human Experts* (Oxford, Oxford University Press, 2015).

[48] See Ross (n 29) 161 ff; Climate Corporation's Climate FieldView.

[49] See *Bayer/Monsanto* (Case M.8084) Commission Decision of 21 March 2018 [2018] OJ C456/10; US Department of Justice, 'Justice Department Secures Largest Negotiated Merger Divestiture Ever to Preserve Competition Threatened by Bayer's Acquisition of Monsanto' (29 May 2018).

[50] Ross (n 29) 28 ff.

[51] See section II.C in ch 3.

[52] Larrey (n 40) 43; Ross (n 29) 32.

[53] Ross (n 29) 30.

[54] MS Gal, '3D Challenges: Ensuring Competition and Innovation in 3D Printing' (SSRN abstract nr 3356891, 20 March 2019).

[55] Ross (n 29) 79.

with more traditional credit card transactions, bank transfers and money lending? When these services evolve, how can market delineation take this into account? For instance, TransferWise is now offering borderless bank accounts in addition to its cross-border money transfers, while Square was initially focused on payment services and has now moved into lending as well. Could digital currencies or cryptocurrencies such as Bitcoin and Libra compete – in some previously unknown sense – with state currencies? Blockchain technologies more generally may also lead to a paradigm shift in antitrust law, allowing for new kinds of engaging in anti-competitive conduct.

Genomics, medical diagnosis via software, personalised medicine and the life sciences more generally are areas in which innovation is pushing frontiers in many respects,[56] leading to entirely new antitrust markets or replacing and complementing established ones. In addition, mobile technologies enable a much broader geographical spread of these important advances in health care,[57] thus impacting the relevant geographical market. A great many antitrust cases have also arisen in the pharmaceutical sector, where innovation and the safeguarding of intellectual property play a particularly significant role. The many cases discussed further below are witness to the manifold antitrust issues – and antitrust market definition issues – that can arise in connection with pharmaceuticals, ranging from active pharmaceutical ingredients being regarded as the relevant product market to a branded pharmaceutical defined as an antitrust market separate from the same generic pharmaceutical, based on customer response.[58] Where pharmaceuticals are differentiated, eg, in the case of non-bioequivalent medicines for treatment of the same ailment, one needs to particularly take customer response into account.[59] Personalised medicine is another field in which antitrust might need to rethink demand substitutability, perhaps resorting to partial substitutability and market boundaries that are not the straight lines competition lawyers would like to see. We might find very narrow antitrust markets for personalised medicine and corresponding augmented antitrust responsibility resulting therefrom.

Data and information play a central role in today's economy and are vital to virtually all innovations that were touched upon in this section. This, of course, encompasses big data analytics to process this data. Based on data's importance, competition laws may need to consider information as a product market in its own right, thereby opening the door to antitrust scrutiny in these highly diverse markets – with examples including credit ratings, crowd-based review systems for travel and restaurants, data lockers, loan rate indicators, online price comparison tools and online search.[60]

[56] See Hiltunen and Hiltunen (n 40) 181 ff; Ross (n 29) 47 ff.
[57] Ross (n 29) 69.
[58] See sections III and IV in ch 6.
[59] DA Crane, 'Market Power without Market Definition' (2014) 90 *Notre Dame Law Review* 31, 78.
[60] Patterson (n 26) 2 f, 14 f.

D. Defining Innovative Product Markets: Conceptual Issues

Antitrust law has developed methods of market definition that primarily cater to static market environments with homogeneous products in which quality and price are the main competitive parameters.[61] The issue, of course, is that one will not find good indicators for innovation occurring in the present or near future by looking at a statically defined market.[62] As was seen in the previous sections, innovation in a great number of areas presents challenging scenarios for antitrust market delineation. Where innovation plays an important role in a market, it can be difficult to uncover demand- and supply-side substitutability so as to enable a traditional market definition for competition law purposes.[63] Under such circumstances, it becomes ever more important to understand how competition works in innovative market environments,[64] thus focusing on the market characterisation function of market definition rather than on its market power function. Only then can the 'boundaries of competition'[65] be properly made out. Section I.A attempted to build such an understanding of some of today's most challenging innovative, dynamic product markets, with a view to navigating possible substitutability patterns. This enterprise is complicated by the fact that traditional market definition looks towards the past to inform itself,[66] while innovation is future-oriented and will often only have effects in the near or perhaps even more distant future. At the same time, data might also be a solution to some of the market definition conundrums in which competition law enforcers find themselves. With its ability to 'nowcast' (ie, to predict the imminent future), big data analytics may become able to point towards competitive relationships in real time, thus allowing competition law insights into market boundaries for antitrust purposes that are unprecedented because they are not based on the (distant) past or the frequently unpredictable future, but on real-time data.

A detailed market characterisation can greatly increase our insight into competition in a market, and into demand and supply substitutability. EU competition law has long emphasised that it is vital to look at an agreement's economic and legal context in order to allow for an insightful competition law analysis.[67] Contextuality is particularly important for innovative market environments, where the market

[61] BR Kern, 'Innovation Markets, Future Markets, or Potential Competition: How Should Competition Authorities Account for Innovation Competition in Merger Reviews?' (2014) 37 *World Competition* 173, 174.

[62] DH Ginsburg and JD Wright, 'Dynamic Analysis and the Limits of Antitrust Institutions' (2012) 78 *Antitrust Law Journal* 1, 4.

[63] DJ Teece and M Coleman, 'The Meaning of Monopoly: Antitrust Analysis in High-Technology Industries' (1998) 43 *Antitrust Bulletin* 801, 826.

[64] MB Coate and JJ Simons, 'In Defense of Market Definition' (2012) 57 *Antitrust Bulletin* 667, 708.

[65] European Commission, Notice on the definition of relevant market for the purposes of Community competition law [1997] OJ C372/5 (EU Market Definition Notice 1997) para 2.

[66] Teece and Coleman (n 63) 826.

[67] Case 23/67 *Brasserie de Haecht v Wilkin* EU:C:1967:54, 415. See already above, section II.B in ch 2.

characteristics require particular attention when delineating the relevant market and when assessing behaviour thought to be anti-competitive. By anchoring a market definition in its specific competitive context, antitrust enforcers can gain a great deal of insight into the realities of the market under scrutiny, thus allowing a more realistic, fact-grounded delineation of the relevant product market.

Market definition concerns questions of fact as well as questions of law.[68] The necessary facts will need to be carefully established and interpreted in innovation-intensive industries so as to obtain a realistic picture of the competitive situation. At the same time, the law may need to evolve with these innovative markets. This is in line with the US Supreme Court's view that antitrust law can and should adapt to new economic insights.[69]

Innovation can have a disruptive impact not only on products but also on processes, on the way in which business is conducted and competition is shaped. 'New technologies are changing the dynamics of competition as we know it' based on data harvesting, big analytics, the increasing use of self-learning algorithms and the subsequent change of market dynamics.[70] One of the possible consequences is a rise in behavioural discrimination based on our user profile or the market circumstances prevailing at a particular moment in time, eg, through price discrimination, dynamic pricing or price steering (nudging).[71] Eventually, such dynamic pricing may indeed lead to the disappearance of a conventional market price.[72] Where such behavioural discrimination becomes part of everyday business, the question for antitrust market definition becomes a very fundamental one: can traditional, price-based tools for market definition apply in such emerging scenarios that lack a uniform price? Two remarks shall be made in this respect: first of all, many more traditional markets do not display a uniform market price, eg, where posted prices are merely the basis for price negotiations; and, second, the lack of a price as the focus of antitrust attention does not mean that the market mechanism is no longer important. Instead, competition law will need to focus on other parameters of competition. While the latter often take second place in current antitrust analysis because price is more readily measurable, this change in competition dynamics may mean that parameters such as quality or innovation finally receive the attention they deserve.

Where markets are asymmetric, ie, characterised by one-way substitution, it becomes particularly important to clearly set out which is the focal product for the antitrust market delineation under way. In an asymmetric market, there may be substantial substitution from product A to product B, so that product B is in

[68] IB Stewart, 'Mergers and Competition: An Analysis of Section 50 of the Trade Practices Act' (2000) 74 *Australian Law Journal* 533, 535.

[69] *Kimble v Marvel Entertainment*, 576 US ___ (2015), 135 SCt 2401, 2412 f (2015).

[70] Ezrachi and Stucke (n 7) 27 (direct quote), 71 ff, 233.

[71] JA Hausman, GK Leonard and CA Vellturo, 'Market Definition under Price Discrimination' (1996) 64 *Antitrust Law Journal* 367, 369 ff; A Lindsay and E McCarthy, 'Do We Need to Prevent Pricing Algorithms Cooking up Markets?' (2017) 38 *European Competition Law Review* 533.

[72] Ezrachi and Stucke (n 7) 212. See also EU Market Definition Notice 1997 (n 65) para 43.

product A's market. Vice versa, however, there may be no substantial substitution from product B to product A, meaning that product A is not in product B's market.[73] Therefore, it will make a difference for the relevant market definition whether product A is being investigated under the competition laws or whether it is product B. A somewhat similar pattern can be observed in multi-sided markets, something that will be discussed further in Chapter 8. While customers are the natural focus of any inquiry into demand substitutability, they have an additional importance for innovative markets, as their customer response can influence how these markets will develop in the future,[74] ie, which products will be further developed, for which products accessories will come to the market and which innovations will exit the market again.

Supply substitution at the stage of market definition has a bigger role to play under EU competition law than under US antitrust law, but this imbalance might be remedied in the US during the substantive assessment.[75] For innovative market environments, a challenge in realising possibilities for supply substitution is the fact that innovation is usually not carried out in the open, and this secretive element makes any such assessment error-prone.

One option in order to better account for innovation is to reject the US Horizontal Merger Guidelines' smallest market principle and instead adopt a 'broad market test' in dynamic markets, which would focus not on single products for which substitutability is established, but rather on the industry concerned. Thereby, partial substitutability as well as potential competition could more easily be taken into account.[76] At the same time, market shares or concentration ratios could then no longer be applied to these broad markets. In addition, the majority of indicia put forward in *Brown Shoe* cannot be meaningfully applied in innovation-driven industries. For instance, the peculiar characteristics and uses of a product that the Supreme Court postulated in that case are not very useful in innovative environments with their heterogeneous products.[77] The data that is gathered in many digital markets and new processes such as 3D printing can be used for multiple purposes and may have effects in an unforeseeable number of markets.

A remarkable feature of companies that can be referred to as big tech is their diversification: Google has expanded from being a renowned online search engine provider to many other markets, such as mobile software, online flight search,

[73] G Niels, H Jenkins and J Kavanagh, *Economics for Competition Lawyers*, 2nd edn (Oxford, Oxford University Press, 2016) § 2.57.

[74] PA Geroski, 'Thinking Creatively about Markets' (1998) 16 *International Journal of Industrial Organization* 677, 677.

[75] RL Smith, 'Defining and Proving Markets and Market Power' in J Duns, A Duke and BJ Sweeney (eds), *Comparative Competition Law* (Cheltenham, Edward Elgar, 2015) 30 f. See sections I.A.ii and II.A.i in ch 2.

[76] AC Hruska, 'A Broad Market Approach to Antitrust Product Market Definition in Innovative Industries' (1992) 102 *Yale Law Journal* 305, 311.

[77] C Pleatsikas and DJ Teece, 'The Analysis of Market Definition and Market Power in the Context of Rapid Innovation' (2001) 19 *International Journal of Industrial Organization* 665, 673 ff. On *Brown Shoe*, see section I.A.iii in ch 2.

online maps, online shopping comparison, video streaming through its acquisition of YouTube, web browser software etc. Somewhat further removed from the public eye, Google's mother company Alphabet is also heavily involved in self-driving car technology (Waymo) and life sciences (Calico and DeepMind Health). Amazon started out as a popular online bookseller and is now present on many other markets, such as online shopping in general, e-books, video streaming, digital assistants, as a platform for retailers etc. A little further removed from the public eye, Amazon is also testing innovative delivery methods. This makes the digital giants appear more and more like the conglomerates of the digital age. For market definition, these developments open up questions on how synergies between these markets affect competition, particularly when it comes to the data accumulation mentioned above.

II. Product Differentiation Based on Innovation

Innovation and technology may lead to considerable product differentiation that challenges antitrust market definition. After some general thoughts concerning innovative product differentiation, we will turn to three areas of particular interest: the online/offline paradigm; substitutability between free and ad-free (freemium) services in the digital environment; and functionalities as sources of product differentiation. IPRs constitute an additional dimension of product differentiation related to innovation, and are discussed in Chapter 6.

A. General Issues

Demand-side and supply-side substitutability are the basic pillars of antitrust market definition, as was amply discussed above (see sections I.A and I.B in Chapter 2). Where products are differentiated (heterogeneous) rather than homogeneous, the question arises as to how substitutability needs to be conceptualised in such market environments.[78] In differentiated products, even less than in homogeneous products, there are rarely clear breaks in the market continuum that allow for an unequivocal finding of an antitrust market.[79] This makes the assessment of product substitutability a particularly delicate enterprise in

[78] R Brenkers and F Verboven, 'Market Definition with Differentiated Products: Lessons from the Car Market' in JP Choi (ed), *Recent Developments in Antitrust: Theory and Evidence* (Cambridge, MA, MIT Press, 2007) 154; J Farrell and C Shapiro, 'Antitrust Evaluation of Horizontal Mergers: An Economic Alternative to Market Definition' (2010) 10 *BE Journal of Theoretical Economics* 1, 1.

[79] JB Baker and C Shapiro, 'Reinvigorating Horizontal Merger Enforcement' in R Pitofsky (ed), *How the Chicago School Overshot the Mark: The Effect of Conservative Economic Analysis on U.S. Antitrust* (Oxford, Oxford University Press, 2008) 264.

differentiated products.[80] At the same time, a reasonable antitrust analysis requires reasonable markets, so it is often advised not to rely on 'atomized mini-markets'.[81] Whether or not products are differentiated is best assessed 'at the point of competition ... where different products compete for a customer's eye'.[82] An observation of the competitive relationships among companies offering differentiated products also provides important insights.[83]

Linking back to the short product cycles in innovative markets, one can further observe that '[p]roduct innovation ... creat[es] goods differentiated over time'.[84] Several generations of a good may be present on the market at the same time – for instance, the original iPhone introduced in 2007, today's newest models, the iPhone 11 and 11 Pro, and all the models in between. Which of these different product generations are substitutable with each other? This aspect of product differentiation is closely connected to market definition's time horizon[85] and is also discussed below.

Product differentiation can emanate from a variety of sources, such as the product's design or physical properties and the seller's differentiation endeavours through advertising, packaging, branding and add-on services.[86] The phenomenon of product differentiation is not unique to the digital environment. Personalised products as such are nothing new either – one merely needs to think of pens with name engravings and tailored suits. From the 1930s to the 1950s, the focus on each supplier's product differentiation led to the emergence of the concept of 'monopolistic competition'.[87] This concept, it seems, receives a new application with the particular ease of product differentiation that e-commerce and new production technologies have enabled. While the copyright laws in the EU and the US may already be based on a model of differentiated products,[88] the drive of digitisation may provide an important impetus in this direction for many fields outside of IP protection. What innovation has achieved in recent years is an unprecedented ease

[80] See already JA Keyte, 'Market Definition and Differentiated Products: The Need for a Workable Standard' (1995) 63 *Antitrust Law Journal* 697, 698.

[81] FJ Säcker, *The Concept of the Relevant Product Market: Between Demand-Side Substitutability and Supply-Side Substitutability in Competition Law* (Frankfurt am Main, Peter Lang, 2008) 23 (with multiple examples at 16 ff).

[82] G Goeteyn, P Smith and S Ashall, 'Away from Market Shares? The Increasing Importance of Contestability in EU Competition Law Cases' (2015) 6 *Journal of European Competition Law & Practice* 197, 198.

[83] R Schmalensee, 'Another Look at Market Power' (1982) 95 *Harvard Law Review* 1789, 1800.

[84] JB Baker, 'Product Differentiation through Space and Time: Some Antitrust Policy Issues' (1997) 42 *Antitrust Bulletin* 177, 178.

[85] See also Office of Fair Trading, 'Market Definition' (OFT 403, December 2004) para 5.1.

[86] JS Bain, *Barriers to New Competition* (Cambridge, MA, Harvard University Press, 1956) 114.

[87] GJ Werden, 'The History of Antitrust Market Delineation' (1992) 76 *Marquette Law Review* 123, 126, citing J Robinson, *The Economics of Imperfect Competition* (London, Macmillan, 1954) and EH Chamberlin, *The Theory of Monopolistic Competition: A Re-orientation of the Theory of Value*, 8th edn (Cambridge, MA, Harvard University Press, 1969). Both works were initially published in the 1930s.

[88] DJ Gifford and RT Kudrle, *The Atlantic Divide in Antitrust: An Examination of US and EU Competition Policy* (Chicago, University of Chicago Press, 2015) 162.

and speed with which products can now be personalised – ranging from books being printed on demand, children's books bearing a particular child's name, personalised clothing labels in all shapes and sizes, software made to a particular company's specifications,[89] personalised marketing and much more. This may lead to differentiated products and limited interchangeability from the demand side. However, if one agrees that supply substitutability has an important role to play in analysing these markets, this differentiation can, to some degree, be countered by supply-side substitution.[90] In this way, innovation may be able to thwart market power that is based on product differentiation by coming up with new ways of producing and delivering differentiated products.[91]

While economists have pointed out long ago that product substitutability is a matter of degree, it is especially apt to call this insight to mind in differentiated products.[92] From a legal perspective, market definition needs to delineate the boundaries of competition, an ever-increasing challenge in differentiated and fast-moving markets.

B. The Online/Offline Paradigm

In the arena of digital markets, a particular aspect of product differentiation that can be referred to as the online/offline paradigm[93] comes into play. It asks whether products that are provided on the internet (online/digitally) compete with identical or comparable products that are offered in the real world (offline/in an analogue way). The question whether online and offline product offerings compete with each other – or whether they are differentiated to such a degree that they do not – is a central one.[94] While the online/offline paradigm may sometimes be confined to issues of distribution channels (eg, the sale of the same branded T-shirt in a brick-and-mortar shop or through a website), some services lead to more intricate questions as they can be directly provided online (eg, a meeting with a tax consultant or using a tax returns online portal).

Advertising is an area in which the online/offline paradigm has been debated with particular intensity. Despite empirical data showing considerable substitution

[89] However, see the Commission's view on enterprise communications services in *Cisco/Tandberg* (Case COMP/M.5669) Commission Decision of 29 March 2010, para 23.

[90] Brenkers and Verboven focus on demand-side substitutability instead. See Brenkers and Verboven (n 78) 158.

[91] VHSE Robertson, 'Delineating Digital Markets under EU Competition Law: Challenging or Futile?' (2017) 12 *Competition Law Review* 131, 151.

[92] JS Bain, *Price Theory* (New York, Holt, Rinehart and Winston, 1952) 25, 52; F Machlup, *The Economics of Sellers' Competition: Model Analysis of Sellers' Conduct*, 5th edn (Baltimore, Johns Hopkins Press, 1952) 213; R Pitofsky, 'New Definitions of Relevant Market and the Assault on Antitrust' (1990) 90 *Columbia Law Review* 1805, 1807.

[93] Robertson, 'Delineating Digital Markets' (n 91) 146.

[94] Similarly, see SW Waller, 'Antitrust and Social Networking' (2012) 90 *North Carolina Law Review* 1771, 1779.

between online and offline advertising,[95] the assumption prevails in both the EU and the US that these two types of advertising constitute separate relevant product markets for the purposes of competition law.[96] In the EU, a number of cases have focused on the question of substitutability between online and offline advertising. For instance, in the early case of *Telia/Telenor/Schibstedt* (1998), the European Commission did not contest the parties' view that online advertising constituted one of the relevant product markets, but ultimately left the market definition open.[97] Shortly afterwards, in *Vodafone/Vivendi/Canal Plus* (2000), the Commission held that online advertising was a separate relevant product market and was likely to be national from a geographical point of view for language and other cultural reasons.[98] In *Google/DoubleClick* (2008), the Commission highlighted several features that distinguished online advertising from more conventional offline advertising. It found that online advertising constituted a separate relevant market for antitrust purposes, based on the industry's perception that it was in fact a separate market, online advertising's capability of precisely targeting its audience and afterwards measuring this effectiveness, and the pricing system based on an advertisement's actual reach.[99] Later cases related to online advertising reached a similar conclusion, such as *Microsoft/Yahoo! Search Business* (2010), *News Corp/BSkyB* (2010) and *Facebook/WhatsApp* (2014).[100] It is interesting to note that these cases did not refer to the two-sidedness of these online markets.[101]

The online/offline paradigm has received less attention in US case law. The FTC investigated the *Google/DoubleClick* merger (2007) on the premise that online advertisements were not in the same relevant market as offline advertisements, without further elaborating on its assumption.[102] The case also gave the FTC a chance to weigh in on the interchangeability of various forms of online

[95] Listing a number of empirical studies to that effect, see A Goldfarb and C Tucker, 'Substitution between Offline and Online Advertising' (2011) 7 *Journal of Competition Law & Economics* 37.

[96] See also ibid.

[97] *Telia/Telenor/Schibstedt* (Case IV/JV.1) Commission Decision of 27 May 1998 [1998] OJ C220/28, paras 13 ff, particularly para 18. See also *Telia/Telenor* (Case COMP/M.1439) Commission Decision 2001/98/EC [2001] OJ L40/1, para 107 (referring to its previous case).

[98] *Vodafone/Vivendi/Canal Plus* (Case COMP/JV.48) Commission Decision of 20 July 2000 [2003] OJ C118/25, paras 43 f.

[99] *Google/DoubleClick* (Case COMP/M.4731) Commission Decision of 11 March 2008 [2008] OJ C184/10, paras 45 f. This product market definition is strongly anchored in online advertising's product characteristics rather than in actual substitutability; see JD Ratliff and DL Rubinfeld, 'Online Advertising: Defining Relevant Markets' (2010) 6 *Journal of Competition Law & Economics* 653, 675 fn 72; Monopolkommission, 'Wettbewerbspolitik: Herausforderung digitale Märkte' (Sondergutachten 68, 2015) para 138. DoubleClick has in the meantime become the Google Marketing Platform.

[100] *Microsoft/Yahoo! Search Business* (Case COMP/M.5727) Commission Decision of 18 February 2010, para 61; *News Corp/BSkyB* (Case COMP/M.5932) Commission Decision of 21 December 2010, paras 262, 268; *Facebook/WhatsApp* (Case COMP/M.7217) Commission Decision of 3 October 2014 [2014] OJ C417/4, paras 74, 75, 79.

[101] See Robertson, 'Delineating Digital Markets' (n 91) 145.

[102] Federal Trade Commission, 'Statement Concerning *Google/DoubleClick*, Case 071-0170' (20 December 2007) 1. The FTC never publicly stated this assumption; see Waller (n 94) 1783.

advertising. It found that online advertising space sold by search engines and online display advertising space sold by content providers ('publishers') was not interchangeable.[103] In addition, it was established that publishers could either directly sell display advertising to advertisers or do so through an intermediary. The FTC held that ad intermediation was not interchangeable with direct advertising contracts between publishers and advertisers,[104] thus effectively finding that a multitude of relevant antitrust markets existed that provided various differentiated, non-substitutable advertising services.

FTC Commissioner Pamela Jones Harbour dissented in *Google/DoubleClick*. She found that there were – or soon would be – significant overlaps in the relevant markets that the majority relied upon.[105] She held that the merger could have an important impact on the development of the entire online advertising market. Even though the majority did not regard search and display advertising as interchangeable, they were closely related and did 'compete for advertising dollars'.[106] She also held that the merger would lead to a faster convergence of search and display advertising by combining the data sets held by Google and DoubleClick, respectively.[107] Thus, she ultimately advocated a broader market definition in the face of the innovation-driven, data-intensive digital marketplace – despite the perceived differentiation.

As the foregoing discussion has shown, there is no agreement on whether the various forms of online advertising constitute one or several antitrust markets, let alone on whether online and offline advertising form part of the same relevant product market. This holds true not only for the US but also for the EU. In *Microsoft/Yahoo!Search Business*, the European Commission held that the notified acquisition related to internet search and search advertising.[108] While internet search is usually provided free of charge, it is financed by advertising revenue generated through search advertising and thus constitutes a multi-sided platform.[109] The Commission found online advertising to constitute a relevant market, but left open the issue of whether segments of that market – such as online search advertising services – constituted their own product market, and did not conclude on internet search as a possible relevant market either.[110] These market segments would be reminiscent of the sub-markets in *Brown Shoe*. In its competitive assessment, the Commission acknowledged that the companies' ability to innovate was

[103] Federal Trade Commission, '*Google/DoubleClick*' (n 102) 3.

[104] ibid 4.

[105] P Jones Harbour, 'Dissenting Statement, In the Matter of Google/DoubleClick' (20 December 2007) 5 f.

[106] ibid 4, 5 (direct quote).

[107] ibid 7.

[108] *Microsoft/Yahoo! Search Business* (Case COMP/M.5727) Commission Decision of 18 February 2010, para 29.

[109] ibid paras 33, 47, 100.

[110] ibid para 87.

an important competitive parameter for search engines, a fact that was underscored by the innovation that the market had exhibited in previous years and by the high barriers to entry.[111] In *Verizon/Yahoo* (2016), the Commission held online and offline advertising to constitute separate relevant markets, but again left open the question of whether online advertising could be further segmented.[112]

In the US District Court case of *KinderStart.com v Google* (2007), the judge held that nothing distinguished search advertising from the broader market of online advertising that would justify the finding of a separate antitrust market. Search-based and non-search advertising were in fact substitutable.[113] In a 2008 investigation of an advertising agreement envisaged between Google and Yahoo, the DoJ concluded that online search advertising and online search syndication constituted their own relevant antitrust markets, with Google by far the biggest market participant in both markets and Yahoo its most important competitor at the time.[114] The DoJ thus differentiated between online search ads that appear when an online search engine is being used and online search ads that appear on partner websites ('syndication partners') on which a search function, eg from Google, is included. This market definition implies that non-search ads are outside the scope of this relevant market. Product differentiation certainly played an important role in the definition of those antitrust markets. However, the DoJ did not discuss whether online and offline advertising could be in the same relevant antitrust market, nor did it when closing its investigation into the Microsoft and Yahoo advertising agreement in 2010.[115]

The relationship between online and offline markets for the purposes of market definition is also highly relevant in areas outside of advertisements. Concerning the provision of online music, for instance, the European Commission decided in the controversial *AOL/Time Warner* merger (2001) that online music was in a separate relevant market from offline music,[116] while in *Telefónica/Terra/Amadeus* (2000), it held that online and offline travel agency services competed (for now) in the same relevant market.[117] This Commission approach was criticised for its

[111] ibid paras 109–11.

[112] *Verizon/Yahoo* (Case M.8180) Commission Decision of 21 December 2016, paras 9–12, 25.

[113] *KinderStart.com v Google*, Case C 06-2057 JF (RS) (ND Cal 16 March 2007) § III.1.a.i.

[114] US Department of Justice, 'Yahoo! Inc. and Google Inc. Abandon Their Advertising Agreement: Resolves Justice Department's Antitrust Concerns, Competition is Preserved in Markets for Internet Search Advertising' (5 November 2008).

[115] US Department of Justice, 'Statement of the Department of Justice Antitrust Division on its Decision to Close its Investigation of the Internet Search and Paid Search Advertising Agreement between Microsoft Corporation and Yahoo! Inc.: Investigation Shows that Agreement Not Likely to Reduce Competition' (18 February 2010).

[116] *AOL/Time Warner* (Case COMP/M.1845) Commission Decision 2001/718/EC [2001] OJ L268/28, para 26 (calling this market 'emerging').

[117] *Telefónica/Terra/Amadeus* (Case COMP/M.1812) Commission Decision of 27 April 2000 [2000] OJ C235/6, para 12. See also M Colangelo and V Zeno-Zencovich, 'Online Platforms, Competition Rules and Consumer Protection in Travel Industry' (2016) 2 *Journal of European Consumer and Market Law* 75, 77.

subjectivity and its reliance on product characteristics.[118] In its 2011 investigation of the Google/ITA acquisition, the DoJ held that brick-and-mortar travel agents were not 'a reasonable substitute for comparative flight search services online' because of the different characteristics these services offered users.[119] Arguably, with recent innovations in this sector, this conclusion might be different today. Similar issues arise when attempting to define the relevant market for the sale of e-books: can Google Play's Book Store, Apple's iBooks, Amazon's Kindle Store and their competitors be said to operate in a market for e-books, ie, are they sufficiently differentiated from physical ('offline') books to constitute a separate relevant antitrust market?[120] In the commitments that the European Commission accepted in the *Amazon E-Books* case (2017), for instance, it considered that the retail distribution of e-books was not in the same relevant market as the retail distribution of print books, based on insufficient interchangeability between the two.[121] If that is the case, can we really assume that there is a combined relevant market for e-books rather than for single platforms? E-book readers provide the platform through which publishers make their e-books available, and readers access e-books by buying them or sometimes borrowing them from a library. An e-book bought through one platform, such as Amazon's Kindle Store, cannot simply be read on Apple's iBooks, but will be accessible through the Kindle app for iOS.[122] One can also find e-book converter apps that change an e-book's format to make it accessible on various platforms, but this is not a standard application that smartphones or e-book readers come equipped with. Here, one might refer back to the conceptualisation of digital platforms for the purposes of market definition noted above (see section I.B.ii). The question of substitutability in e-books, of course, not only involves issues of product differentiation but also links back to IPRs and touches upon the issue of multi-sided platform markets, both of which will be discussed below (see Chapters 6 and 8). The example shows how the multiple characteristics of innovative markets (see section II in Chapter 3) lead to a combination of issues arising in one and the same market, thereby posing enhanced challenges to market delineation.

Sometimes, cases involving innovative products may need to be assessed (slightly) differently, depending on whether they relate to offline markets (such as the *Microsoft* cases involving software) or online markets (such as the European

[118] G Monti, 'Article 82 EC and New Economy Markets' in C Graham and F Smith (eds), *Competition, Regulation, and the New Economy* (Oxford, Hart Publishing, 2004) 26.

[119] *United States v Google/ITA Software*, Case 1:11-cv-00688 (Complaint, DDC 8 April 2011) para 22; this case was subsequently decided on 5 October 2011.

[120] The search within Google Books' electronic repository is not identical to the Play Book Store in which electronic copies of books can be bought. For a discussion of e-books as multi-sided markets, see section III.A in ch 8.

[121] *E-Book MFNs and Related Matters (Amazon)* (Case AT.40153) Commission Decision of 4 May 2017 [2017] OJ C264/7, paras 42 f (further sub-dividing the markets by language).

[122] There is currently no equivalent iBooks app for Android phones.

Commission's *Google* cases).[123] However, it should be highlighted that these online and offline markets are converging. With conventional offline software becoming more and more integrated into the online world, we might today strongly question whether software is actually an offline market; the days where software was bought on a floppy disk are long gone, and software now regularly includes integrated online features and online help. In many cases, a clear-cut online/offline distinction may no longer be possible, not only for software but also in many other domains. This is also true for pricing: while one might still find different prices offered by online and offline retailers, this differentiation may disappear thanks to algorithm-driven pricing mechanisms and enabled by digital price tags in shops that can be centrally administrated.[124] Online price comparison sites may also transport price competition from the online realm to the offline realm, for customers may rely on such sites in order to haggle a better deal out of brick-and-mortar retailers, or brick-and-mortar retailers may issue 'price guarantees' based on such price comparison sites. One therefore finds multiple indicators for a gradual convergence of many online/offline markets.[125]

During its e-commerce sector inquiry, the European Commission observed that in retailing, six out of 10 retailers now rely on a multi-channel approach, selling their goods online and offline.[126] At the same time, consumers are increasingly switching between online and offline sales channels, leading to increased price transparency and price competition amongst those two distribution channels.[127] The importance of distribution channels as a criterion for market delineation is repeatedly addressed in European cases. In its 2015 *HRS* judgment on parity clauses in a hotel booking portal, a German court argued that whether products were distributed online or offline merely constituted a strategic decision and should not lead to the finding of separate antitrust markets.[128] And the French Autorité de la concurrence, in its *Fnac/Darty* decision (2016), held that online and offline retailing had become sufficiently substitutable in the case before it so as to constitute one antitrust market, with the result that the merger was cleared subject to very few commitments.[129] In the EU, cases on restrictions on online sales, such

[123] G Surblytė, 'Competition Law at the Crossroads in the Digital Economy: Is it All about Google?' (2015) 1 *Journal of European Consumer and Market Law* 170, 172.

[124] Ezrachi and Stucke (n 7) 21.

[125] Arguing that the online/offline dimension may be assessed differently depending on the type of business concerned, see J Kagan, 'Bricks, Mortar, and Google: Defining the Relevant Antitrust Market for Internet-Based Companies' (2011) 55 *New York Law School Law Review* 271, 284.

[126] European Commission, 'Preliminary Report on the E-Commerce Sector Inquiry' (15 September 2016) SWD(2016) 312 final, 65.

[127] European Commission, 'Final Report on the E-Commerce Sector Inquiry' (10 May 2017) COM(2017) 229 final, paras 11 f.

[128] Higher Regional Court Düsseldorf, Case VI – Kart 1/14 (V) *HRS* (9 January 2015) paras 47 f.

[129] Autorité de la concurrence, Case 16-DCC-111 *Fnac/Darty* (27 July 2016). On this case and the authority's shift in the market definition approach, see S Genevaz and J Vidal, 'Going Digital: How Online Competition Changed Market Definition and Swayed Competition Analysis in *Fnac/Darty*' (2017) 8 *Journal of European Competition Law & Practice* 30.

as *Pierre Fabre* (2011) and *Coty v Parfümerie Akzente* (2017), implicitly broached the online/offline paradigm.[130]

In the hotel merger *Marriott/Starwood* (2016), the European Commission concluded that hotel accommodation services constituted the relevant market without determining whether short-stay residences formed part of this market.[131] Arguably, it thereby overlooked the fact that traditional hotel accommodation is increasingly competing with digital platforms labelled the sharing economy, such as Airbnb, which allow short-letting private residences.[132] In *Randstad/Monster Worldwide* (2016), the Commission considered that online job board services constituted a separate antitrust market from offline markets, although ultimately leaving that market definition open.[133]

Where online and brick-and-mortar shops offer different services in relation to goods, such as advice, customisation, fitting, maintenance or updates, identical goods may no longer be seen as substitutable from a customer perspective.[134] The question then shifts to whether supply substitutability could broaden this narrow market. Another issue is the question of customer migration, ie, scenarios in which customers flock to online offerings while abandoning equivalent brick-and-mortar offerings. In the UK merger case of *Ladbrokes/Coral* (2016), the Competition and Markets Authority concluded that customer migration from brick-and-mortar betting shops to online betting may be such that it does not amount to substitutability and 'cannot be equated with diversion (which is relevant for market definition purposes)'.[135] The question is whether this type of migration, which is brought about by the digitisation of markets, should not suffice to understand it as substitutability from an antitrust perspective.[136]

The online/offline paradigm highlights that competitive relationships amongst products that are encountered both online and offline must be carefully assessed. From a comparative perspective, it can be said that at present, competition authorities and courts alike are still very much focusing on what divides online and offline markets rather than on what unites them. This results in markets being delineated

[130] Case C-439/09 *Pierre Fabre Dermo-Cosmétique* EU:C:2011:649, para 2; Case C-230/16 *Coty Germany v Parfümerie Akzente* EU:C:2017:941, paras 37–58; Robertson, 'Delineating Digital Markets' (n 91) 148.

[131] *Marriott/Starwood* (Case M.7902) Commission Decision of 27 June 2016 [2016] OJ C411/1, paras 64–69.

[132] ibid paras 66, 201; AG Yaşar, 'Achieving Symbiosis between Disruptive Innovation and Merger Control: Challenges and Remedies' (SSRN abstract nr 3015007, 1 June 2017).

[133] *Randstad/Monster Worldwide* (Case M.8201) Commission Decision of 26 October 2016, paras 7(d), 19 f. For online/offline recruitment services and job advertisements, see also *Randstad/VNU/JV* (Case COMP/M.2057) Commission Decision of 30 August 2000 [2000] OJ C368/5, paras 11–15.

[134] MR Baye, 'Market Definition and Unilateral Competitive Effects in Online Retail Markets' (2008) 4 *Journal of Competition Law & Economics* 639, 642.

[135] Competition and Markets Authority, *Ladbrokes/Coral* (26 July 2016) para 6.25.

[136] S MacMahon Baldwin, 'Common Sense or Non-sense? The "Migration vs Substitution" Paradigm in Market Definition' *Kluwer Competition Law Blog* (3 June 2019), competitionlawblog.kluwercompetitionlaw.com/2019/06/03/common-sense-or-non-sense-the-migration-vs-substitution-paradigm-in-market-definition.

more narrowly and stricter antitrust rules being applied. With the integration of markets also increasingly encompassing these market dimensions, the online/offline divide may soon be a thing of the past.

C. Product Differentiation in the Digital World: Free versus Ad-Free Services

In the world of digital platforms, which will be discussed in more detail below (see Chapter 8), a phenomenon can be observed which raises interesting questions as to interchangeability: some online platforms rely on advertisements in order to generate revenue, while others charge all or some of their users for their services. Yet other platforms have found a middle ground in that users can choose whether to use the free, ad-based service or pay so as to remove the ads or obtain more functionalities ('freemium' business models).

One example of this kind of product differentiation is the popular video streaming service YouTube. YouTube Premium (previously YouTube Red) is a streaming service that Google first launched in the US in the autumn of 2015.[137] It is an advertisement-free version of YouTube, which was until then exclusively financed through ads. Millions of YouTube users are now given a choice between receiving a service for free – but enduring the endless stream of ads that accompanies this service – or paying for the ad-free version of the same service. YouTube Premium will spare them ads before and during a video, banner ads, search ads and video overlay ads, according to the terms of service. When delineating the antitrust market for YouTube Premium, interesting questions arise from the point of view of product differentiation, but also regarding digital platforms and allegedly free services: is YouTube Premium in the same relevant market as its ad-sponsored counterpart? Or is the latter an ad-sponsored digital platform with three sides to it (free user side, advertiser side and content side) and the former limited to two market sides (paying user side and content side)? Is YouTube Premium only partially in the same market as YouTube, due to the lack of advertising on this digital platform? Does a significant number of users regard the two YouTube versions as interchangeable?

Another such example is the music streaming service Spotify, which operates a free, ad-sponsored version (Spotify Free) as well as the ad-free Spotify Premium, for which music aficionados need to pay a monthly subscription. The same applies to other music streaming services, such as Google Play Music and Deezer. Yet other music streaming services operate fee-based music streaming services, eg, Tidal. In multimedia, there were already discussions about how to define markets

[137] By June 2019, the service was available in a whole range of countries, including Austria, Australia, France, Germany, Mexico, New Zealand and South Korea.

related to pay-TV, free-access TV and the like in the late 1990s.[138] Today, one might have to include streaming services in those markets, providing an illustration of how technological progress can reshape relevant product markets.

The previous examples demonstrate the close link that free-user sides in platform markets share with revenue-generating market sides, inevitably leading to questions of product differentiation and interchangeability. In each case, customer behaviour as regards interchangeability will need to be taken into account when delineating an antitrust market. Market delineation might be complicated by the fact that in multi-sided markets, one market side may regard a platform as interchangeable, while another may not. For instance, users may regard video streaming on YouTube or on YouTube Premium as interchangeable, while for the advertisers, the ad-free YouTube Premium will not represent a substitute.

D. Functionalities and Product Differentiation

A product's functionalities – be it music streaming services, communication services or software more generally – are sometimes used as an anchor for defining the relevant antitrust market. This practice needs to be questioned against the background of innovative markets that seem quite differentiated, but may evolve quickly, sometimes converging in functionalities or continuously developing new functionalities.

The music streaming services referred to above provide various functionalities, such as downloading options in addition to the streaming function, music videos enhancing the experience, personalised or off-the-shelf playlists, search etc.[139] The same is true for subscription video streaming services such as Amazon Prime and Netflix, which offer an array of functionalities and have now moved into content production as well. Their relationship with on-demand video streaming services that rely on a pay-per-view pricing model needs to be assessed from a substitutability point of view. As these are all highly innovative markets that keep developing further functionalities, this form of product differentiation certainly raises the question of limited substitutability.

Another example is word processing software. While plaintiffs might insist on narrow product markets being delineated for this software based on current or past functionalities, software producers will argue that functionalities are constantly being developed in this very dynamic market, with the consequence

[138] See PD Camesasca, 'Mayday or Heyday? Dynamic Competition Meets Media Ownership Rules after Premiere' (2000) 21 *European Competition Law Review* 76, 78–80.

[139] From a competition law perspective, it is also interesting to note that all of these services – apart from Amazon Prime Music – amount to about €10 or $10 per month, raising the question whether this is due to the cost of licensing fees that all these services incur, conscious parallelism or actual competition.

of ever-shifting market boundaries.[140] It requires a great deal of insight into how competition works in a specific market in order to be able to come to a workable conclusion on this matter.

The perils of focusing a product market definition on an innovative product's functionalities have largely been recognised. The merger between the software giants Oracle and PeopleSoft in 2004, which was investigated both by the DoJ and the European Commission, contained a number of interesting issues on market definition and its relationship with functionalities.[141] Most notably, and although the analysis focused on unilateral effects, market definition played a decisive role in the case, leading to the rejection of the proposed simulation model in both jurisdictions.[142] The markets that the investigations relied upon both in the US and the EU were sub-categories of enterprise application software, namely high-function software for human resource management and for financial management services.[143] The US definition was perhaps even narrower, as the Commission ultimately broadened its market definition to include additional providers.[144] On both sides of the Atlantic, the authorities therefore found very narrow markets consisting only of 'certain high-function software packages sold to large customers'.[145] In the US, the DoJ's market definition was ultimately rejected. In its final order, the District Court of Northern California characterised the industry under investigation in some detail, highlighting how the DoJ delineated this narrow market by focusing on very limited sub-categories, before finding that the DoJ had not successfully proved many of its assumptions that were necessary to arrive at such a narrow market, for instance relating to competitive relationships amongst companies.[146] The Court regarded it as vital to look beyond functionalities and assess competitive relationships, as these might easily transcend certain functionalities, particularly where the latter can evolve. The European Commission laid out its own difficulties in delineating the relevant market in its merger decision and stated that, based on additional evidence, it had decided to slightly broaden the market it had initially arrived at.[147]

[140] R Schmalensee, 'Antitrust Issues in Schumpeterian Industries' (2000) 90 *AEA Papers and Proceedings* 192, 193.

[141] *United States DoJ and Plaintiff States v Oracle*, 331 FSupp2d 1098 (ND Cal 2004); *Oracle/PeopleSoft* (Case COMP/M.3216) Commission Decision 2005/621/EC [2005] OJ L218/6.

[142] O Budzinski and A Christiansen, 'The Oracle/PeopleSoft Case: Unilateral Effects, Simulation Models and Econometrics in Contemporary Merger Control' (2007) 34 *Legal Issues of Economic Integration* 133, 155.

[143] *United States DoJ and Plaintiff States v Oracle*, 331 FSupp2d 1098 (ND Cal 2004, Complaint dated 26 February 2004) para 23; *Oracle/PeopleSoft* (Case COMP/M.3216) Commission Decision 2005/621/EC [2005] OJ L218/6, paras 55 ff.

[144] *Oracle/PeopleSoft* (Case COMP/M.3216) Commission Decision 2005/621/EC [2005] OJ L218/6, para 57.

[145] Budzinski and Christiansen, 'Oracle/PeopleSoft' (n 142) 152.

[146] *United States DoJ and Plaintiff States v Oracle*, 331 FSupp2d 1098, 1101 ff, especially 1108 f (ND Cal 2004) (for what the DoJ did not successfully prove).

[147] *Oracle/PeopleSoft* (Case COMP/M.3216) Commission Decision 2005/621/EC [2005] OJ L218/6, para 57.

In *LiveUniverse v MySpace* (2007), the question before the US District Court of the Central District of California was what constituted the relevant antitrust market in a case involving social networking. The Court reasoned that, also in this case, markets should be defined based on substitutability. LiveUniverse argued – and the District Court ultimately accepted – that internet-based social networking sites constituted the relevant product market. MySpace, on the other hand, submitted that market definition should account for social networking sites' different products, qualities and prices; in short, it argued that based on different features in social networking sites, they could not all be regarded as interchangeable for antitrust purposes. The Court also discussed the interchangeability between online dating sites and social networking sites, finding that online dating sites were not interchangeable with social networking sites, even if some might use the latter for dating.[148] This showcases an important feature of many differentiated products: one-way substitutability.

In the 2011 case of *Microsoft/Skype*, the European Commission did not decide whether the two communication services at issue should be further segmented by functionality, despite having a clear preference against such segmentation.[149] Only two years later, in *Microsoft/Nokia* (2013), it made clear that further segmentation of consumer communication services by functionality should not be undertaken because 'consumers increasingly demand a user experience that integrates a range of communication functionalities [but also because] most providers offer the whole range of functionalities'.[150] In *Facebook/WhatsApp* (2014), the Commission pointed to further functionalities in consumer communication services, such as one-to-one or group communications through voice or multimedia messaging, video chats, group chats, voice calls and the sharing of locations.[151]

In the *TomTom/Tele Atlas* merger of 2008, it was not so much functionalities, but rather attributes that were at issue: the market definition question boiled down to whether narrowly construed navigable digital map databases or more broadly construed map databases constituted the relevant market. While the European Commission opted for the narrower market delineation,[152] the continuing sophistication in map databases shows that attributes can evolve – and, with them, the market definition.[153]

[148] *LiveUniverse v MySpace*, Case CV 06-6994 AHM (RZx) (CD Cal 4 June 2007).

[149] *Microsoft/Skype* (Case COMP/M.6281) Commission Decision of 7 October 2011, paras 10 (noting that the main functionalities were instant messaging, voice and video calls), 28, 29, 51. However, the differentiation between consumer and enterprise communication services itself was already partially based on functionalities; see paras 14, 46.

[150] *Microsoft/Nokia* (Case COMP/M.7047) Commission Decision of 4 December 2013 [2014] OJ C44/1, para 38.

[151] *Facebook/WhatsApp* (Case COMP/M.7217) Commission Decision of 3 October 2014 [2014] OJ C417/4, para 16. It considered that segmentation based on functionalities was not appropriate in this case (para 22).

[152] *TomTom/Tele Atlas* (Case COMP/M.4854) Commission Decision of 14 May 2008 [2008] OJ C237/8, paras 17 ff.

[153] On this case, see ME Stucke and AP Grunes, *Big Data and Competition Policy* (Oxford, Oxford University Press, 2016) §§ 6.03 ff.

When Google sought to acquire the navigation app Waze in 2013, this equally touched upon the turn-by-turn navigation market. Waze had a number of functionalities that went beyond turn-by-turn navigation, such as real-time traffic information.[154] The UK's Office of Fair Trading did not challenge the *Google/Waze* merger and based its analysis on the relevant market of 'turn-by-turn navigation applications for mobile devices'.[155] Nor was the merger challenged by the FTC.[156]

From a comparative perspective, it is interesting to note how both jurisdictions have repeatedly tried not to fall into the market delineation trap that evolving functionalities pose. A strong focus on functionalities can mean that the broader picture of competitive relationships is lost from sight. In many ways, market definition based on functionalities links back the issues of sub-markets that were raised in the 1962 US Supreme Court case of *Brown Shoe*, already discussed above.[157] With the overly narrow sub-markets delineated in *Brown Shoe* in mind, one should be wary of delineating increasingly narrow markets for innovative products based on particular characteristics or functionalities. Functionalities may poorly depict the competitive situation in which products find themselves.

When thinking along the lines of product functionalities, one can also draw a parallel to cluster markets as they were relied upon in *Philadelphia National Bank* in the US or in *Lombard Club* in the EU.[158] When taken together, do various functionalities of an innovative product make up a new product? And can various innovative products with different and partially overlapping functionalities be characterised as a cluster market?

E. Differentiated Products and Innovation: Some Conceptual Issues

Under US antitrust law, relevant antitrust markets were initially conceptualised as inherently homogeneous, with relevant markets relatively isolated from each other and thus from each other's competition.[159] Many markets today are quite the opposite, namely heterogeneous with partial or one-way substitutability, thus

[154] ibid § 6.77.

[155] Office of Fair Trading, Case ME/6167/13 *Motorola Mobility (Google)/Waze* (17 December 2013) para 13.

[156] Noting the absence of a closing statement, see Stucke and Grunes (n 153) § 6.79.

[157] *Brown Shoe v United States*, 370 US 294, 325 (1962). See section I.A.iii in ch 2.

[158] *United States v Philadelphia National Bank*, 374 US 321 (1963); Joined Cases C-125, C-133, C-135 and C-137/07 P *Raiffeisen Zentralbank Österreich and Others v Commission (Lombard Club)* EU:C:2009:576; JA Newberg, 'Antitrust for the Economy of Ideas: The Logic of Technology Markets' (2000) 14 *Harvard Journal of Law & Technology* 83, 125. On these cases, see above, sections I.A.iv and II.A.iv in ch 2.

[159] RS Markovits, 'Predicting the Competitive Impact of Horizontal Mergers in a Monopolistically Competitive World: A Non-market-Oriented Proposal and Critique of the Market Definition-Market Share-Market Concentration Approach' (1978) 56 *Texas Law Review* 587, 595 f.

making it increasingly difficult to conceive of a reasonably well-defined antitrust market. One suggestion has been to move away from the focus on substitutability at the product level and instead opt for focusing on the end use of differentiated products and the substitutability for that end use. The intention is to thereby avoid the delineation of overly narrow product markets.[160]

Although it has been argued that the hypothetical monopolist test was 'expressly designed to assess competition between differentiated offerings',[161] the Horizontal Merger Guidelines are not easily applied when product markets are highly differentiated.[162] This similarly applies to innovative markets that produce a chain of differentiated products in succession, some of which compete with each other, while others do not.

III. Innovation and the Time Horizon for Market Definition

The question of which time horizon should be applied to a market definition is particularly significant for dynamic markets and depends on a number of parameters.[163] Traditional market definition is 'relatively static [as it is] concerned primarily with competitive effects in the relatively short term'.[164] A central question when delineating product markets in innovation-intensive industries is therefore which time horizon should be applied,[165] eg, how quickly does a company need to be able to supply product X in order to be considered a supply-side substitute or a potential competitor, or how fast do customers switch to another product when faced with a price increase or a decrease in quality.

Based on the prevailing market conditions, the time horizon will differ on a case-by-case basis. As the guidance issued by competition authorities does not explicitly refer to the time horizon in delineating antitrust markets, those applying competition law are not provided with any bright line tests to rely on. At the same time, each industry has its own peculiarities and development cycles, thus making it impossible to provide generalised yet meaningful guidance on the time horizon

[160] See Keyte, 'Market Definition' (n 80) 746.

[161] S Bishop and M Walker, *The Economics of EC Competition Law: Concepts, Application and Measurement*, 3rd edn (London, Sweet & Maxwell, 2010) para 4-007.

[162] With reference to the cereal market, see DL Rubinfeld, 'Market Definition with Differentiated Products: The Post/Nabisco Cereal Merger' (2000) 68 *Antitrust Law Journal* 163, 177.

[163] Distinguishing discrete from coherent temporal market boundaries, see A Gerasymenko and S Afendikova, 'The Relevant Temporal Market Definition in Antitrust Analysis' (2018) 4 *Baltic Journal of Economic Studies* 68, 75.

[164] Newberg, 'Economy of Ideas' (n 158) 91.

[165] Similarly, see T Madiéga, 'Innovation and Market Definition under the EU Regulatory Framework for Electronic Communications' (2006) 29 *World Competition* 55, 57; Bishop and Walker (n 161) para 4-029.

for market delineation. Bearing this in mind, it becomes even more important to thoroughly characterise and understand a particular market environment, including for gaining an insight into the issue of timing for the purposes of market definition.

A short-term, static analysis of the market is naturally at odds with the long-term benefits to consumer welfare that innovation is thought to bring about.[166] Frequently, substitution that is considered in the very short term only produces relevant markets that are quite narrow.[167] The argument can be made that for innovative market environments, a preference is to be given to a longer time horizon if the competition law framework regards the promotion of innovation as a worthy goal. The implication of this is a possible trade-off between allocative efficiency that reduces prices in the short run and dynamic efficiency that increases innovation in the long run.[168] The short-term or long-term goals which a certain competition policy pursues therefore also influence the preference that is given to a short-term or long-term view as regards substitutability for the purposes of market delineation.

Another aspect of the time horizon is considering at what point in time a product is bought or consumed[169] – examples are peak/off-peak train travel, cocktail happy hours at less busy times of day and half-price sushi starting from half an hour before the shop closes. All these goods or services are essentially the same – a train ride, a cocktail, some sushi. But the different points in time at which they are bought or consumed lead to considerably different prices and the question to what extent these products are in the same relevant antitrust market. The same question might also be transferred to the innovation sphere in that different generations of innovative goods are being marketed simultaneously, with one generation (eg, the iPhone 7s) losing in value upon the marketing of a new generation (eg, the iPhone 11).[170]

Time also matters insofar as some antitrust analyses might be directed at the past (eg, anti-competitive agreements, abuse of dominance/monopolisation), while others are typically directed at the future (eg, mergers).[171] While not justifying a different approach to market delineation, it is a factor that can lead to a different perspective on the market definition enquiry. The delineation of future markets – particularly in merger control – is looked at more closely in the following section.

[166] R Hartman et al, 'Assessing Market Power in Regimes of Rapid Technological Change' (1993) 2 *Industrial and Corporate Change* 317, 324.

[167] Bishop and Walker, *Economics* (n 161) para 4-029.

[168] V Kathuria, 'A Conceptual Framework to Identify Dynamic Efficiency' (2015) 11 *European Competition Journal* 319, 319 f.

[169] Europe Economics, 'The Development of Analytical Tools for Assessing Market Dynamics in the Knowledge Based Economy' (12 September 2003) 44; Niels, Jenkins and Kavanagh (n 73) § 2.12.

[170] On this differentiation, based on generations of the same products, see section II.A.

[171] Niels, Jenkins and Kavanagh (n 73) § 2.13.

IV. The Definition of Future Markets

From a basic conception, there are two possibilities: either one can analyse innovation as an aspect of a product market; or innovation can be analysed outside of existing product markets.[172] The definition of future markets covers some middle ground between these two options, incorporating innovation considerations into markets that have yet to fully emerge. It requires a prediction of market developments that even companies themselves can often not reliably foresee, in order to assess the competitive impact of companies' current or foreseeable market conduct on these future markets. One has every right to be sceptical towards such crystal ball gazing, especially where legal consequences are attached to this exercise. Nevertheless, antitrust law often requires that we anticipate certain market developments in order to prevent anti-competitive harm from arising. As the case law both from the EU and the US shows, future markets are only relied upon where their coming into existence is relatively certain. In cases of uncertainty, other concepts are relied upon (see Chapter 5).

The definition of future markets allows for a prospective analysis of markets based on the further development of current markets or the emergence of entirely new markets.[173] However, innovation's uncertainty bestows a great amount of unpredictability on the definition of future markets. As with technology forecasting,[174] we can only assess whether a market definition was correct ex post, and possibly only after antitrust intervention has taken place and perhaps influenced market dynamics.

In the realm of innovation, the definition of future markets is closely related to the R&D market concept discussed – and largely dismissed – below (see section II in Chapter 5). Indeed, where a market for a certain innovation can be made out, the next step will be the manufacturing and subsequent marketing of that innovation, thus leading to a (future) market. However, the reliance on the future market concept does require that R&D is at a relatively advanced and observable stage, implying that regulated markets, and specifically the pharmaceutical sector, are particularly apt for such market definitions. Other market environments may not be as suitable for such a future market analysis.

A. Future Markets in Pharmaceuticals

In a number of cases in the pharmaceutical industry, mergers between competitors have raised the question of whether competition on future markets was at risk.

[172] Similarly, see M Glader, *Innovation Markets and Competition Analysis: EU Competition Law and US Antitrust Law* (Cheltenham, Edward Elgar, 2006) 92, 94.
[173] Europe Economics (n 169) 41.
[174] Hiltunen and Hiltunen (n 40) 21.

As befits this globalised market environment, many of the mergers in this field were analysed in parallel in the EU and the US.

In *Upjohn/Pharmacia* (1996), the FTC defined the relevant market as 'the research, development, manufacture and sale of topoisomerase I inhibitors for the treatment of colorectal cancer'.[175] The manufacture and sale of the drug were necessarily future markets, as no such drug had been marketed yet. Upjohn's drug for treating colorectal cancer was allegedly fairly advanced in the market approval process, while the drug that had been developed by Pharmacia was lagging behind by a few years. The FTC's concern was that once the two pharmaceutical companies merged, their incentive to also bring the drug developed by Pharmacia on the market as quickly as possible would be diminished. The consent order established that the merging companies would divest all of Pharmacia's assets relating to the R&D of this particular drug.[176] The intention of this divestiture was to bring another company into the competitive environment that would continue the innovation process and ultimately compete with the merged entity's drug, thus safeguarding competition on this future market. The merger of *Upjohn/Pharmacia* (1995) was also investigated by the European Commission, which referred not only to 78 current pharmaceutical product markets, but equally to areas in which both merging parties were conducting R&D on active compounds.[177] Some of the R&D considerations related to possible future markets. Concerning active compounds for treating Parkinson's disease, for instance, the Commission noted that both merging parties had compounds in phase III of development, with an expected launch within two years. In addition, it noted that a dozen competing products were being developed by competitors.[178]

The FTC's assessment of the *Glaxo/Wellcome* merger (1995) regarded R&D of certain agonists for the treatment of migraine as the relevant market,[179] thereby relating to a future pharmaceutical market. In its own *Glaxo/Wellcome* merger assessment of the same year, the European Commission stated that at least '[i]n the pharmaceutical sector, in order to be complete a competition assessment will require scrutiny of products which are not yet on the market but which are at an advanced stage of development'.[180] Several other EU cases also illustrate how the European Commission relied on future markets, at least as part of its analysis. In *Hoechst/Rhône-Poulenc* (1999), the Commission reiterated its statement from

[175] *Upjohn and Others*, 121 FTC 44, 45 (1996).

[176] ibid 50. See also S DeSanti and W Cohen, 'Competition to Innovate: Strategies for Proper Antitrust Assessments' in R Cooper Dreyfuss, D Leenheer Zimmerman and H First (eds), *Expanding the Boundaries of Intellectual Property: Innovation Policy for the Knowledge Society* (Oxford, Oxford University Press, 2001) 329 f.

[177] *Upjohn/Pharmacia* (Case IV/M.631) Commission Decision of 28 September 1995 [1995] OJ C294/9, paras 5, 7.

[178] ibid paras 30 f.

[179] *Glaxo*, 119 FTC 815, 816 (1995).

[180] *Glaxo/Wellcome* (Case IV/M.555) Commission Decision of 28 February 1995 [1995] OJ C65/3, para 9.

Glaxo/Wellcome and added that it had 'to look at R&D potential in terms of its importance for existing markets, but also for future market situations'.[181]

In *Ciba-Geigy/Sandoz* (1996), the European Commission said that pharmaceuticals had to be assessed based on their characteristics and intended therapeutic use, and also in relation to future markets.[182] Although that case pre-dated the Commission's policy documents on innovation competition, it already relied on this concept when it looked at whether the involved companies' 'pipeline products' would have a competitive effect on existing or future markets.[183] However, the Commission did not properly delineate these future markets on which a merger might have an impact.[184] Instead, it stated that where R&D 'must be assessed in terms of its importance for future markets, the relevant product market must, by its very nature, be defined in a less clear-cut manner than in the case of existing markets'.[185] Where no future market can be ascertained, the Commission might look at whether competition in innovation is harmed through a merger.[186] The future technology market of gene therapy technology was at issue in the parallel US *Ciba-Geigy/Sandoz* merger investigation (1997).[187]

The Commission, with reference to pharmaceuticals, has stressed that while it may be perceived as insufficient to only look at the current market situation, as it is prone to rapid change, it is also 'difficult if not impossible to predict the economic success of future products, especially if they are in early stages of development'.[188] Therefore, R&D needs to have reached a certain stage before it can be relied upon for delineating a future market. As R&D is conducted at a global level, the analysis should equally have a wide geographical swath.[189]

B. Non-pharmaceuticals and Future Market Definition

Outside of the pharmaceuticals context, the EU and US antitrust authorities have relied on future markets as the analytical point of reference in a range of cases. For instance, in the joint venture case of *Shell/Montecatini* (1994), the European Commission defined a downstream market, namely the production and sale of

[181] *Hoechst/Rhône-Poulenc* (Case IV/M.1378) Commission Decision of 9 August 1999 [1999] OJ C254/5, para 26.

[182] *Ciba-Geigy/Sandoz* (Case IV/M.737) Commission Decision 97/469/EC [1997] OJ L201/1, para 42.

[183] L Gyselen, 'Competition in Innovation: A Novel Concept? The Case Law on Pharmaceuticals' in P Lugard and L Hancher (eds), *On the Merits: Current Issues in Competition Law and Policy – Liber Amicorum Peter Plompen* (Cambridge, Intersentia, 2005) 38 (direct quote).

[184] ibid 44.

[185] *Ciba-Geigy/Sandoz* (Case IV/M.737) Commission Decision 97/469/EC [1997] OJ L201/1, para 44.

[186] Gyselen (n 183) 47.

[187] *Ciba-Geigy/Sandoz*, 123 FTC 842, 844 (1997). On this characterisation, see Newberg (n 158) 115.

[188] European Commission in OECD (ed), *Policy Roundtable: Competition and Regulation Issues in the Pharmaceutical Industry* (2000) DAFFE/CLP(2000)29, 345.

[189] *Ciba-Geigy/Sandoz* (Case IV/M.737) Commission Decision 97/469/EC [1997] OJ L201/1, para 51.

polypropylene, and an upstream market consisting of the licensing of polypropylene technology.[190] In the latter, the Commission assessed which research was currently being carried out and concluded that some of that research might only lead to marketable innovations in over 10 years' time, while other research efforts were highly uncertain.[191] Concerning the same joint venture, the FTC issued a consent order in *Montedison* (1995), in which it discerned five relevant lines of commerce, amongst which polypropylene technology as well as the licensing of polypropylene technology. Particularly for the former, it highlighted the importance of innovation through R&D competition.[192]

In *Sensormatic* (1995), the two merging companies were independently developing electronic article surveillance labels, so-called anti-shoplifting labels, that would not be compatible. The FTC delineated the relevant markets as 'the research and development of disposable labels developed or used for source labelling and the research and development of processes to manufacture disposable labels'.[193] Its main concern was that, post-merger, the company would abandon one of the R&D efforts and thereby decrease the number of R&D tracks in this already concentrated market.[194] To address this concern, the FTC ordered that Sensormatic should not have sole control over the technology of the other merging company.[195] In its analysis, it found future goods markets for the next generation of anti-shoplifting labels to be of relevance.[196]

In the *Microsoft/Intuit* (1995) merger case, the DoJ analysed current as well as future markets for personal finance software.[197] In *Intel* (1998), the FTC was concerned not only about current product markets, but also about future product and technology markets, as evidenced by the complaint in that case.[198]

In a merger case on pay-TV, *Bertelsmann/Kirch/Premiere* (1998), the European Commission discussed the possible future convergence of pay-TV and free TV, but ultimately concluded that 'this possible future development is not enough now to justify the acceptance of a common market for pay and free TV'.[199] This demonstrates that the Commission is cautious when assessing future markets, only resorting to this option when the emergence of such a future market is relatively

[190] *Shell/Montecatini* (Case IV/M.269) Commission Decision 94/811/EC [1994] OJ L332/48, para 44.

[191] ibid para 30.

[192] *Montedison*, 119 FTC 676, 678 f (1995).

[193] *Sensormatic Electronics*, 119 FTC 520, 522 (1995).

[194] ibid 520.

[195] ibid 526.

[196] Federal Trade Commission, 'Anticipating the 21st Century: Competition Policy in the New High-Tech, Global Marketplace: Staff Report', vol 1 (May 1996) § 7.II.A; LB Landman, 'Competitiveness, Innovation Policy, and the Innovation Market Myth: A Reply to Tom and Newberg on Innovation Markets as the "Centerpiece" of "New Thinking" on Innovation' (1998) 13 *St John's Journal of Legal Commentary* 223, 264, 266.

[197] Landman, 'Innovation Market Myth' (n 196) 277; *United States v Microsoft*, 56 F3d 1448 (DC Cir 1995).

[198] *Intel*, 128 FTC 213, 214 (1999); Newberg (n 158) 117 f.

[199] *Bertelsmann/Kirch/Premiere* (Case IV/M.993) Commission Decision 1999/153/EC [1999] OJ L53/1, para 18.

certain. By doing so, the Commission essentially handles future markets in a similar fashion to potential competition.[200]

In its 2011 Horizontal Cooperation Guidelines, the European Commission emphasised that the effects that cooperation between competitors has on future product markets should be factored into the analysis. It also noted that the analysis of innovation competition may often entail an assessment relating to future markets.[201]

C. Innovation and the Reliance on Future Markets

Future markets have been labelled the 'necessary extension' of potential competition.[202] However, under a different characterisation, future markets provide an analytical tool that grasps a competitive dimension that is different from that of potential competition, rather than extending the latter. Future markets are concerned with markets that are thought to emerge in the future – either based on current products or on products that are yet to be marketed. Potential competition, on the other hand, relates to possible market entry either in a current or in a future market.

Because of the uncertainty surrounding the emergence of future markets, it is only sensible to define a future market under two circumstances.[203] The first is where market developments are well understood and the future market would build on a current market yet differ from it in important ways. In such a case, antitrust analysis can use the current market as a starting point in order to analyse what the future market would look like. This can be understood as a future market which grows out of an established current market. The second is a situation in which the connection between a current market and a likely future market is weak, and they would probably not compete with each other in the future. In this case, an entirely separate future market should be delineated to depict this competitive situation. This can be seen as a novel future market which has no strong ties to any current market. In cases in which a future market cannot be delineated with sufficient certainty, either because it is not yet clear what the future market might encompass or because market developments are too little understood to develop any meaningful alternatives for possible future markets, antitrust enforcers should refrain from delineating a future market.[204] Scenarios within which a future market analysis is useful are thus limited, particularly in dynamic market environments.

[200] Camesasca (n 138) 77.

[201] European Commission, Guidelines on the applicability of Article 101 of the Treaty on the Functioning of the European Union to horizontal co-operation agreements [2011] OJ C11/1, para 121.

[202] Kern (n 61) 178.

[203] Europe Economics (n 169) 43.

[204] ibid.

As was argued above, in innovative markets antitrust law analysis might have to focus more on competition for the (future) market than on competition in the current (product or technology) market.[205] Such an approach assigns a much less important role to the definition of current markets,[206] and instead emphasises the future development of the market, for instance, based on observable R&D efforts. The delineation of future markets can prove fruitful where R&D efforts are fairly advanced so that future goods or services can already be singled out.[207] It also requires a certain transparency in the R&D environment, as future competitors need to be made out. This might be possible in pharmaceuticals, where a certain medicinal product is moving along the mandatory approval process, but less so in other areas in which companies or individuals are quietly working on their innovations.[208]

US courts and antitrust agencies rarely try to predict future market developments in dynamic industries.[209] Despite the examples provided in this section, it appears that the European Commission is even less willing to look at future market developments than the FTC.[210] Innovators that are active both in the EU and the US, of course, may welcome a cautious approach to future markets because of the possibility that the authorities or courts may reach conflicting conclusions as to the future markets on which they base their antitrust analysis, perhaps entailing contradictory legal consequences for the innovators. While administrative cooperation amongst the antitrust authorities will usually ensure that no such contradictory outcome is reached,[211] such a scenario would indeed be most unfortunate, leading to a legal divergence based on differing predictions.

An example of a market environment in which future markets could be relied upon is data, where the fact that data is used in-house rather than being traded means that it does not constitute a current relevant market in the antitrust sense. However, one might argue that such data already constitutes a potential market – similar to potential or hypothetical markets that are sometimes relied upon in refusal to license cases.[212] At what point in time companies with decision-making

[205] See section II.C in ch 3.

[206] J Drexl, 'Anticompetitive Stumbling Stones on the Way to a Cleaner World: Protecting Competition in Innovation without a Market' (2012) 8 *Journal of Competition Law & Economics* 507, 510.

[207] Similarly, see Glader, *Innovation Markets* (n 172) 237, 248 (distinguishing between distant and imminent future markets).

[208] Kern (n 61) 179.

[209] Ginsburg and Wright (n 62) 11 f, 21.

[210] J Temple Lang, 'European Community Antitrust Law: Innovation Markets and High Technology Industries' (1997) 20 *Fordham International Law Journal* 717, 769; LB Landman, 'Innovation Markets in Europe' (1998) 19 *European Competition Law Review* 21, 23.

[211] See, for instance, Agreement between the European Communities and the Government of the United States of America on the application of positive comity principles in the enforcement of their competition laws [1998] OJ L173/28.

[212] See V Korah, *Intellectual Property Rights and the EC Competition Rules* (Oxford, Hart Publishing, 2006) 147; J Temple Lang, '"Potential" Downstream Markets in European Antitrust Law: A Concept in Need of Limiting Principles' (2011) 7 *Competition Policy International* 106.

power over such data decide to enter into a market relationship with other companies lies exclusively in their own hands.

Market definition in the context of anti-competitive agreements, (attempted) monopolisation and abuse of dominance generally focuses on the past or present, while in mergers it regularly concerns future developments, explaining why it has primarily been merger cases that have featured future market considerations. However, where innovation comes into play, one needs to consider to what degree future – and perhaps non-predictable – market developments can and should be taken into consideration at the stage of market delineation.[213] In a great many cases, the answer may be that they should not be factored into the market definition because they simply cannot be reliably predicted, but appropriate weight needs to be given to such considerations at a later stage in the competition assessment, going beyond mere tokenism. An approach that relies on future markets rather than on innovation competition may protect competition in future product markets and would thereby implicitly favour price competition over more dynamic competitive parameters.[214] In addition, future markets also bear the danger of extending antitrust liability into areas that are still developing – and in which anti-competitive conduct or even the market itself might not actually materialise. Chapter 5 sets out how antitrust analysis can approach more uncertain market developments related to innovation through instruments such as potential competition, innovation markets and innovation competition.

V. Conclusion

The product dimension is the all-important starting point for antitrust market delineation. In dynamic markets, it is challenging to grasp this point of departure. Similarly, substitutability becomes more and more difficult to get a hold of in innovative environments. The advent of digital technologies has led to a convergence of markets based on integrated systems, making it hard to draw (sometimes rather artificial) market boundaries that traditional market delineation requires. Super-platforms and their increasingly integrated product portfolios increasingly resemble 'old world' conglomerates, but with the added pinch of salt of markets being digitally interconnected. Digital assistants and IoT functionalities may lead to a narrowing of product substitutability from the consumer perspective based on data relied upon, as well as a possible broadening where multi-task algorithms come into play. The market conditions that led to this state of affairs were already highlighted in Chapter 3, ie, interoperability concerns, network effects and the use of big data in digital platforms. Based on the digitisation of the economy, techno-

[213] See also TE Kauper, 'The Problem of Market Definition under EC Competition Law' (1997) 20 *Fordham International Law Journal* 1682, 1726 f.

[214] See also Kern (n 61) 193.

logical innovation may sooner or later lead to the emergence of new substitutability patterns in all imaginable sectors, with far-reaching effects on the relevant market delineation. Product differentiation, for instance in the form of the online/offline paradigm, product functionalities or freemium versus paid-for services, are areas to which heightened attention must be paid in innovative markets. With regard to product functionalities, ongoing innovation can expand functionalities, and functionalities may be partially overlapping. This means that market delineation along the lines of current or anticipated functionalities may be particularly error-prone.

The time horizon is a challenging aspect of market definition in innovative industries, as it links back to uncertainty in innovation and the question of how far into the future market developments should be anticipated. While a case-by-case analysis of the temporal market dimension is required, a focus on dynamic parameters of competition might require a more long-term view. As regards future markets, these can primarily be discerned in the context of pharmaceuticals and other products which, through regulatory processes, allow for an early awareness of an innovation. For those innovations that remain secret until market entry, the delineation of future markets needs to be restricted to those scenarios in which the emergence of a future market is relatively certain.

Going forward, it will be important to recognise demand and supply substitutability where it exists, even if this means bridging the online/offline divide or taking one-way substitutability into account. Product functionalities, particularly where they are still constantly developing, should not determine the relevant product market. Finally, a thorough understanding – ie, characterisation – of dynamic markets and the way in which competition operates on those markets is vital in order to adequately apply competition law in those markets.

5

Beyond Market Definition

Potential Competition, R&D Markets and Innovation Competition

In dynamic market contexts, antitrust law needs to take into account market behaviour that occurs before a relevant product market has properly emerged, as this might seriously affect both competition and innovation.[1] The question is within which framework these market developments can properly be assessed. In order to capture antitrust markets before they are properly established, several concepts have been applied. One of these is the notion of future markets, which has been discussed earlier (see section IV in Chapter 4). Further options are explored in the following, including potential competition as a concept to grasp the restraining effect that looming competition might have, the innovation (now R&D) market concept and the concept of innovation competition. Neither of these concepts relates to an actual market as such; they relate to a market in the making or competitive conditions that can influence how a relevant market is competitively constrained.

I. Potential Competition in Innovative Markets

The cut-off point between the more immediate competitive constraint of supply-side substitution, which directly filters into market delineation, and the more removed competitive constraint of potential competition, which does not, is not easily established. Criteria to rely on include the quick re-utilisation of assets as well as the time dimension. The cut-off point is particularly difficult to establish in highly innovative industries,[2] as '[t]he dynamic nature of modern competition

[1] Similarly, see B Lundqvist, *Standardization under EU Competition Rules and US Antitrust Laws: The Rise and Limits of Self-Regulation* (Cheltenham, Edward Elgar, 2014) 115.

[2] AJ Padilla, 'The Role of Supply-Side Substitution in the Definition of the Relevant Market in Merger Control' (Madrid, June 2001) 76. On that economist's distinction between supply-side substitution and potential entry along the three dimensions of time, uncommitted/committed entry and impact, see ibid 19.

affects the nature of potential competition'.[3] The following sets out what EU competition and US antitrust law understand by the term 'potential competition' and what role it plays in innovative markets. It will be argued that potential competition is an important factor in the competitive analysis, but should only be considered after the relevant market has been delineated. Especially in innovative markets, potential competition is too far removed from actual markets to be considered at an earlier stage.[4] Nevertheless, it is an important force in innovative markets and should be taken into account where 'certain conditions of probability and immediacy of entry'[5] are fulfilled.

A. Contrasting the EU and US Doctrines of Potential Competition

The concept of potential competition has been particularly relevant to merger cases in the US, where companies frequently sought to acquire potential competitors. In the EU, potential competition has mainly been an issue in cases on anti-competitive agreements, where companies sought to keep potential competitors out of their market. The starting point for the potential competition analysis in the EU and the US was thus a very different one, with one being prospective (the US) and the other oriented towards past behaviour (the EU). This focus makes a difference to how the potential competition doctrine is applied. The development of the potential competition doctrines in these two jurisdictions provides valuable comparative insights, as it allows us to see a legal doctrine taking two very different trajectories based on the environment in which the doctrine was applied.

The US doctrine of potential competition was developed within the merger sphere, in cases in which an acquirer wanted to buy a company with which it could have competed – thus eliminating the prospect of competition ('potential competition') from that market. When investigating mergers, potential competition can aid in discerning which competition the merging companies could face in the future and which competition is being eliminated through the merger. US antitrust law differentiates between perceived potential competition and actual potential competition: perceived potential competition means that one of the merging companies could plausibly enter the market, while actual potential competition is found when one of the merging parties is likely to enter the market in the future. In both cases, this entry would increase competition on the relevant market.

[3] TA Hemphill, 'Role of Competition Policy in the US Innovation System' (2003) 30 *Science and Public Policy* 285, 289.

[4] See European Commission, Notice on the definition of relevant market for the purposes of Community competition law [1997] OJ C372/5 (EU Market Definition Notice 1997) para 14.

[5] JG Sidak and DJ Teece, 'Dynamic Competition in Antitrust Law' (2009) 5 *Journal of Competition Law & Economics* 581, 614.

However, this theory is only applied to concentrated markets in which companies with market power can control output and prices,[6] and only where market entry is complicated or only few market entries are expected.[7]

Already in the 1960s, the US Supreme Court emphasised that potential competition can have the power to restrain anti-competitive harm under certain circumstances, just like actual competition does.[8] The elimination of potential competition, on the other hand, might lead to anti-competitive effects.[9] In *Falstaff Brewing*, the US Supreme Court held that it would be appropriate to consider whether the acquiring company 'was a potential competitor in the sense that it was so positioned on the edge of the market that it exerted beneficial influence on competitive conditions in that market'.[10] In *Siemens*, the Court of Appeals for the Second Circuit had to deal with the question of whether an acquisition by Siemens would eliminate potential competition on the relevant market by Siemens itself under section 7 of the Clayton Act. It observed that '[t]he "perceived" potential competition doctrine differs from the "actual" potential competition doctrine in that it is concerned with the present effect that a company not in the oligopolistic market is having on competitors within that market.'[11]

The elements of the US doctrine of perceived potential competition can be summarised as follows:

> (1) [T]he relevant market is substantially concentrated; (2) the acquiring firm has the capabilities and incentives to render it a perceived potential entrant, and is one of only a few firms with such characteristics; and (3) the firm's presence on the fringe of the relevant market in fact tempered oligopolistic behavior on the part of market participants.[12]

An FTC Report of 1996 considered how the actual potential competition doctrine might apply to market definition, finding that such an application would result in the question of 'whether a merger might eliminate the "actual potential competition" of an existing innovation effort directed toward producing a product to compete in a current or future goods market'. In such an analysis, the direction of the existing innovation effort would be determinative of the market definition.[13]

[6] *United States v Marine Bancorporation*, 418 US 602, 630 (1974). See also BR Kern, 'Innovation Markets, Future Markets, or Potential Competition: How Should Competition Authorities Account for Innovation Competition in Merger Reviews?' (2014) 37 *World Competition* 173, 176; P Areeda and HJ Hovenkamp, *Antitrust Law: An Analysis of Antitrust Principles and Their Application*, 4th edn (Wolters Kluwer, 2017) § 423.

[7] RT Rapp, 'The Misapplication of the Innovation Market Approach to Merger Analysis' (1995) 64 *Antitrust Law Journal* 19, 39 (with reference to further case law in fn 70).

[8] *United States v Penn-Olin Chemical*, 378 US 158, 172–74 (1964). Similarly, see *United States v El Paso Natural Gas*, 376 US 651, 659 (1964).

[9] *FTC v Procter & Gamble*, 386 US 568, 578 (1967).

[10] *United States v Falstaff Brewing*, 410 US 526, 532 f (1973).

[11] *United States v Siemens*, 621 F2d 499, 504 (2d Cir 1980).

[12] MD Whitener, 'Potential Competition Theory: Forgotten But Not Gone' (1991) 5 *Antitrust* 17, 18 (citing case law).

[13] Federal Trade Commission, 'Anticipating the 21st Century: Competition Policy in the New High-Tech, Global Marketplace: Staff Report', vol 1 (May 1996) § 7.V.B.1.

In the EU, the doctrine of potential competition was primarily an issue when anti-competitive agreements between (potential) competitors were at issue. In *European Night Services* (1998), the General Court acknowledged that potential competition needed to be examined in addition to competition in the current relevant market in order to obtain a full picture of the conditions of competition.[14] In its earlier *Delimitis* judgment (1991), the Court of Justice had underlined the importance of analysing whether there were 'real concrete possibilities' of future market entry.[15] However, in both of these cases, potential competition was only referred to when discussing anti-competitive effects on the market rather than in the context of delineating the relevant market.

In *Visa* (2011), the General Court found that whether a company could be regarded as a potential competitor depended on whether 'there would have been real concrete possibilities for it to enter the [relevant] market and to compete with established undertakings'.[16] On the time horizon, the General Court argued that potential entry needed to 'take place with sufficient speed',[17] but underlined that the timeframe of one year that could be found in a Commission policy document was merely illustrative.[18] Relying on *European Night Services*, it pointed out that mere hypothetical entry was not enough and that potential competitors needed to have the actual ability for market entry.[19] That same year, in *Hitachi*, the General Court held that potential competition was established where market participants 'regarded' or 'perceived' other companies to be 'potential credible competitors'.[20] Linguistically, this diction is reminiscent of the US doctrine of perceived potential competition.

In *Toshiba* (2014), the General Court reiterated its previous findings. It also held that a market-sharing agreement between companies indicated that the parties considered each other 'to be at least potential competitors'.[21] Upon appeal, the Court of Justice ruled that whether or not there was potential competition between the parties at issue was a question of fact, not of law; therefore, it did not need to assess this question.[22]

[14] Joined Cases T-374, T-375, T-384 and T-388/94 *European Night Services and Others v Commission* EU:T:1998:198, para 137.

[15] Case C-234/89 *Stergios Delimitis v Henninger Bräu* EU:C:1991:91, para 21.

[16] Case T-461/07 *Visa v Commission* EU:T:2011:181, para 166. For a near-identical wording, see Case T-360/09 *E.ON v Commission* EU:T:2012:332, para 86.

[17] Case T-461/07 *Visa v Commission* EU:T:2011:181, para 189.

[18] Here, the Court referred to European Commission, Guidelines on the applicability of Article 81 of the EC Treaty to horizontal cooperation agreements [2001] OJ C3/2, para 9, fn 9. For the current version of these guidelines, see European Commission, Guidelines on the applicability of Article 101 of the Treaty on the Functioning of the European Union to horizontal co-operation agreements [2011] OJ C11/1.

[19] Case T-461/07 *Visa v Commission* EU:T:2011:181, paras 167 f.

[20] Case T-112/07 *Hitachi v Commission* EU:T:2011:342, paras 90, 226, 319. Equally referring to these passages, see *Lundbeck* (Case AT.39226) Commission Decision of 19 June 2013 [2015] OJ C80/13, para 613, fn 1100.

[21] Case T-519/09 *Toshiba v Commission* EU:T:2014:263, para 231.

[22] Case C-373/14 P *Toshiba v Commission* EU:C:2016:26, para 35.

More recently, the question of potential competition reached the General Court in the pay-for-delay case of *Lundbeck* (2016).[23] In that case, the European Commission relied on *European Night Services* and *Visa* when arguing that generic pharmaceutical companies had been perceived as potential competitors by the originator pharmaceutical company Lundbeck before it entered into pay-for-delay agreements with them, agreements that would postpone the generic companies' imminent market entry once Lundbeck's patents for citalopram expired.[24] Upon appeal, the General Court recalled the importance not only of current competition but also of potential competition in the competition law assessment. One should take into account whether 'there are real concrete possibilities for the undertakings concerned to compete among themselves or for a new competitor to enter the relevant market and compete with established undertakings'.[25] In this particular case, the Court viewed potential competition as relating to the current relevant product market. The Court's position was criticised insofar as it considered the possibility of potential competition, even where entering the market would have required a patent infringement on the part of the potential competitor.[26]

The Commission's Horizontal Co-operation Guidelines (2011) regard a company as another's potential competitor where it would realistically enter the relevant market within a short period of time. Market entry has to be plausible, while 'the mere theoretical possibility to enter a market is not sufficient'.[27] Whether a period of time is considered short depends on the particular circumstances of a case.[28]

B. Potential Competition and Innovation

Potential competition and future markets are complementary concepts that relate to different dimensions of the relevant market. While future markets are an

[23] For the US merger case involving Lundbeck, see *FTC v Lundbeck*, 650 F3d 1236 (8th Cir 2011) (a case which the FTC lost based on the court's rejection of its narrow market definition); HJ Hovenkamp, 'Mergers with Dominant Firms: The *Lundbeck* Case' [2011] *Competition Policy International Antitrust Chronicle* 1, 2.

[24] *Lundbeck* (Case AT.39226) Commission Decision of 19 June 2013 [2015] OJ C80/13, paras 621–23.

[25] Case T-472/13 *Lundbeck v Commission* EU:T:2016:449, para 99 (citations to cases omitted; amongst other things, the General Court referred to *European Night Services* and *Visa*). This case is currently on appeal as Case C-591/16 P *Lundbeck v Commission* [2017] OJ C30/25 (appeal pending). Similar questions on potential competition amongst originators and generics manufacturers in pharmaceuticals are currently before the Court of Justice in Case C-307/18 *Generics and Others v Competition and Markets Authority (Paroxetine)* [2018] OJ C35/29 (preliminary ruling pending).

[26] S Marco Colino et al, 'The *Lundbeck* Case and the Concept of Potential Competition' [2017] *Concurrences* 24, 26 fn 13; with reference to P Ibáñez Colomo, 'GC Judgment in Case T-472/13, Lundbeck v Commission: On Patents and Schrödinger's Cat' *Chillin' Competition* (13 September 2016), chillingcompetition.com/2016/09/13/gc-judgment-in-case-t-47213-lundbeck-v-commission-on-patents-and-schrodingers-cat.

[27] European Commission, Guidelines on the applicability of Article 101 of the Treaty on the Functioning of the European Union to horizontal co-operation agreements (n 18) para 10.

[28] ibid para 10, fn 3. The Commission noted that 'both the R&D and the Specialisation Block Exemption Regulations consider a period of not more than three years a "short period of time"'.

integral part of market definition, potential competition is an analytical dimension that goes beyond market definition proper. Potential competition may relate to current or future markets.[29]

As potential competition in the EU only takes credible market entry into account, the concept caters more to product markets on which the company concerned is already active. This can be likened to actual potential competition in the US. This version of potential competition would not take into account innovation that is thought to have an influence on product markets that have yet to emerge.[30] The US concept of perceived potential competition, on the other hand, can be likened to innovation competition in highly concentrated market settings (on this, see below). However, where an innovation is at a stage at which it will reach the market with a certain predictability, it may well be considered as a future market. Potential competition may then relate to that market, as evidenced in the European *Hitachi* case (see section I.A above).

Just as it is highly relevant for supply-side substitutability, the time horizon also plays an important role for potential competition. It provides the dividing line between current product markets or future product markets and possible entry into those markets. In dynamic markets, this dividing line is particularly difficult to draw where potential competitors are not yet operating on the relevant product market.[31] From a behavioural antitrust perspective, potential market entry is quite unpredictable, occurring when least expected and perhaps not occurring when most expected.[32] Unless it is very close to actual market entry – as is suggested by the European case law – it cannot and perhaps should not be taken into account at the market definition stage. This is why the enforcers also take into account companies' statements on whom they perceive their (future) competitors to be.

In the EU, it is recognised that potential competition has a special role to play in technology markets.[33] Furthermore, the Commission acknowledges that the dynamic nature of competition implies that it is not only the prevailing market situation that needs to be taken into account, but also the possibility that competitors might expand into or newly enter the incumbent's market.[34]

[29] As such, the future market approach does not represent an extension of the potential competition doctrine into the future (see section IV in ch 4). But see Kern (n 6) 178.

[30] Kern (n 6) 176–78 (who draws on the US distinction between perceived and actual potential competition).

[31] T Madiéga, 'Innovation and Market Definition under the EU Regulatory Framework for Electronic Communications' (2006) 29 *World Competition* 55, 61, 64.

[32] A Tor, 'The Market, the Firm, and Behavioral Antitrust' in E Zamîr and D Teichman (eds), *The Oxford Handbook of Behavioral Economics and the Law* (Oxford, Oxford University Press, 2014) 553.

[33] European Commission, Guidelines on the applicability of Article 101 of the Treaty on the Functioning of the European Union to horizontal co-operation agreements (n 18) paras 116 f.

[34] European Commission, Guidance on the Commission's enforcement priorities in applying Article 82 of the Treaty to abusive exclusionary conduct by dominant undertakings [2009] OJ C45/7 (Guidance Paper) para 16.

Such (potential) expansion or entry can restrain the currently dominant company's market power. This approach very much reflects Schumpeter's view on creative destruction.[35]

In the US, an FTC Report from 1996 highlighted that the potential competition doctrine was sometimes regarded as an alternative to the innovation market concept and that it could be applied to current or future product markets.[36] In the testimony before the FTC at the time, some preferred the potential competition approach because it related to the more tangible effects on price or output, not innovation.[37] However, potential competition considerations are not identical to the US innovation market approach put forward by the 1995 IP Guidelines:[38] while potential competition is anchored in current or future markets, innovation markets relate to R&D developments that are still very much up in the air. Some have argued that one of the achievements of the innovation market approach was to extend the timeframe within which potential competition can be considered by the antitrust agencies.[39]

As has been emphasised above, potential competition is not considered at the stage of market definition itself, but might be taken into account at a later stage. The European Commission normally only takes potential competition into account once a relevant market has been identified.[40] The US agencies will scrutinise mergers with potential competitors with special care.[41] Their Horizontal Merger Guidelines count 'rapid entrants' amongst the market participants on a relevant market; they are defined as producers 'that would very likely provide rapid supply responses with direct competitive impact in the event of a SSNIP, without incurring significant sunk costs'.[42] While some prefer taking potential competition into account when defining the relevant market,[43] this becomes particularly difficult in dynamic markets in which market definition itself is short-lived, making an analysis of credible market entry by potential competitors even more uncertain. Nevertheless, potential competition can act as a powerful competitive restraint[44] and, as such, it should certainly be considered during the competitive assessment.

[35] See section III.A in ch 3.

[36] Federal Trade Commission, 'Anticipating the 21st Century' (n 13) § 7.V.

[37] ibid § 7.V.A.

[38] NA Widnell, 'The Crystal Ball of Innovation Market Analysis in Merger Review: An Appropriate Means of Predicting the Future?' (1996) 4 *George Mason Law Review* 369, 380.

[39] M Glader, *Innovation Markets and Competition Analysis: EU Competition Law and US Antitrust Law* (Cheltenham, Edward Elgar, 2006) 129.

[40] EU Market Definition Notice 1997 (n 4) para 24.

[41] US Department of Justice and Federal Trade Commission, Horizontal Merger Guidelines (19 August 2010) (US Horizontal Merger Guidelines 2010) § 5.3.

[42] ibid § 5.1.

[43] Sidak and Teece, 'Dynamic Competition' (n 5) 614.

[44] See already JA Schumpeter, *Capitalism, Socialism and Democracy*, 5th edn (London, Allen & Unwin, 1976) 85.

II. The US Innovation Market Approach

Innovation is 'intangible, uncertain, unmeasurable, and often even unobservable, except in retrospect. There are no market transactions in innovation, only in the inputs … and the outputs'.[45] What we frequently target when trying to define an *innovative* market is not so much the current R&D, but rather the fruits of previous innovative efforts. In doing so, we are always one step behind. The *innovation* market approach reflected the desire to overcome this limitation by incorporating innovation considerations into the antitrust law framework at an early stage, based on the conceptual framework of an innovation 'market'. It targeted R&D efforts before they culminated in marketable products or in marketable intellectual property. With the help of this analytical tool, the US agencies wanted to analyse which effects a certain behaviour (eg, merger) had on innovation that had not yet led to a conventional product.[46] This first attempt is today widely regarded as non-satisfactory, as it is only workable in a limited set of cases with particular characteristics. Its position within antitrust law analysis remains disputed.

The EU did not endorse the US-style innovation market approach as such,[47] and its brief flirtation with the innovation market concept and its later focus on competition in innovation are discussed below.

A. Policy Developments in Innovation Markets

In 1984, the US Congress passed the National Cooperative Research Act (NCRA), which urged antitrust authorities to take into account 'properly defined, relevant research and development markets' when assessing the legality of joint ventures.[48] That same year, a former head of the DoJ's Antitrust Division pointed out that competition concerns relating to dynamic industries not only arose in current product markets, but also in future markets and in the R&D process. He referred to the latter as antitrust markets in their own right.[49]

From the NCRA, the idea of an R&D market was picked up by the antitrust agencies.[50] The theory of innovation markets was then formally introduced into antitrust policy by the IP Guidelines of 1995, jointly issued by the FTC and the

[45] Rapp (n 7) 27.

[46] S DeSanti and W Cohen, 'Competition to Innovate: Strategies for Proper Antitrust Assessments' in R Cooper Dreyfuss, D Leenheer Zimmerman and H First (eds), *Expanding the Boundaries of Intellectual Property: Innovation Policy for the Knowledge Society* (Oxford, Oxford University Press, 2001) 327.

[47] Kern (n 6) 183.

[48] National Cooperative Research Act (1984), 15 USC §§ 4301–05, § 4302.

[49] WF Baxter, 'The Definition and Measurement of Market Power in Industries Characterized by Rapidly Developing and Changing Technologies' (1984) 53 *Antitrust Law Journal* 717, 717 f.

[50] See US Department of Justice, Antitrust Enforcement Guidelines for International Operations, 53 Fed Reg 21584, 21586 (1988).

DoJ.[51] This policy change greatly went beyond what the NRCA had anticipated.[52] The innovation market approach was again reaffirmed as good policy in the Competitor Collaboration Guidelines of 2000.[53] In their 2017 update to the IP Guidelines, the agencies renamed innovation markets R&D markets, but without changing their substance.[54]

An R&D market as envisioned by the IP Guidelines 'consists of the assets comprising research and development related to the identification of a commercialisable product, or directed to particular new or improved goods or processes, and the close substitutes for that research and development'.[55] The R&D market approach assumes that the R&D process itself can be regarded as its own antitrust market.[56] This will be delineated in an analogy to the hypothetical monopolist test: while the conventional SSNIP test looks for a small but significant non-transitory increase in prices, in the realm of R&D markets, the Guidelines look for a small but significant non-transitory *decrease* in R&D.[57]

In their CC Guidelines, the antitrust agencies refer to the relevant section of their 1995 IP Guidelines in order to define innovation markets,[58] but also elaborate on the innovation market concept when outlining antitrust safety zones and how they are applied in the context of innovation markets. They emphasise that several factors are relevant when deciding whether or not certain R&D is substitutable and thus forms part of the same innovation market, namely the nature, scope and magnitude of R&D efforts, the company's access to financial support, IPRs, skilled personnel or other specialised assets, the timing of R&D efforts and the company's ability to successfully commercialise innovations.[59] In their 2017 update to the IP Guidelines, the US agencies include these same factors in order to demonstrate which factors they would look to when assessing close substitutes for R&D activities.[60]

[51] US Department of Justice and Federal Trade Commission, Antitrust Guidelines for the Licensing of Intellectual Property (6 April 1995) § 3.2.3.

[52] LB Landman, 'Innovation Markets in Europe' (1998) 19 *European Competition Law Review* 21, 22.

[53] Federal Trade Commission and US Department of Justice, Antitrust Guidelines for Collaborations among Competitors (April 2000) (US CC Guidelines).

[54] US Department of Justice and Federal Trade Commission, Antitrust Guidelines for the Licensing of Intellectual Property (14 January 2017) (US IP Guidelines 2017) § 3.2.3; VHSE Robertson, 'A Brief Comment on the 2017 Update of the US Intellectual Property Licensing Guidelines' (2018) 39 *European Competition Law Review* 461, 462.

[55] US IP Guidelines 2017 (n 54) § 3.2.3.

[56] LB Landman, 'Competitiveness, Innovation Policy, and the Innovation Market Myth: A Reply to Tom and Newberg on Innovation Markets as the "Centerpiece" of "New Thinking" on Innovation' (1998) 13 *St John's Journal of Legal Commentary* 223, 236.

[57] US IP Guidelines 2017 (n 54) § 3.2.3. See also RJ Gilbert and SC Sunshine, 'Incorporating Dynamic Efficiency Concerns in Merger Analysis: The Use of Innovation Markets' (1995) 63 *Antitrust Law Journal* 569, 594; Rapp (n 7) 32.

[58] US CC Guidelines 2000 (n 53) § 3.32(c).

[59] ibid § 4.3.

[60] US IP Guidelines 2017 (n 54) § 4.3.

The IP Guidelines distinguish between an R&D market and a technology market, as the former relates to ongoing R&D that has not yet culminated in IPRs or product innovation. Importantly, the agencies will only define an R&D market 'when the capabilities to engage in the relevant research and development can be associated with specialized assets or characteristics of specific firms'.[61] This is so because only the limited access to these assets makes the innovation inaccessible to third parties. If one can determine who has access to these or similar specialised assets, then one can also single out actual or potential competitors.[62] Where no specialised assets or characteristics are involved, anybody with the necessary means could engage in said R&D or replicate it. Although it is essential to combine the R&D market approach with access to such tangible assets, this criterion does not receive the necessary attention in practice.[63]

B. Cases Involving Innovation Markets in the US

The FTC has repeatedly relied on the innovation or R&D market concept in its investigations, as has the DoJ.[64] As none of these cases was tried in court,[65] the innovation market approach very much remains a child of antitrust policy rather than antitrust law. Richard Rapp has suggested that the innovation market framework has been applied in three categories of cases: (a) in most cases in which innovation market considerations were applied, there were also competition considerations relating to actual product markets, meaning that the DoJ could have relied on the latter alone in order to decide these cases.[66] The DoJ perhaps decided to add these innovation market aspects in order to further its nascent doctrine. (b) Some cases were actually potential competition cases; and (c) merely a few cases were, as Rapp terms it, 'R&D only'.[67] In this light, the importance of the innovation market concept for competition law practice dwindles.

While the innovation market seems to lend itself to the pharmaceutical industry, where ongoing and culminating R&D efforts can more easily be

[61] ibid § 3.2.3. See also US CC Guidelines 2000 (n 53) § 3.32(c).

[62] Gilbert and Sunshine, 'Incorporating Dynamic Efficiency Concerns' (n 57) 588.

[63] Rapp (n 7) 37; K Bernard, 'Innovation Market Theory and Practice: An Analysis and Proposal for Reform' (2011) 7 *Competition Policy International* 159, 162, 166.

[64] For an overview, see RJ Hoerner, 'Innovation Markets: New Wine in Old Bottles?' (1995) 64 *Antitrust Law Journal* 49, 70–73; BR Kern, R Dewenter and W Kerber, 'Empirical Analysis of the Assessment of Innovation Effects in US Merger Cases' (2016) 16 *Journal of Industry, Competition and Trade* 373, 385. While the mere mentioning of innovation-related aspects does not automatically translate into the use of an innovation market, it is a good indicator of the continued relevance of innovation in US market definition.

[65] See also JT Rosch, 'Antitrust Regulation of Innovation Markets' (ABA Antitrust Intellectual Property Conference, Berkeley, 5 February 2009) 13 f.

[66] Rapp (n 7) 22.

[67] ibid 40.

observed, it is not the only one in which the agencies have applied that concept. Beyond pharmaceuticals, cases have included computer software and medical equipment.

The *Genzyme/Novazyme* investigation, closed in 2001, has been referred to as 'the poster child for pure innovation market analysis'.[68] In this case, both Genzyme and Novazyme were engaged in R&D towards developing a therapy for Pompe, a rare metabolic disorder. No drug had been approved for this rare disease yet, and the merging companies were the only ones engaged in R&D to that effect. The investigation looked at whether the merger could harm innovative efforts and found that it would not.[69] However, the analysis could just as well have been based on a future market that, in view of the early stages of the R&D in the drug approval process, was very uncertain and antitrust enforcement would thus have considerably increased uncertainty for the companies involved.

In its case regarding Amgen's acquisition of Immunex (2002), the FTC held that the relevant product markets related to three different pharmaceutical products. It considered the R&D, manufacture and sales relating to each of these.[70] As these products had already reached the market, the innovation market approach was merely an add-on.[71] The FTC discussed the structure of these markets, emphasising the little competition that existed in them, already at the R&D stage.[72] It also noted that the acquisition would reduce innovation competition in three markets as well as eliminating potential competition in two markets.[73] The investigation was resolved through a consent order, with Amgen agreeing to divest certain of Immunex's assets and license certain of the latter's IPRs.

Further insight into the FTC's application of the innovation market concept can be gained from its *Johnson & Johnson* case of 2005, in which the FTC investigated Johnson & Johnson's proposed acquisition of Guidant. Both current and future/innovation markets were at issue. The FTC regarded the relevant market to be the R&D, manufacture and/or sale of three specific interventional cardiology products and cardiac surgery devices.[74] These were markets in which both parties to the acquisition were operating.[75] For one of these products, Johnson & Johnson was one of only two companies producing these in the US, while at least three companies were engaged in R&D in that area and were set to obtain FDA approval

[68] Bernard (n 63) 176.

[69] Federal Trade Commission, 'FTC Closes its Investigation of Genzyme Corporation's 2001 Acquisition of Novazyme Pharmaceuticals, Inc' (13 January 2004).

[70] *Amgen and Immunex*, 134 FTC 333, 337 (2002).

[71] With a similar finding for a range of merger cases, see RJ Gilbert and WK Tom, 'Is Innovation King at the Antitrust Agencies? The Intellectual Property Guidelines Five Years Later' (2001) 69 *Antitrust Law Journal* 43, 83.

[72] *Amgen and Immunex*, 134 FTC 333, 337–39, (2002).

[73] ibid 340.

[74] *Johnson & Johnson*, 140 FTC 1062, 1065 f (2005).

[75] ibid 1064 f.

for their products within the next two to three years. Guidant was one of these three companies.[76] The FTC was concerned that the proposed acquisition would substantially lessen competition by, amongst other things, eliminating actual, direct and substantial competition between Johnson & Johnson and Guidant in the markets for the R&D, marketing and sale of certain cardiac surgery devices and, more generally, reducing R&D in the relevant markets.[77] Although the FTC did not explicitly refer to the innovation market concept in its analysis, its market definition clearly took R&D of a particular nature into account and, as such, can be construed to fit within the innovation market rationale, coupled with an already-existing product market. The innovation market approach allowed the FTC to incorporate expected market developments into its analysis. However, this case could just as well have relied on a current product market and potential competition in this market. Where market entry was more certain, it could have relied on the delineation of a future market. Also, it should be noted that the foreseeable market developments at issue are unique to regulated markets in which companies must obtain regulatory approval before marketing their goods or services, particularly in pharmaceuticals in which lengthy approval procedures are the norm. Wherever possible, innovators try to keep their pipeline products secret until they (nearly) reach the market, thereby making innovations unobservable and the innovation market approach inapplicable.

The European Commission investigated many of the same mergers as its US counterparts. A survey of innovation-intense cases that were investigated in parallel by both US and EU competition authorities – eg, *Shell/Montedison*,[78] *Upjohn/Pharmacia*[79] and *Glaxo/Wellcome*[80] – revealed that the Commission based its analysis on current or future product or technology markets in these cases, embedding the innovation analysis within this analytical framework.[81] An FTC Staff Report of 1996 made clear that 'it seems inevitable that an innovation market will be defined with respect to an ultimate goods market, such as "R&D directed at [a class of products]"'.[82] In the end, this is nothing else than the future goods markets that the European Commission has been relying on.[83]

[76] ibid 1066.

[77] ibid 1067. Eventually, Johnson & Johnson did not acquire Guidant, leading to the re-opening and setting aside of the order; *Johnson & Johnson*, 141 FTC 487 (2006).

[78] *Shell/Montecatini* (Case IV/M.269) Commission Decision 94/811/EC [1994] OJ L332/48; *Montedison*, 119 FTC 676 (1995).

[79] *Upjohn/Pharmacia* (Case IV/M.631) Commission Decision of 28 September 1995 [1995] OJ C294/9; *Upjohn and Others*, 121 FTC 44 (1996).

[80] *Glaxo/Wellcome* (Case IV/M.555) Commission Decision of 28 February 1995 [1995] OJ C65/3; *Glaxo*, 119 FTC 815 (1995).

[81] Landman, 'Innovation Market Myth' (n 56) 23 ff, especially 29 f for the conclusion.

[82] Federal Trade Commission, 'Anticipating the 21st Century' (n 13) § 7.VI.A.1.

[83] This is also noted by Landman, 'Innovation Market Myth' (n 56) 248.

C. Criticism of the Innovation Market Approach

The innovation market approach provides one possibility for factoring innovation into antitrust analysis. It applies traditional product market definition to R&D efforts. The US agencies adopted it as a third type of market, in addition to product and technology markets. However, the approach was met by severe criticism after the IP Guidelines were issued in 1995. Following the 'Pitofsky Hearings', in which the FTC under Chairman Robert Pitofsky held hearings on the IP Guidelines and other issues,[84] the FTC was reproached for not retracting its innovation market approach in the light of the outcomes of these hearings.[85] The fact that a number of FTC and DoJ officials published articles and held speeches promoting the innovation market approach[86] was denounced as the propagation of an 'innovation market myth'.[87] Several commentators regarded the innovation market concept as an instrument created in order to be able to challenge certain mergers.[88] Richard Gilbert and Steven Sunshine, themselves DoJ proponents of the approach, identified the major criticisms of the innovation market concept as (a) the weak economic evidence on the connection between market structure and innovation, (b) the non-fulfilment of Clayton Act prerequisites by innovation markets, and (c) the aptness of conventional antitrust analysis to deal with innovation-related issues.[89] To this criticism, Michael Carrier added that (d) innovation is speculative and includes unidentifiable market participants and that (e) the relationship between R&D and innovation is unclear.[90] We will now look at these and a number of additional criticisms in turn.

Where the innovation market approach is used as an add-on to current product market cases, future product market cases or potential competition cases, it is not actually needed.[91] Cases with pure R&D are 'the only setting where the innovation market approach is arguably necessary',[92] but at the same time this 'is also the most potentially dangerous application'[93] because the effects of thus tampering

[84] Federal Trade Commission, 'Anticipating the 21st Century' (n 13).

[85] Widnell (n 38) 369 (citing a number of economists and lawyers in fn 3 who openly spoke out against the innovation market approach at the Pitofsky Hearings); JR Eiszner, 'Innovation Markets and Automatic Transmissions: A Shift in the Wrong Direction?' (1998) 43 *Antitrust Bulletin* 297, 329 ff.

[86] Eg, Baxter (n 49); AK Bingaman, 'Innovation and Antitrust' (Commonwealth Club of California, San Francisco, 29 July 1994); RJ Gilbert and SC Sunshine, 'The Use of Innovation Markets: A Reply to Hay, Rapp, and Hoerner' (1995) 63 *Antitrust Law Journal* 75; WK Tom and JA Newberg, 'Antitrust and Intellectual Property: From Separate Spheres to Unified Field' (1997) 66 *Antitrust Law Journal* 167.

[87] Landman, 'Innovation Market Myth' (n 56) 234 f.

[88] Hoerner (n 64) 50, 56; AR Chin, 'The Misapplication of Innovation Market Analysis to Biotechnology Mergers' (1997) 3 *Boston University Journal of Science and Technology Law* 6, § 48.

[89] Gilbert and Sunshine, 'A Reply' (n 86) 75.

[90] MA Carrier, *Innovation for the 21st Century: Harnessing the Power of Intellectual Property and Antitrust Law* (Oxford, Oxford University Press, 2009) 297.

[91] For a contrary opinion, see Glader, *Innovation Markets* (n 39) 197 ff.

[92] Rapp (n 7) 43.

[93] ibid 44.

with the innovation process are unknown. In the end, the innovation market concept may do more harm than good.[94]

The notion of an innovation market may in itself be misleading, as it uses the term 'market' to describe an approach that tries to identify competitive effects on innovation[95] rather than providing a new way for defining antitrust relevant markets. In the absence of market transactions, no real market lies at the heart of an innovation market; instead, Josef Drexl suggests that the original proponents of the approach 'tried to capture problems related to competition that occur outside existing markets'.[96] As R&D – just like human resources or raw materials – is simply an input into a company's business, it might be too far-fetched to actually construe a relevant antitrust market around this input.[97] Only when a company is in the business of selling R&D services could one speak of a market that satisfies the criteria of section 7 of the Clayton Act.[98] A similar discussion is currently taking place with reference to data as an input or as a standalone antitrust market.[99]

The Second Circuit has held that the acquisition of an IPR before the relevant market even came into existence, and particularly before the commercialisation of the patented art, should never lead to antitrust liability.[100] Temporally speaking, the innovation market is relied upon too early in the lifecycle of product development, at a time at which all there is are predictions about which current R&D avenues might prove relevant in the future. Talking of 'markets' so early on in the innovation process may simply be premature.

The US Supreme Court understands 'commerce' within the meaning of the Clayton Act to relate to 'the practical, economic continuity in the generation of goods and services for interstate markets and their transport and distribution to the consumer'.[101] This criterion is arguably not met by the innovation market concept.[102] In addition, as highlighted above, no actual transactions take place in an innovation market.

The ex ante analysis, which is the defining feature of the innovation market concept,[103] might prove to be its greatest weakness. It can lead to misguided

[94] J Temple Lang, 'European Community Antitrust Law: Innovation Markets and High Technology Industries' (1997) 20 *Fordham International Law Journal* 717, 767; Eiszner (n 85) 298, 334.

[95] Gilbert and Sunshine, 'A Reply' (n 86) 82 (suggesting that 'the innovation market analysis is simply a tool to aid in the analysis of competitive effects').

[96] J Drexl, 'Anticompetitive Stumbling Stones on the Way to a Cleaner World: Protecting Competition in Innovation without a Market' (2012) 8 *Journal of Competition Law & Economics* 507, 517.

[97] Hoerner (n 64) 51.

[98] ibid 53. For the EU, see Temple Lang, 'Innovation Markets' (n 94) 764 f.

[99] See section III.D.ii in ch 8.

[100] *Scm Corporation v Xerox Corporation*, 645 F2d 1195, 1206 (2d Cir 1981). See also a District Court's holding that 'the absence of a relevant market in [certain] products at the time of patent acquisition precludes the applicability of Section 7' of the Clayton Act; *Crucible v Stora Kopparbergs Bergslags*, 701 FSupp 1157, 1162 f (WD Penn 1988).

[101] *Gulf Oil Corp v Copp Paving Co*, 419 US 186, 195 (1974).

[102] Eiszner, 'Innovation Markets' (n 85) 341; however, see Gilbert and Sunshine, 'Incorporating Dynamic Efficiency Concerns' (n 57) 600.

[103] HJ Hovenkamp et al, *IP and Antitrust: An Analysis of Antitrust Principles Applied to Intellectual Property Law*, 3rd edn, vol I (New York, Wolters Kluwer, 2018) § 4.03[E].

predictions, imposing antitrust liability that can result in harm to competition and innovation rather than in their promotion. Where initial expectations regarding a certain R&D endeavour are flawed, antitrust enforcement based on such expectations would necessarily lead to counter-productive results.

The innovation market approach relies on the premise that there is a 'deterministic relationship between R&D expenditure and innovation'[104] even though there is no such established relationship.[105] This criticism has been acknowledged by the agencies,[106] but has not led to any substantive change in the 2017 update to the IP Guidelines.

The innovation market approach is not needed in order to incorporate innovation considerations into antitrust analysis. Product and technology markets, future markets and innovation competition already provide the analytical framework for doing so. Indeed, innovation is valued mainly because it leads to a new or improved good or service, which in turn can be described through a conventional (future) antitrust product market.[107] The possibility of relying on potential competition, be it perceived or actual (on this, see section I.A above), enables antitrust analysis to capture companies that may have no market share in any current markets, but are likely to be important innovators in the future.[108]

The usefulness of the innovation market, but also of technology markets, is generally seen in their ability to forecast the effect of company behaviour on prices and output[109] rather than on innovation. From an innovation perspective, this is rather frustrating, for one might expect that the purpose of engaging in the innovation market analysis will yield some benefit for innovation as a dynamic competitive parameter, not for the static competitive parameters of prices and output.

The innovation market approach is notably absent from the 2010 Horizontal Merger Guidelines, which subscribe to a market definition approach that is different from the 1995 IP Guidelines.[110] The 2010 Merger Guidelines emphasise that a merger's effect on innovation shall be analysed, albeit without ever resorting to the notion of an innovation market analysis.[111] However, the 2017 update of the IP Guidelines made no changes to the innovation market approach, apart from renaming it an R&D market.

[104] Gilbert and Sunshine, 'Incorporating Dynamic Efficiency Concerns' (n 57) 579.

[105] Eiszner (n 85) 317.

[106] United States in OECD (ed), *Policy Roundtable: Merger Review in Emerging High Innovation Markets* (2002) DAFFE/COMP(2002)20, 149.

[107] Bernard (n 63) 163.

[108] Rapp (n 7) 37 ff; GA Hay, 'Innovations in Antitrust Enforcement' (1995) 64 *Antitrust Law Journal* 7, 13 f.

[109] RJ Gilbert, 'Competition and Innovation' in ABA Section of Antitrust Law (ed), *Issues in Competition Law and Policy* (Chicago, ABA Publishing, 2008) 581.

[110] K Feng, 'Patent-Related Mergers and Market Definition under the 2010 Horizontal Merger Guidelines: The Need to Consider Technology and Innovation Markets' (2012) 34 *Thomas Jefferson Law Review* 197, 207.

[111] US Horizontal Merger Guidelines 2010 (n 41) § 6.4. See section III.B above.

D. Lessons from the US Innovation Market Concept

Before we briefly explore the short-lived emergence of innovation market analysis in the EU, it is worth reflecting on the lessons that can be learned from the agencies' attempt of establishing the innovation market concept. The first certainly is that it is quite hard to conceptually grasp innovation, it being the unpredictable animal it is. Therefore, any estimate of how, when and why innovation will ensue needs to be taken with a pinch of salt, and antitrust concepts incorporating predictions about innovation need to be crafted very carefully.

Second, basing policy recommendations on a fragile theoretical background is doomed to fail. Where antitrust authorities – as the FTC did during the Pitofsky Hearings – ask experts to contribute to their policy-making, they are well-advised to not only record their arguments but also incorporate them rather than propose policies that perfectly contradict the expert advice that has been received. It remains to be seen what the current FTC leadership makes of the most recent FTC Hearings.[112]

Third, an important lesson to take away from the innovation market approach is that even if that approach is not as successful as it was meant to be, we must not stop our attempts at meaningfully incorporating dynamic competition and innovation considerations into our antitrust framework. Innovation competition (see section III below) can be regarded as the most recent development in this respect.

E. Some Reflections on the Innovation Market Concept in the EU

The EU has never relied on an innovation market approach identical to the one set out in the US IP Guidelines, even if some authors have tried to liken the European approach to that of the US.[113] While the US regarded innovation as a separate market, the EU preferred to frame innovation considerations within given product markets.[114] In Europe, attention has been drawn to the serious conceptual problems with defining an innovation market outside of actual products or technology,[115] and there is widespread scepticism towards the innovation market concept based

[112] Federal Trade Commission, 'Hearings on Competition and Consumer Protection in the 21st Century' (2018–19), www.ftc.gov/policy/hearings-competition-consumer-protection.

[113] See, eg, Glader, *Innovation Markets* (n 39).

[114] European Commission, Guidelines on the application of Article 81 of the EC Treaty to technology transfer agreements [2004] OJ C101/2, para 25; European Commission, Guidelines on the application of Article 101 of the Treaty on the Functioning of the European Union to technology transfer agreements [2014] OJ C89/3, para 26; LB Landman, 'The Economics of Future Goods Markets' (1997) 21 *World Competition* 63, 64.

[115] RC Lind et al, 'Report on Multiparty Licensing' CRA Report (22 April 2003) 47.

on its input-relatedness.[116] It is generally acknowledged that R&D expenditure does not translate well into future market power and that basing an approach on ongoing R&D expenditure necessarily leads to a high margin of error.[117]

This is not to say that the European Commission did not use the term 'innovation market' in the past. In fact, in its 2004 Technology Transfer Guidelines, it discussed innovation markets as a third type of relevant market – next to product and technology markets – within the realm of market definition.[118] Some viewed this as the Commission's 'final step [to] include innovation markets' within its analytical framework.[119] While the Commission emphasised that it preferred to analyse innovation considerations within the framework of actual markets, or perhaps within the framework of potential competition, it also stated that in rare cases it can be 'useful and necessary to also define innovation markets', particularly where R&D poles can readily be made out at an early stage.[120] This, one could add, only applies to pharmaceuticals or biotechnology and little else. The Commission's reticence in this respect was clearly palpable. Yet, it acknowledged that something outside of product and technology markets may be at stake that requires antitrust attention.

This language slightly changed in the Commission's 2014 Technology Transfer Guidelines, in which the Commission no longer uses the term 'innovation market'. Instead, the Commission points out that it may rely on innovation competition where this dimension of competition is affected, rather than competition in products.[121] This language relates back to the Commission's 2001 Guidelines on Horizontal Cooperation, in which it also discussed these issues within the framework of competition in innovation.[122] This remained unchanged in its 2011 Guidelines on Horizontal Cooperation.[123] While these competitive effects are highly relevant for market definition, the Commission acknowledges that they 'can in some cases not be sufficiently assessed by analysing actual or potential competition in existing product/technology markets'.[124] In these cases, the Commission will focus on innovation competition, as is further detailed below.

[116] Europe Economics, 'The Development of Analytical Tools for Assessing Market Dynamics in the Knowledge Based Economy' (12 September 2003) 74.

[117] Temple Lang, 'Innovation Markets' (n 94) 765 f.

[118] Guidelines on the application of Article 81 of the EC Treaty to technology transfer agreements (n 114) para 25.

[119] Glader, *Innovation Markets* (n 39) 3.

[120] Guidelines on the application of Article 81 of the EC Treaty to technology transfer agreements (n 114) para 25.

[121] Guidelines on the application of Article 101 of the Treaty on the Functioning of the European Union to technology transfer agreements (n 114) para 26.

[122] Guidelines on the applicability of Article 81 of the EC Treaty to horizontal cooperation agreements (n 18) paras 50 f.

[123] Guidelines on the applicability of Article 101 of the Treaty on the Functioning of the European Union to horizontal co-operation agreements (n 18) para 119.

[124] ibid.

III. The Emergence of Innovation Competition

Innovation competition – or competition in innovation – is the competitive process between companies that are striving to bring interchangeable new products or processes on to the market.[125] This competition takes place before a proper antitrust market emerges and is characterised by the kind of high uncertainty that accompanies all innovation enterprises. US antitrust agencies now increasingly focus on this dimension of competition. The European Commission has also incorporated innovation competition into its competition policy. Thereby, innovation competition is seen as a dimension that needs to be taken into account outside of market definition. However, it is still in search of a coherent approach.[126] It particularly remains uncertain how the adoption of the concept of innovation competition relates to more traditional antitrust market definition.

A. Competition in Innovation in the EU

A number of current Commission policy documents refer to competition in innovation, such as the Technology Transfer Guidelines[127] and the Guidelines on Horizontal Cooperation.[128] The Commission understands that 'R&D co-operation may not only affect competition in existing markets, but also competition in innovation and new product markets'.[129] While these effects are highly relevant for market definition, the Commission acknowledges that they 'can in some cases not be sufficiently assessed by analysing actual or potential competition in existing product/technology markets'.[130]

The Commission spells out two scenarios that might help in identifying an agreement's effect on competition in innovation. The first is an innovative environment in which competing R&D efforts or poles can be identified early on, eg, in the case of pharmaceuticals. In order to assess whether other R&D efforts represent credible competition, the Commission will look at criteria such as their 'nature, scope and size', 'their access to financial and human resources, know-how/patents, or other specialized assets' and their 'timing and their capability to exploit possible results'.[131] The Commission also draws attention to the fact that R&D cooperation

[125] JB Baker, 'Beyond Schumpeter vs Arrow: How Antitrust Fosters Innovation' (2007) 74 *Antitrust Law Journal* 575, 579.

[126] Kern (n 6) 175.

[127] Guidelines on the application of Article 101 of the Treaty on the Functioning of the European Union to technology transfer agreements (n 114) para 26.

[128] Guidelines on the applicability of Article 101 of the Treaty on the Functioning of the European Union to horizontal co-operation agreements (n 18) paras 119 ff.

[129] ibid para 119.

[130] ibid.

[131] ibid para 120. Here, the dynamic capabilities discussed by David Teece can also come into play; see DJ Teece, *Dynamic Capabilities and Strategic Management: Organizing for Innovation and Growth* (Oxford, Oxford University Press, 2009).

might affect future product markets. This effect might be 'implicitly incorporated in the analysis of competition in innovation'.[132] The second scenario is an innovative environment in which competing R&D efforts cannot be identified early on. In such a case, the Commission will normally not be able to evaluate what kind of effect an R&D cooperation will have on innovation.[133] Of these two scenarios, the first is highly reminiscent of the innovation market approach and it appears to be the only one in which innovation competition can be properly assessed under the Commission's current framework. The limitations that the innovation market approach was confronted with also constrain the Commission's analytical framework for innovation competition. Regarding the relationship between innovation competition and antitrust market definition, it remains an open question as to whether the market within which such competition in innovation occurs or will occur needs to be defined at all. As was pointed out above when discussing the characteristics of innovative markets (see section II.C in Chapter 3), competition in such environments is frequently found to be for the market rather than in the market.[134] Competition for the market may exert less – or at least a different kind of – pressure than competition from within the market.[135]

The Commission's approach to innovation competition in its Horizontal Cooperation Guidelines has been criticised because it is 'trying to assess innovation-related competition law cases in light of existing technology and product markets to the extent possible', and such an approach is naturally biased towards static competition.[136] Competition in innovation as an antitrust tool might need to be particularly carefully applied.[137] Especially where R&D poles cannot be readily identified, it would seem that the Commission does not currently have an analytical framework that allows it to take competition in innovation into account.

Two merger proceedings involving Microsoft have shed light on how the Commission might take innovation competition into account in practice when R&D poles cannot be readily identified. In *Microsoft/Yahoo! Search Business* (2010), the Commission relied on online advertising services as the relevant product market, but left open whether online search advertising, mobile search advertising or intermediation in online advertising constituted their own

[132] Guidelines on the applicability of Article 101 of the Treaty on the Functioning of the European Union to horizontal co-operation agreements (n 18) para 121.

[133] ibid para 122.

[134] See DS Evans and R Schmalensee, 'Some Economic Aspects of Antitrust Analysis in Dynamically Competitive Industries' in AB Jaffe, J Lerner and S Stern (eds), *Innovation Policy and the Economy*, vol 2 (Cambridge, MA, MIT Press, 2002) 1; PA Geroski, 'Competition in Markets and Competition for Markets' (2003) 3 *Journal of Industry, Competition and Trade* 151, 152.

[135] Geroski, 'Competition in Markets' (n 134) 162. Geroski also warns that antitrust law's standard analytical tools are only to a limited degree apt for this analysis of competition *for* a market (ibid 165).

[136] Drexl, 'Anticompetitive Stumbling Stones' (n 96) 522 f.

[137] L Gyselen, 'Competition in Innovation: A Novel Concept? The Case Law on Pharmaceuticals' in P Lugard and L Hancher (eds), *On the Merits: Current Issues in Competition Law and Policy – Liber Amicorum Peter Plompen* (Cambridge, Intersentia, 2005) 48.

relevant markets.[138] Having concluded its market delineation, the Commission turned to the competitive assessment and to characterising the market. In this respect, it took particular account of innovation, stressing that search engines' ability to innovate both on the user side and on the advertiser side of their market constituted '[a]nother important dimension of competition'.[139] The Commission also underlined the importance of incremental and leap-frog innovation in this industry.[140] While explicitly recognising the importance of innovation competition in the industry under investigation, the Commission did so outside of market definition. This is in line with the view that innovation competition necessarily takes place outside of antitrust product markets.[141]

A further merger investigation concerned *Microsoft/Skype* (2011). The Commission assessed this case based on the narrowest possible product market, without deciding on the actual relevant market.[142] It then went on to characterise the market for consumer communication services. It noted the dynamic nature of this market, which went hand in hand with the Commission's finding that market shares were only 'a preliminary indication of the competitive situation in these dynamic markets'.[143] Upon appeal, the General Court agreed with the Commission that '[i]n such a dynamic context, high market shares are not necessarily indicative of market power.'[144] This case can be seen as another example of the Commission taking innovation competition into account, again outside the framework of market definition. A further instance was the *General Electric/Alstom* (2015) merger, in which the Commission also referred to the transaction's effects on innovation competition.[145]

In the *Dow/DuPont* (2017) merger, the European Commission found that while there could be no market in innovation as such, this should not keep competition authorities from investigating a transaction's 'impact … at the level of innovation efforts'.[146] It set out to define 'innovation spaces', which in this particular case it understood to be 'those spaces in which innovation competition occurs in the crop protection industry'.[147] It then explained that 'in order to assess innovation competition, the Commission will both consider metrics of innovation taking place at industry level, as well as innovation taking place in spaces consisting of groupings of crop/pest combinations'.[148] In order to assess innovation competition, the

[138] *Microsoft/Yahoo! Search Business* (Case COMP/M.5727) Commission Decision of 18 February 2010, paras 60, 75, 81, 83, 87.

[139] ibid para 109.

[140] ibid.

[141] Drexl, 'Anticompetitive Stumbling Stones' (n 96) 540.

[142] *Microsoft/Skype* (Case COMP/M.6281) Commission Decision of 7 October 2011, para 63.

[143] ibid paras 78, 99 (direct quote).

[144] Case T-79/12 *Cisco Systems & Messagenet v Commission* EU:T:2013:635, para 69.

[145] *General Electric/Alstom* (Case M.7278) Commission Decision of 8 September 2015, paras 1392, 1418.

[146] *Dow/DuPont* (Case M.7932) Commission Decision of 27 March 2017 [2017] OJ C353/9, para 348.

[147] ibid para 350.

[148] ibid para 352.

authority 'will focus on measures of innovation output'.[149] One could argue that the innovation spaces referred to by the Commission are those R&D efforts which might at a later stage give rise to a proper antitrust market. Before this occurs, the Commission will take into account innovation competition occurring in these spaces in the realm of the substantive analysis.

The Commission carried out a similar analysis in *Bayer/Monsanto* (2018), where it defined innovation spaces as

> spaces in which innovation competition occurs (be it in the crop protection sector or in the traits sector). R&D players do not innovate for all the product markets composing a sector at the same time. They also do not innovate randomly without targeting specific spaces within that sector. When setting up their innovation capabilities and conducting their research, R&D players have specific research targets.[150]

It found that the merging parties had been and would probably continue to be innovation competitors in relation to established product markets, new product markets and product markets that were still at the R&D stage and were considered to constitute important innovation spaces.[151]

In the *Apple/Shazam* (2018) merger decision, the Commission would have been prepared to take innovation competition into account, yet found that the market for dedicated music recognition services on which Shazam was active did not represent a market that was characterised by disruptive innovation, but was a relatively mature market.[152]

When the European Commission prohibited the merger of *Siemens/Alstom* (2019), it did so because it was concerned that the merger would significantly impede competition on the markets of rail signalling systems and very high-speed trains. A clear nexus to innovation competition was palpable in this case, as the Commission and stakeholders were concerned that the merger would diminish innovation.[153] Similarly, in the ongoing *Car Emissions* case, the European Commission holds the preliminary view that several car manufacturers restricted competition by agreeing to limit the development and marketing of emission cleaning technology for cars (Article 101 TFEU).[154] The Commission appears to be pursuing a theory of harm that is closely linked to innovation competition – or, rather, an agreement to limit innovation competition that may have, in due course, hindered the development of more environmentally friendly technology.

[149] ibid para 379.

[150] *Bayer/Monsanto* (Case M.8084) Commission Decision of 21 March 2018 [2018] OJ C456/10, para 80, fn 23.

[151] ibid para 81.

[152] *Apple/Shazam* (Case M.8788) Commission Decision of 6 September 2018 [2018] OJ C417/4, paras 162 f.

[153] *Siemens/Alstom* (Case M.8677) Commission Decision of 6 February 2019 [2019] OJ C300/14; European Commission, 'Mergers: Commission Prohibits Siemens' Proposed Acquisition of Alstom' (IP/19/881, 6 February 2019).

[154] *Car Emissions* (Case AT.40178) (Commission Decision pending); European Commission, 'Antitrust: Commission Sends Statement of Objections to BMW, Daimler and VW for Restricting Competition on Emission Cleaning Technology' (IP/19/2008, 5 April 2019).

Again, the dimension of innovation competition relates to the substantive assessment rather than to market delineation.

B. Innovation Competition in the US

The Horizontal Merger Guidelines of 2010 acknowledge that a merger's impact on innovation competition ranges amongst the possible unilateral effects of mergers. A merger might lead to lower levels of innovation on the part of the merging companies, be it because (1) existing products are not further developed or (2) new products are not being developed, in each case because it would harm one of the merging firms' revenue.[155] The concerns are thus related to innovation competition in both existing markets and possible future markets.

When analysing a merger in innovative markets, the agencies will look closely at the merger's possible harm for innovation by focusing on whether or not the merged company will reduce R&D below the level of R&D as it would exist were it not for the merger (but-for analysis).[156] This approach to analysing innovation competition is reminiscent of the IP Guidelines' innovation market approach, which heavily concentrates on a lessening of R&D expenditure. As such, innovation competition as envisaged by the Horizontal Merger Guidelines 2010 does not appear to be aligned with competition in innovation as formulated by the European Commission (discussed above). Of course, a focus on R&D expenditure is simpler in that it is more easily ascertainable than innovative outcome. However, R&D expenditure alone does not appear to be a reliable indicator in this respect, as it is not correlated with innovation.[157]

The agencies' focus on innovation competition could entail that even a relatively small merger will be closely scrutinised if the companies involved have important innovative potential.[158] The Guidelines make these policy considerations within the realm of the unilateral effects analysis (§ 6) rather than in the section on market definition (§ 4). This is an indication that innovation competition will not form part of the market delineation as such, but enters the antitrust analysis during the substantive assessment.

C. The Move from Innovation Markets to Innovation Competition

The innovation market approach (see section II above) was a first attempt to incorporate considerations relating to innovation competition into the analytical

[155] US Horizontal Merger Guidelines 2010 (n 41) § 6.4.
[156] ibid.
[157] Similarly, see Eiszner (n 85) 317.
[158] US Horizontal Merger Guidelines 2010 (n 41) § 6.4.

framework of antitrust. However, by remaining within the borders of market definition, this approach was not very successful. More recent attempts now refer to 'innovation competition'[159] or 'competition in innovation'.[160] In order to capture the innovative environment surrounding R&D, the European Commission now appears to favour the term 'innovation spaces'.[161] It seems that the antitrust authorities thereby try to avoid the difficulties associated with the innovation market concept.[162] From a comparative perspective, it is interesting to note that the discourse on innovation competition is a transatlantic one and the jurisdictions should not be regarded in isolation – as evidenced by the introduction of innovation markets in the US and the appearance of the same terminology a few years later in the EU. The same has more recently happened with innovation competition, this time with its point of departure in the EU and making its way across the Atlantic. Here, the same caveat that applies to all legal transplants needs to be taken into account.[163]

If innovation competition is seen as competition outside of defined antitrust markets,[164] then one needs to acknowledge that it cannot be included within the analytical framework for defining antitrust markets. Indeed, the shift away from innovation markets and towards innovation competition reflects the realisation that current innovative activities cannot be properly framed as antitrust markets before they have culminated in something resembling a market. While market definition continues to be one of the central pillars of antitrust analysis (see section III in Chapter 2), it would be misguided to try to incorporate innovation competition considerations into it at any cost. While it is certainly possible to make the traditionally static antitrust market definition exercise more dynamic, innovation competition is too uncertain, too future-oriented and not market-oriented enough to be able to use it as a market definition criterion. Instead, it should constitute a central element of the substantive competitive assessment.[165] This is precisely what occurred in the European Commission's merger decisions discussed above, where the innovation considerations were not analysed within market definition proper. Instead, the market characterisation in those cases was able to inform the competitive assessment, including the transactions' effects on innovation. In *Microsoft/Skype*, the characterisation of the dynamic market at issue was also able

[159] ibid.

[160] Guidelines on the applicability of Article 101 of the Treaty on the Functioning of the European Union to horizontal co-operation agreements (n 18) para 126.

[161] *Dow/DuPont* (Case M.7932) Commission Decision of 27 March 2017 [2017] OJ C353/9, para 350; *Bayer/Monsanto* (Case M.8084) Commission Decision of 21 March 2018 [2018] OJ C456/10, para 80 fn 23.

[162] Drexl, 'Anticompetitive Stumbling Stones' (n 96) 520–22.

[163] See generally A Watson, *Legal Transplants: An Approach to Comparative Law*, 2nd edn (Athens, GA, University of Georgia Press, 1993).

[164] See Drexl, 'Anticompetitive Stumbling Stones' (n 96) 540.

[165] See VHSE Robertson, 'Delineating Digital Markets under EU Competition Law: Challenging or Futile?' (2017) 12 *Competition Law Review* 131, 151.

to counter the EU's general tendency to heavily rely on market shares, from which the Commission and the General Court readily departed in this case.

Interestingly, the criteria that the US Competitor Collaboration Guidelines and the US Intellectual Property Guidelines 2017 cite as being relevant to define an innovation or R&D market (the nature, scope and magnitude of R&D efforts; the company's access to financial support, IPRs, skilled personnel or other specialised assets; the timing of R&D efforts and the company's ability to successfully commercialise innovations)[166] are almost identical to the factors that the EU Horizontal Cooperation Guidelines refer to in order to establish competing R&D poles when considering innovation competition (the nature, scope and size of R&D efforts; the company's access to financial and human resources, know-how/patents or other specialised assets; the timing of R&D efforts and the company's capability to exploit possible results).[167] This shows that the antitrust authorities have converging views in terms of the factors that need to be taken into account when considering innovation in an antitrust analysis. The weighing of these factors, of course, is a separate issue. Nevertheless, this parallelism should assist the authorities in reaching transatlantic agreement on questions related to innovation competition.

As of now, there is no coherent approach to the practical application of considerations relating to innovation competition. Innovation competition continues to be a fleeting concept, urging antitrust agencies to take innovation into account even where no current or future product or technology market can be properly defined. This fluid concept needs to be further brought into focus in order to deliver useful and predictable results. The difficulty lies in the fact that innovative environments are very diverse, making it hard to devise a one-fits-all analytical framework. At the theoretical level, however, there appears to be transatlantic convergence. It would seem that the criticism with which the innovation market concept was met has led to a process of rethinking the analytical framework for antitrust markets in dynamic contexts.

Even if one acknowledges that innovation competition itself needs to be accounted for in the substantive analysis rather than in the course of market definition,[168] there also remain analytical tools within market definition that can accommodate some innovation competition-type analysis. One of the tools that can be relied upon where the emergence of a market is fairly certain is the delineation of future product markets (see section IV in Chapter 4), while another option that should be further explored in innovative markets is the acknowledgement of

[166] US CC Guidelines 2000 (n 53) § 4.3; US IP Guidelines 2017 (n 54) § 4.3.

[167] Guidelines on the applicability of Article 101 of the Treaty on the Functioning of the European Union to horizontal co-operation agreements (n 18) para 120.

[168] Arguing against the analytical separation of innovation competition and product market competition, see J Crémer, Y-A de Montjoye and H Schweitzer, 'Competition Policy for the Digital Era' (Report for DG Competition, 3 April 2019) 120.

potential competition as an important factor (see section I). In both cases, this is only possible where actual or future product markets can be distinguished.

IV. Conclusion

Going beyond the definition of markets in a strict sense, potential competition represents a useful approach to taking market entry into account. US and EU approaches to this question are quite different. The US focuses on potential competition in the context of mergers and distinguishes between actual and perceived potential competition, depending on whether or not there is already an innovation effort directed at entry into the relevant market by the potential competitor. In the EU, potential competition is primarily employed in the context of anti-competitive agreements that aim at keeping a potential competitor out of the market. Here, the potential competitor's impending market entry must be credible and occur within a short time. Potential competition plays an important role in innovative market environments, as it can exert competitive pressure either on current or on future markets. Rather than being incorporated at the stage of market definition, it should play a role during the substantive assessment.

The US innovation market concept as developed in the 1990s is a theory in decline. While it was central in drawing attention to the issue of innovation in antitrust, the innovation market concept is problematic in that it does not relate to an actual market and is thus easily misconceived. Today, both in the EU and the US, the focus is on innovation competition as a dimension that is explicitly situated outside of market delineation proper, but that is highly relevant as it takes up where market definition has to leave off – eg, for lack of a discernible market. A coherent approach to innovation competition is still being developed. Here, it will be important to recognise that innovation competition is not always observable and that not all companies that are in possession of the necessary assets to compete on innovation will have an incentive to join this competition. This means that there is uncertainty not only in innovation, but also in innovation competition.

Together, current product markets, future markets, potential competition and innovation competition are able to capture large parts of the innovation dimension in dynamic markets, either at the stage of market definition or at the subsequent stage of the substantive analysis.

6

Intellectual Property Rights

I. Market Definition and Intellectual Property Rights

Intellectual property rights (IPRs) include, amongst other things, copyright, database rights, design rights, patents and trademarks. These rights allow an intellectual property (IP) owner to control the use (application, copying etc) of the protected database, design, innovation, sign or work, and also to license the IPR.[1] The relationship between competition law and IP law is not a straightforward relationship. While the granting of IPRs is reserved to IP law, competition law is only concerned with the way in which IP owners exercise their IPRs.[2] IPRs can be understood as 'a distortion of free-market principles'[3] because they allocate temporary exclusive rights to creators. Competition law needs to ensure that IP owners refrain from 'extending market power beyond the legitimate scope of the property right a patent [or other IPR] generates'.[4] As such, it adds another layer of regulation on IPRs.[5] At the same time, both competition law and IP law have an innovation agenda.[6] A dilemma can arise when a competition authority must ensure that the incentive to innovate, which IP laws want to promote, is not thwarted by the application of the antitrust laws.[7]

A potential source of misunderstanding between competition law and IP law lies in the fact that they both use the term 'monopoly', with competition law understanding it as market power and IP law understanding it as a bundle of rights over

[1] D Geradin, 'Pricing Abuses by Essential Patent Holders in a Standard-Setting Context: A View from Europe' ('The Remedies for Dominant Firm Misconduct' Conference, University of Virginia, June 2008) 1.

[2] AF Abbott, 'Intellectual Property Licensing and Antitrust Policy: A Comparative Perspective' (2003) 34 *Law & Policy in International Business* 801, 801.

[3] JH Barton, 'The Balance between Intellectual Property Rights and Competition: Paradigms in the Information Sector' (1997) 18 *European Competition Law Review* 440, 441.

[4] AF Abbott, S Michel and A Irizarry, 'The Right Balance of Competition Policy and Intellectual Property Law: A Federal Trade Commission Perspective' in P Marsden (ed), *Handbook of Research in Trans-Atlantic Antitrust* (Cheltenham, Edward Elgar, 2006) 357.

[5] SD Anderman, 'The Interface between Intellectual Property Rights and EU Competition Law' in A Ohly (ed), *Common Principles of European Intellectual Property Law* (Tübingen, Mohr Siebeck, 2012) 242.

[6] US Department of Justice, 'Antitrust Enforcement and Intellectual Property Rights: Promoting Innovation and Competition' (April 2007) 2.

[7] ibid 2.

a (de)sign, invention or work.[8] While IPRs confer some legal exclusivity on their owner, it needs to be determined whether or not this translates into economic power within the meaning of the antitrust laws.[9] Antitrust scholars might too readily equate this IP-type legal monopoly with significant market power in an economic sense and subsequently in an antitrust sense.[10] A monopoly over an IPR will only translate into market power in an antitrust sense if there are effectively no substitutes for the IP-protected work and none can emerge.[11] There are a number of ways in which the market power attributed to an IPR can quickly dwindle: through substitution by other innovations in the same market or in neighbouring markets, work-arounds, unsuccessful commercialisation, a change in consumer preferences evidenced by cross-market elasticity, or simply by the IP-protected technology or work turning out to be unsuccessful.[12]

Several important questions arise for market definition in the context of IPRs. Can possession of an IPR be directly linked to market power without any further need to define this market? Do IPRs delineate an antitrust market and, if so, under what circumstances? Are IPRs but a special case of product differentiation or does their relevance to antitrust market definition go beyond this aspect? Do standard-essential patents need to be treated differently? And how can technology markets be applied in antitrust law analysis? We will now turn to these issues.

II. The Move Away from Intellectual Property Rights Seen as Conferring Market Power

As an IPR grants its owner an exclusive right over a certain design, invention, sign, work or other creation, one could easily come to the conclusion that few IP-protected creations are perfectly interchangeable with any other. This might then feed into an assumption that any IP-protected subject matter delineates its

[8] K Coates, *Competition Law and Regulation of Technology Markets* (Oxford, Oxford University Press, 2011) para 5.217; HJ Hovenkamp, 'Competition for Innovation' [2012] *Columbia Business Law Review* 799, 800.

[9] HJ Hovenkamp et al, *IP and Antitrust: An Analysis of Antitrust Principles Applied to Intellectual Property Law*, 3rd edn, vol I (New York, Wolters Kluwer, 2018) § 4.02[A]; J Drexl, 'The Relationship between the Legal Exclusivity and Economic Market Power: Links and Limits' in I Govaere and H Ullrich (eds), *Intellectual Property, Market Power and the Public Interest* (Brussels, Peter Lang, 2008) 14, 16.

[10] T Heide, 'Trade Marks and Competition Law after Davidoff' (2003) 25 *European Intellectual Property Review* 163, 165 fn 21.

[11] CS Yoo, 'Copyright and Product Differentiation' (2004) 79 *New York University Law Review* 212, 217 f.

[12] See A Heinemann, 'The Contestability of IP-Protected Markets' in J Drexl (ed), *Research Handbook on Intellectual Property and Competition Law* (Cheltenham, Edward Elgar, 2008) 54; EF Sherry and DJ Teece, 'Royalties, Evolving Patent Rights, and the Value of Innovation' (2004) 33 *Research Policy* 179, 179; R Feldman, 'Patent and Antitrust: Differing Shades of Meaning' (2008) 13 *Virginia Journal of Law & Technology* 1, 10.

own relevant antitrust product market. In due course, an IP owner holding a 100 per cent market share in an IPR market would easily be found to be dominant (under EU competition law) or to hold monopoly power (under US antitrust law). However, both in the EU and US, the move has clearly been away from any such assumption, even if recent cases on standard-essential patents in both jurisdictions have given this popular assumption something of a renaissance (see section V below).

The well-established view in the EU is that an IPR does not automatically delineate a relevant antitrust market and thus an IPR holder does not automatically have a dominant position on such a product market.[13] This understanding was developed very early on in the case law. Roughly half a century ago, the Court of Justice stated in *Deutsche Grammophon* that while EU law is not concerned with the granting of IPRs,[14] it can affect their exercise. As a case in point, the Court highlighted that Article 36 TFEU foresees very narrow exceptions to the free movement of goods in order to accommodate industrial and commercial property, ie, IPRs.[15] However, the Court also made it clear that where a company holds a copyright, this IP ownership does not automatically put it in a dominant position as required under Article 102 TFEU.[16] This case law was further reinforced two decades later, when the Court of Justice commented on the relationship between IPRs and market power in *Magill*. In this landmark ruling, which concerned the refusal to license an IPR, the Court again held that 'mere ownership of an intellectual property right cannot confer such a [dominant] position'.[17] More recently, the Court of Justice pointed out in *AstraZeneca* that 'although the mere possession of intellectual property rights cannot be considered to confer such a [dominant] position, their possession is none the less capable, in certain circumstances, of creating a dominant position'.[18] A dominant position, of course, will be more easily found if the relevant market to which that dominant position relates is correspondingly narrow, eg, delimited by the IPR itself. Factors that will lead to the finding of a narrower market include the absence of technologies or works that are substitutable with the one protected by the IPR, and the likelihood of actual or potential competition in those technologies or works.[19]

[13] See also L Zhang, 'Refusal to License Intellectual Property Rights under Article 82 EC in Light of Standardisation Context' (2010) 32 *European Intellectual Property Review* 402, 402. Sometimes, the Court is too reticent in admitting that IPRs can also confer dominance; see DT Keeling, *Intellectual Property Rights in EU Law – Volume I: Free Movement and Competition Law* (Oxford, Oxford University Press, 2003) 371.

[14] The fact that national IP laws grant IPRs, while EU law may regulate the exercise of IPRs, continues to lead to friction; see D Hickman, 'Patents: Competition Law a Defence to Patent Infringement Claims?' (2003) 25 *European Intellectual Property Review* N114, N116.

[15] Case 78/70 *Deutsche Grammophon Gesellschaft v Metro* EU:C:1971:59, para 11.

[16] ibid para 16.

[17] Joined Cases C-241 and C-242/91 P *RTE and ITP v Commission (Magill)* EU:C:1995:98, para 46.

[18] Case C-457/10 P *AstraZeneca v Commission* EU:C:2012:770, para 186.

[19] E Cortés, A Dawson and C Hatton, 'Squaring the Circle: The EU's Quest for Balance between Antitrust and Intellectual Property' [2015] *European Antitrust Review* 16.

In the US, 'the natural monopoly every manufacturer has in the production and sale of its own product cannot be the basis for antitrust liability'.[20] By analogy, this reasoning can be applied to the sphere of IPRs, which award something akin to a 'natural monopoly' to the IP owner: simple ownership of an IPR is not a sufficient basis for antitrust liability. Instead, the plaintiff must clearly demonstrate why a given IPR – be it a copyright, a patent or a trademark (brand) – delineates the relevant product market in each specific case.

In 1995, the DoJ and the FTC first issued their IP Guidelines (see section II.A in Chapter 5). In these Guidelines, they underline innovation and consumer welfare as the common goals of IP law and antitrust law.[21] In the context of antitrust analysis, the agencies will not equate the possession of an IPR with the possession of market power.[22] This conclusion is today strongly supported by case law, although for many decades, IP owners were considered to have market power for certain antitrust actions.

In a number of older cases, the US Supreme Court had held that market power could be derived from an IPR.[23] Despite this precedent, the Court of Appeals for the Federal Circuit found in 2000 that '[a] patent alone does not demonstrate market power'.[24] Only shortly afterwards, in *Independent Ink* (2006), the Supreme Court was confronted with the question of 'whether the presumption of market power in a patented product should survive as a matter of antitrust law'[25] for tying cases tried under section 1 of the Sherman Act. It reversed its own case law in this respect and held that 'the mere fact that a tying product is patented does not support such a presumption'.[26] In order to conclude that a tying arrangement is unlawful, this finding must be substantiated 'by proof of power in the relevant market rather than by a mere presumption thereof'.[27] In reaching its conclusion, the Supreme Court also cited the IP Guidelines of 1995.[28] The *Independent Ink* rationale can be extended to copyright and other IPRs, as these can be seen as less potent than patents, thus negating any presumption of market power based on an IPR.[29]

[20] *Belfiore v New York Times*, 654 FSupp 842, 846 (D Conn 1986) (relating to the market for general interest daily newspapers directed primarily at upscale readers).

[21] US Department of Justice and Federal Trade Commission, Antitrust Guidelines for the Licensing of Intellectual Property (14 January 2017) (US IP Guidelines 2017) § 1.0.

[22] ibid §§ 2.0, 2.2.

[23] *United States v Lowe's*, 371 US 38, 45 (1962). See also *International Salt v United States*, 332 US 392 (1947) (with the tying product being patented machinery); *United States v Paramount Pictures*, 334 US 131 (1948) (extending the rationale of the antitrust treatment of patents to copyright); *Standard Oil v United States*, 337 US 293, 307 (1949) (regarding a patent as 'at least prima facie evidence of [market] control'); *Jefferson Parish Hospital District v Hyde*, 466 US 2, 16 (1984) ('if the Government has granted the seller a patent … the inability to buy the product elsewhere gives the seller market power').

[24] *CSU and Others v Xerox Corp*, 203 F3d 1322, 1325 (Fed Cir 2000), also referring to earlier case law.

[25] *Illinois Tool Works and Others v Independent Ink*, 547 US 28, 31 (2006).

[26] ibid.

[27] ibid 43.

[28] ibid 45.

[29] Hovenkamp et al (n 9) § 4.02[E][7].

Even before *Independent Ink*, the Supreme Court had already made it quite clear that the presumption of market power in patented goods does not extend across the whole antitrust spectrum. In *Walker Process* (1965), a case under section 2 of the Sherman Act, the Supreme Court insisted that a proper definition of the relevant market had to be carried out in order to assess market power. In order to 'establish monopolization … it would then be necessary to appraise the exclusionary power of the illegal patent claim in terms of the relevant market for the product involved'.[30] The Court thus regarded market definition as an indispensable step in the analysis and insisted on assessing the availability of substitutes that did not infringe the patent.

From a comparative point of view, the EU and the US took quite different trajectories in order to arrive at today's conclusion as to the market power implications of IPRs. The US starting point was very much based on the IP laws and their understanding of a monopoly. This view was then only gradually overcome in the antitrust arena. The EU started out with more of an antitrust outlook on IP, perhaps also due to the fact that EU law for a very long time only concerned the exercise of IPRs rather than their granting, and in large part this is still the case.[31] Today, neither under EU competition law nor under US antitrust law is there a presumption that the mere possession of an IPR automatically confers market power on its owner. Instead, market power will need to be proved with reference to the relevant market at issue, which might or might not be delineated by the IPR in question. However, this theoretical clarity clashes with the reality as it presents itself in some cases, where the existence of an IPR continues to feed an intuition that the presence of an IPR must indicate market power.[32]

III. Market Definition in the Presence of Intellectual Property Rights

The following section examines to what extent relevant product markets continue to be delineated along the lines of IPRs, also asking whether this might depend on the nature of the IPR in question or on the type of market behaviour concerned. Then, cases are assessed in which IPRs played a role in the market definition, but without the market being defined along the lines of the IPR. Only if the distinction between these two scenarios is sufficiently clear can companies rely on this analytical framework in order to predict their liability under the antitrust laws.

[30] *Walker Process Equipment v Food Machinery*, 382 US 172, 177 f (1965).

[31] See generally C Geiger (ed), *Constructing European Intellectual Property: Achievements and New Perspectives* (Cheltenham, Edward Elgar, 2013).

[32] See also HKS Schmidt, 'The Influence of IP Rights on Product Definition in Competition Law: The Curious Case of Tying' (2010) 26 *International Company and Commercial Law Review* 224, 224.

A. Delineating the Relevant Market Along the Lines of an Intellectual Property Right

A number of antitrust cases in both the EU and the US have seen relevant markets delineated along the lines of the IPR in question.[33] In the area of copyright, licensing issues have been at the centre of attention. In three well-known cases on the refusal to license copyrighted works, the EU Courts adopted a narrow view of the relevant market.[34] The *Magill* case (1995) concerned listings of TV programmes in Ireland. The General Court had held that two relevant markets could be distinguished: a relevant market for the copyrighted weekly television listings and a separate relevant market for magazines which publish television listings in Ireland and Northern Ireland.[35] The Court of Justice, siding with the Commission and the General Court, found that the incumbent held a dominant position on the market for weekly television magazines.[36] However, it did not set out in any more detail how the relevant market it had identified related to the incumbent's copyright in its television listings and merely reiterated its position that 'mere ownership of an intellectual property right cannot confer [a dominant] position'.[37]

Tiercé Ladbroke (1997) was about the transmission of the sound and pictures of horse races, all of which were copyrighted.[38] The copyrighted work constituted the relevant market, and the General Court had to assess whether the racecourse societies could legally refuse to license the transmission of their sound and pictures of horse races.[39]

A copyrighted brick structure for the sales data of pharmaceuticals in Germany was at the centre of *IMS Health* (2004).[40] While the Court of Justice did not state this explicitly in the preliminary ruling, it implicitly held that this copyrighted brick structure delineated the relevant antitrust market in which an abuse had possibly occurred through a refusal to license. However, one must acknowledge that this is how the referring court had posed the question.[41] Interestingly, the Court assessed customers' switching costs and thus, essentially, substitutability when discussing the possibility of economic obstacles to the creation of competing products.[42] This type of analysis usually pertains to the market definition. Indeed,

[33] This is mainly the case if there is market power in the product to which the IPR relates; see HJ Hovenkamp, 'Response: Markets in IP and Antitrust' (2012) 100 *Georgetown Law Journal* 2133, 2138.

[34] On the licensing of IPRs and technology markets, see also section IV below.

[35] Joined Cases C-241 and C-242/91 P *RTE and ITP v Commission (Magill)* EU:C:1995:98, para 24.

[36] ibid para 47.

[37] ibid para 46.

[38] Case T-504/93 *Tiercé Ladbroke v Commission* EU:T:1997:84, para 72.

[39] Ultimately, the Commission decision was partially annulled because the Commission had wrongly concluded that the refusal to license could not be the product of an anti-competitive agreement between the racecourse societies; ibid para 162.

[40] Case C-418/01 *IMS Health v NDC* EU:C:2004:257, paras 4, 17 (first question).

[41] ibid para 17 (first question).

[42] ibid paras 28 f.

if one considers the copyrighted brick structure to constitute its own relevant product market because no substitutable products are available, then the very fact that customers would not turn to a competing product should not be used as the basis for finding an abuse of dominance.

In both *Magill* and *IMS Health*, it is plausible that the weak and sometimes secondary nature of the copyrights in question might have influenced the Court's willingness to accept the existence of an abuse of dominance. While the television listings in *Magill* were copyrighted in Ireland, but not in many other EU Member States,[43] the copyrighted information was also merely a by-product of the broadcaster's main business.[44] In *IMS Health*, various industry members had significantly contributed to the creation of the copyrighted database,[45] perhaps leading the Court to question whether copyright protection was justified in this case.[46] However, it is not clear whether this might have also influenced the acceptance of the very narrow market definitions in these cases, which revolved around the copyright, as did the transmissions in *Tiercé Ladbroke*.

In the US, licensing agreements over copyrighted works were equally subjected to antitrust scrutiny. In *PolyGram* (2005), the FTC held a particular musical album by the Three Tenors to constitute its own antitrust market because consumers would not regard it as interchangeable with any other musical album.[47] PolyGram and Warner had entered into a licensing agreement for distributing the newest album of the Three Tenors, ie, a copyrighted musical work. At the same time, the parties separately agreed not to advertise or discount two earlier albums of the Three Tenors, one of which was distributed by PolyGram and the other by Warner.[48] Another US District Court case revolved around TV shows. In *Paramount Pictures* (2006), the substitutability of copyrighted TV programmes was discussed, based on expert opinions. One expert identified two copyrighted TV programmes – *Judge Judy* and *Judge Joe Brown* – as the relevant tying product market, based on substitutability patterns that included the availability of programmes to be licensed.[49] While the relevant market was thus not exclusively defined based on the IPR, the availability of IP licences did play a decisive role in the market definition.

[43] Schmidt, 'Influence of IP Rights' (n 32) 226 fn 26. It was never doubted that such a copyright existed under Irish law; see DJ Gifford and RT Kudrle, *The Atlantic Divide in Antitrust: An Examination of US and EU Competition Policy* (Chicago, University of Chicago Press, 2015) 174.

[44] A Ezrachi and M Maggiolino, 'European Competition Law, Compulsory Licensing, and Innovation' (2012) 8 *Journal of Competition Law & Economics* 595, 604.

[45] See Case C-418/01 *IMS Health v NDC* EU:C:2004:257, paras 17 (second question), 29.

[46] Ezrachi and Maggiolino (n 44) 605.

[47] The Court of Appeals for the District of Columbia Circuit denied PolyGram's petition for review; *PolyGram Holding v FTC*, 416 F3d 29 (DC Cir 2005). Here, the Three Tenors' album was implicitly regarded as a separate relevant antitrust market; see also on this MA Lemley and MP McKenna, 'Is Pepsi Really a Substitute for Coke? Market Definition in Antitrust and IP' (2012) 100 *Georgetown Law Journal* 2055, 2085 fn 136, 2088 fn 150.

[48] *PolyGram Holding v FTC*, 416 F3d 29, 31 (DC Cir 2005).

[49] *Paramount Pictures v Johnson Broadcasting*, 432 FSupp2d 707, 709 f (SD Tex 2006).

In other antitrust scenarios, copyrighted works by a particular artist have repeatedly been regarded as their own antitrust markets in the US, based on non-substitutability. A US District Court found on two occasions (1994 and 1995) that there was a relevant antitrust (sub-)market for paintings by Jackson Pollock, based on a lack of interchangeability with other contemporary artists' paintings.[50] In *Kramer v Pollock-Krasner Foundation*, the plaintiff had alleged that the defendants had monopolised the market of Jackson Pollock paintings.[51] In a similar case several years later, an art collector claimed that the Warhol Foundation and others had conspired in order to artificially restrict competition in the market for authentic Andy Warhol paintings, in addition to monopolisation and other claims. In the *Warhol* case (2009), the District Court held that there was a relevant antitrust market for art by Andy Warhol.[52]

There is now a whole body of case law on the question of whether trademarks confer market power, with a majority of US cases holding that this is not the case.[53] These cases implicitly reject that trademarks can delineate their own antitrust markets. Nevertheless, a trademark may influence market delineation by reducing demand substitutability. Where 'preferences created by [trademark] information are substantial and rivals cannot readily attain the same status, then a brand might constitute a relevant market'.[54] The case of *Eastman Kodak* is instructive when it comes to a trademark's ability to delineate an antitrust product market. In this case, revolving around spare parts for and the servicing of Kodak's photographic equipment, the US Supreme Court found that only 'in some instances one brand of a product can constitute a separate market'.[55] It also clarified that in *du Pont*, it had only wanted to indicate that 'one brand does not necessarily constitute a relevant market if substitutes are available'.[56] In the majority of cases, single brands will therefore not be seen as separate product markets for the purposes of antitrust law, while in some cases they will. With reference to *Grinnell*, the Court also pointed out that the commercial realities that consumers faced were decisive in finding whether or not a brand delineated a market.[57] Several US courts have insisted

[50] *Kramer v Pollock-Krasner Foundation*, 890 FSupp 250, 254 f (SDNY 1995). See also *Vitale v Marlborough Gallery*, Case 93 Civ 6276 (PKL) (SDNY 5 July 1994); M Weber, 'Liability for the Acquisition of Faked or Wrongly Attributed Works of Art in US Law' in K Odendahl and PJ Weber (eds), *Kulturgüterschutz – Kunstrecht – Kulturrecht* (Baden-Baden, Nomos, 2010) 419–21.

[51] *Kramer v Pollock-Krasner Foundation*, 890 FSupp 250, 253 (SDNY 1995).

[52] *Simon-Whelan v Andy Warhol Foundation for the Visual Arts*, Case 07 Civ 6423 (LTS) (SDNY 26 May 2009); GS Lacy, 'Standardizing Warhol: Antitrust Liability for Denying the Authenticity of Artwork' (2011) 6 *Washington Journal of Law, Technology & Arts* 185, 194 ff.

[53] Hovenkamp et al (n 9) § 4.02[E][7][b] fn 241 (with many references to US case law).

[54] ibid § 4.03[A][1].

[55] *Eastman Kodak v Image Technical Services*, 504 US 451, 482 (1992), referring to a number of previous judgments.

[56] ibid 482 fn 30, referring to *United States v EI du Pont de Nemours & Co (Cellophane)*, 351 US 377, 393 (1956).

[57] *Eastman Kodak v Image Technical Services*, 504 US 451, 482 (1992); *United States v Grinnell Corp*, 384 US 563, 572 (1966).

that '[s]ingle brand markets are, at a minimum, extremely rare'[58] and '[f]or good reason',[59] arguing that one manufacturer's products generally do not constitute their own relevant product market.[60] Ultimately, therefore, a trademark's ability to capture consumers, thus shaping the 'commercial realities' that consumers are facing, is crucial as to whether or not it delineates the antitrust market. At least as far as trademarks are concerned, marketing may become a key factor in this analysis.

The relevance of trademarks – or brands – for antitrust market definition has been discussed in a number of US cases related to pharmaceuticals. A District Court found that 'brand-name pharmaceuticals and their generic counterparts might not always compete in the same markets at all because, based on the higher prices of the brand-name drugs, there is less cross-elasticity of demand than one might expect'.[61] In these cases, while there may be full therapeutic substitutability between the drugs at issue, customers might nevertheless regard them as non-substitutable based on brand perception.[62] In *Geneva Pharmaceuticals* (2004), the Court of Appeals for the Second Circuit relied on a trademark to delineate the relevant antitrust market. It found that, although 'certified by the FDA as therapeutically equivalent',[63] the branded drug produced by the plaintiff was not in the same relevant market as the generic drug produced by the defendant. The branded drug did not face significant competitive pressure from the generic drug, such that a small but significant non-transitory price increase in the branded drug would not push consumers away from it.[64] In another industry, the carbonated soft drink market, a District Court held that where a company would like to enter this market, 'barriers to entry include the substantial time and expense required to build a brand name that may overcome existing consumer preferences'[65] – trademarks are invariably an important part of branding. This possibility was also highlighted in *Henry v Chloride* (1987), where the Court of Appeals for the Eighth Circuit emphasised that barriers to entry counted as circumstantial evidence for predatory intent and discussed the customers' loyalty to the salespersons of one particular brand.[66]

In an ongoing US case, the question is whether the contraceptive Loestrin forms part of a larger market for oral birth control, whether it is in a relevant

[58] *Apple v Psystar*, 586 FSupp2d 1190, 1198 (ND Cal 2008).

[59] *US Ring Binder v World Wide Stationery Manufacturing*, 804 FSupp2d 588, 598 (ND Ohio 2011).

[60] *Green Country Food Market and Others v Bottling Group*, 371 F3d 1275 (10th Cir 2004).

[61] *Ciprofloxacin Hydrochloride Antitrust Litigation*, 363 FSupp2d 514, 522 (EDNY 2005).

[62] On the question of who is the customer in the case of pharmaceuticals, and the market definition issues associated with this question, see A Vaishnav, 'Product Market Definition in Pharmaceutical Antitrust Cases: Evaluating Cross-price Elasticity of Demand' [2011] *Columbia Business Law Review* 586, 596–611, 625.

[63] *Geneva Pharmaceuticals Technology v Barr Laboratories*, 386 F3d 485 (2d Cir 2004).

[64] ibid.

[65] *FTC v Coca-Cola*, 641 FSupp 1128, 1137 (DDC 1986), vacated and remanded; *FTC v Coca-Cola*, 829 F2d 191 (8th Cir 1987).

[66] *Henry v Chloride*, 809 F2d 1334, 1344, 1342 (8th Cir 1987).

market together with its generic or whether, based on the trademark's effect on the market definition, it constitutes its own antitrust market ('single drug market').[67]

Another possibility for thinking about trademarks as constructing an antitrust market is in professional sports teams, which attach their trademark to their merchandise. Would it go too far to find a relevant product market for ice hockey jerseys of the New York Rangers? In professional sports, consumer behaviour (ie, fan loyalty) generally leads to very narrow markets.[68] And it can easily be ascertained that plain ice hockey jerseys sell at a lower price than New York Rangers jerseys. Would a New York Ranger fan happily buy a Vancouver Canucks jersey instead? An interesting aspect in such a scenario, of course, is that supply substitutability may be limited because licensing of trademarks may exclude it to a large extent.

In the EU, there has been less litigation on trademarks delineating their own antitrust market. In the early case of *Consten and Grundig* (1966), the Commission had found that the relevant market at issue was Grundig products, ie, products bearing the Grundig trademark.[69] The defendants argued that the relevant market should not be 'limited to products of a single brand' and that interchangeability between products of different brands should be taken into account.[70] The Court of Justice agreed with the Commission that a restriction of intra-brand competition could satisfy the criterion of 'restriction of competition' under Article 101(1) TFEU.[71] It did not engage in an elaborate market definition analysis as would be the case today, over half a decade later. Nevertheless, the Court did emphasise that the separate agreement on the Grundig trademark was one of the competition infringements, as it served to enable Consten to enforce absolute territorial protection in France.[72]

Several years on, *Nungesser* (1982) was a case on licensing terms analysed under Article 101 TFEU, which centred on plant breeders' rights relating to three different varieties of hybrid maize seeds especially developed for use in temperate climatic conditions.[73] The question which is of interest in the present context is whether and to what extent these rights were each considered to be their own relevant market and whether any suitable substitutes existed for these specific climatic conditions. This question was not directly addressed by the Court of Justice. In its decision, the Commission had held that maize seeds of three categories were

[67] *Loestrin 24 FE Antitrust Litigation*, Case 1:13-md-2472-S-PAS (Memorandum of law, DRI 13 January 2017); 'Only Loestrin Market Matters in Antitrust Row, Retailers Say' *Law360* (26 October 2018), www.law360.com/articles/1096127.

[68] J Winfree, 'Fan Substitution and Market Definition in Professional Sports Leagues' (2009) 54 *Antitrust Bulletin* 801, especially 822.

[69] Joined Cases 56 and 58/64 *Consten and Grundig v Commission* EU:C:1966:41, 320, 325.

[70] ibid 325.

[71] ibid 342.

[72] ibid 344–46.

[73] Case 258/78 *Nungesser v Commission* EU:C:1982:211, p 2018 and para 3.

the 'relevant products',[74] possibly allowing the conclusion that each of these varieties was thought to constitute its own product market delineated by the IPR in question.

The case law on patents within the realm of market definition will be explored when discussing technology markets (see section IV.B below) and standard-essential patents (see section V below).

B. The Role of Intellectual Property Rights in Relevant Markets Not Defined Along the Lines of Intellectual Property Rights

A whole host of cases, in both the EU and the US, does not define the relevant antitrust market along the lines of the IPR in question, but nevertheless acknowledges the importance of the IPR either when delineating the market or in the substantive antitrust analysis. Interestingly, many of the cases were litigated on both sides of the Atlantic, providing us with an opportunity to directly compare the respective influence of IPRs on market definition.

In 1992, the US Court of Appeals for the Tenth Circuit had to decide whether a TV channel could constitute its own relevant antitrust market. It held that antitrust law would not hold a company accountable for 'the natural monopoly it holds over its own product'.[75] As a TV channel consists of a string of copyrighted programmes, one could argue that this reasoning necessarily is of relevance in the realm of IPRs. As was seen above, the issue might be resolved differently where a particular TV show is in question. If we conceive of the TV channel as a multi-sided platform, a parallel can easily be drawn with the contemporary question whether a platform in itself – in addition to its markets sides – may constitute a relevant market (see Chapter 8).

In one of the US *Microsoft* cases (2001), Microsoft was under antitrust scrutiny for its alleged anti-competitive conduct towards Netscape's Internet Navigator and Sun's Java technologies. The relevant market in this case was found to be 'the licensing of all Intel-compatible PC operating systems worldwide'.[76] Microsoft argued that three more products needed to be added to this narrowly defined market, namely non-Intel-compatible operating systems such as Apple's Mac OS, operating systems for non-PC devices (eg, tablets) and middleware, which is software that exposes its own application programming interfaces (APIs). The Court

[74] *Nungesser* (Case IV/28.824) Commission Decision 78/823/EEC [1978] OJ L286/23, para 1.

[75] *TV Communications Network v Turner Network*, 964 F2d 1022, 1025 (10th Cir 1992), also cited by *Compliance Marketing v Drugtest*, Case 09-cv-01241-JLK (D Colo 7 April 2010).

[76] *United States v Microsoft*, 253 F3d 34 (DC Cir 2001). On this market definition and how it differs from the European Commission's decision in this respect, see R Pardolesi and A Renda, 'The European Commission's Case against Microsoft: Kill Bill?' (2004) 27 *World Competition* 513, 515 fn 7; *Microsoft* (Case COMP/C-3/37.792) Commission Decision 2007/53/EC [2007] OJ L32/23, para 326.

of Appeals for the District of Columbia Circuit rejected this, mainly based on a clear lack of interchangeability from a consumer perspective, and in relation to middleware also based on a lack of substitutability between middleware and the Windows OS.[77] Following the judgment by the Court of Appeals in June 2001, Microsoft entered into commitments with the DoJ and nine Attorneys General in November 2001. Several of the negotiated commitments related to interoperability information: Microsoft would draw up specifications of its communication protocols and allow third parties to access this interoperability information, which was necessary to interact with its successful Windows client PC OS, under certain conditions. It would also license its IPRs to third parties if this became necessary. Furthermore, Microsoft would allow original equipment manufacturers (OEMs) to activate or eliminate access to its middleware so that third-party developers could develop middleware (eg, media players) that would work well on Windows.[78] These commitments were approved by the District Court for the District of Columbia in November 2002.[79] They ultimately led to the introduction of the Microsoft Communications Protocol Program, which has since been superseded by other programs relating to technical specifications needed in order to interoperate with Microsoft's products.[80]

In *Microsoft*, the concurrent presence of copyrighted software, APIs protected by trade secrets and network effects led to an 'essentially patent-like' protection afforded to Microsoft,[81] which was viewed sceptically by the authorities. In its defence, Microsoft repeatedly relied on its copyright protection in order to justify its business conduct towards OEMs, and the Court accepted many of these justifications as necessary in the realm of IPRs in order to foster innovation.[82]

The very elaborate European *Microsoft* case (2007) dealt with a number of similar antitrust issues, amongst them the interoperability information for Microsoft's work group servers that was not readily available to competitors. The relevant market was defined as work group server operating systems, ie, 'operating systems designed and marketed to deliver collectively "basic infrastructure services" to relatively small numbers of client PCs connected to small or medium-sized networks'.[83] The Commission qualified the withholding of interoperability information as an abuse under Article 102 TFEU and fined Microsoft.[84] Microsoft argued that the specifications of the communication protocols that it was supposed

[77] *United States v Microsoft*, 253 F3d 34 (DC Cir 2001).

[78] US Department of Justice, 'Department of Justice and Microsoft Corporation Reach Effective Settlement on Antitrust Lawsuit' (2 November 2001). See also Case T-201/04 *Microsoft v Commission* EU:T:2007:289, paras 51–58.

[79] *United States v Microsoft*, Case 98-1232 (CKK) (DDC 12 November 2002), affirmed upon appeal in *Massachusetts v Microsoft*, 373 F3d 1199 (DC Cir 2004).

[80] Case T-201/04 *Microsoft v Commission* EU:T:2007:289, para 58.

[81] Gifford and Kudrle (n 43) 170.

[82] ibid 171. For instance, see *United States v Microsoft*, 253 F3d 34, 62 ff (DC Cir 2001).

[83] Case T-201/04 *Microsoft v Commission* EU:T:2007:289, para 25.

[84] ibid para 36.

to draw up and share were copyrighted, while at the same time also constituting trade secrets.[85] The Commission disputed the IPR nature of the information at issue.[86] The General Court found that whether or not Microsoft's communication protocols were covered by IPRs did not influence the outcome in this case; despite its view to the contrary, the Commission had, in its infringement decision, started from the premise that Microsoft held IPRs in the information concerned, thus adopting 'the strictest legal test and therefore the one most favourable to Microsoft'.[87] In line with this reasoning, the General Court also proceeded on the presumption that Microsoft's interoperability information was covered by IPRs or by trade secrets.[88]

The relevant market in the European *Microsoft* case was not defined based on the interoperability information (which was likened to IPRs), but was found to be the work group server operating systems. The anti-competitive conduct, ie, the competition law abuse, by the super-dominant company concerned IP-like information that was related to the relevant market. In this respect, the European case was remarkably similar to the Intel-compatible PC operating systems at the centre of the US *Microsoft* case, in that IPRs played a role in those operating systems, but were not the all-decisive element. The role that IPRs then played in the subsequent substantive analysis showed that US courts were more receptive to IPR arguments to counter antitrust claims than EU courts. Also, while the US case displayed characteristics of a technology market,[89] the EU case focused on the ultimate product.

Another parallel case concerned Intel. The European Commission's *Intel* decision (2009) revolved around discount schemes that Intel operated in order to obtain exclusivity on the market. The first step that the Commission took in its legal and economic assessment of the case was to delineate the relevant market. For the relevant product market, the Commission first focused on demand-side substitution, finding that central processing units (CPUs) for desktop computers, for laptop computers and for servers were generally not regarded as substitutable by the demand side, despite some interchangeability at the margins.[90] However, there was significant supply-side substitutability.[91] On the other hand, the commercial and consumer segments of CPUs were not found to constitute separate markets, either from the demand or from the supply side.[92]

[85] ibid paras 271, 273.

[86] ibid paras 275–80.

[87] ibid paras 283 f.

[88] ibid para 289. In the US, it was assumed that Microsoft's application programming interfaces were covered by trade secret law; see DJ Gifford and RT Kudrle, 'Antitrust Approaches to Dynamically Competitive Industries in the United States and the European Union' (2011) 7 *Journal of Competition Law & Economics* 695, 708.

[89] JA Newberg, 'Antitrust for the Economy of Ideas: The Logic of Technology Markets' (2000) 14 *Harvard Journal of Law & Technology* 83, 113.

[90] *Intel* (Case COMP/C-3/37.990) Commission Decision of 13 May 2009 [2009] OJ C227/13, para 799.

[91] ibid para 819.

[92] ibid paras 802, 820.

No demand-side or supply-side substitutability was found between non-x86 CPUs and x86 CPUs, or between CPUs for non-computer devices and CPUs for computers.[93] Ultimately, the Commission held that the relevant product market was confined to x86 CPUs, but left open whether these had to be sub-segmented for desktop computers, laptop computers and servers.[94] The Commission emphasised that the market for x86 CPUs had high barriers to entry, which consisted of IPRs, most importantly patents.[95] However, it did not define the market along the lines of those IPRs, instead regarding them as an important feature of that market and the competition on that market. The General Court summarised the Commission's findings regarding product market definition, but did not comment on them, nor did Intel challenge them.[96] When analysing whether Intel had a dominant position at the time of the alleged abuse, the Commission relied on the importance of R&D in the market and particularly on the vital role of IPRs for producing an x86 CPU.[97] A typical feature of cases in innovative markets was present in *Intel* as well: the relevant geographical market was found to be worldwide.[98]

On the US side, the complaint that the FTC lodged against Intel in December 2009 was aimed at stopping Intel's allegedly anti-competitive practices that had started in 1999,[99] and was brought under section 5 of the FTCA. The FTC held that Intel was seeking to maintain its monopoly on the market for CPUs, as well as to gain a monopoly position on the market for graphics processing units (GPUs).[100] The FTC lodged its complaint only seven months after the European Commission issued its *Intel* decision, based on similar reasoning.[101] The FTC delineated two 'sets' of relevant product markets in its complaint. The first were CPUs for use in various kinds of computers, servers etc. Like the European Commission, the FTC understood CPUs to only encompass x86 microprocessors, as this had become an industry standard.[102] The second relevant market set were GPUs for use in various kinds of computers, servers etc.[103] In both cases, the relevant geographical market

[93] ibid paras 808, 813, 824, 830.

[94] ibid para 835.

[95] ibid para 858.

[96] Case T-286/09 *Intel v Commission* EU:T:2014:547, paras 21–23. Therefore, this issue was not discussed by the Court of Justice when it rendered its judgment in September 2017; see Case C-413/14 P *Intel v Commission* EU:C:2017:632, para 7. For AG Wahl's short commentary on the market definition in his 2016 Opinion, see Case C-413/14 P *Intel v Commission* EU:C:2016:788, Opinion of AG Wahl, paras 20–22.

[97] *Intel* (Case COMP/C 3/37.990) Commission Decision of 13 May 2009 [2009] OJ C227/13, paras 129 f, 856–58; Case T-286/09 *Intel v Commission* EU:T:2014:547, para 25.

[98] *Intel* (Case COMP/C 3/37.990) Commission Decision of 13 May 2009 [2009] OJ C227/13, para 836; Case T-286/09 *Intel v Commission* EU:T:2014:547, para 24.

[99] For the *Intel* cases in the 1990s, see the account in JT Soma and KB Davis, 'Network Effects in Technology Markets: Applying the Lessons of *Intel* and *Microsoft* to Future Clashes between Antitrust and Intellectual Property' (2000) 8 *Journal of Intellectual Property Law* 1, 35–44, especially 40 f on market definition.

[100] *Intel*, 150 FTC 420, 421 f (2010).

[101] See also *Intel* (Case COMP/C 3/37.990) Commission Decision of 13 May 2009 [2009] OJ C227/13.

[102] *Intel*, 150 FTC 420, 428 f (direct quote at 428) (2010).

[103] ibid 429 f.

was worldwide.[104] While the FTC did not mention IPRs when delineating the relevant market, it described IPRs as one of the major entry barriers to these two relevant markets.[105] It alleged that Intel had used its IPRs relating to the x86 CPU in order to strengthen its market power on that market.[106] Indemnification from IP litigation was one of the means with which Intel favoured OEMs that exclusively sourced their requirements from Intel.[107]

In the *Intel* cases, it could thus be observed how the two antitrust authorities reached very similar conclusions on market definition, relying on x86 CPUs as the relevant product market and regarding IPRs as an important feature of that market, which also constituted a barrier to entry. The fact that x86 CPUs had become an industry standard, which was protected by IPRs, was considered highly relevant.

The European case of *AstraZeneca* (2010) dealt with that company's alleged abuses in exercising its rights as a patent holder. The General Court described AstraZeneca as a 'pharmaceutical group active, worldwide, in the sector of inventing, developing and marketing innovative products'.[108] The definition of the relevant product market adopted by the Commission in its decision was one of the five issues that AstraZeneca contested upon appeal.[109] The Commission had held that the relevant product market consisted of proton pump inhibitors (PPIs) for the treatment of frequent heartburn.[110] The General Court confirmed that the therapeutic uses for PPIs and antihistamines differed, that the former were increasingly used in treatments while the latter's use declined, and that the therapeutic quality of antihistamines did not account for the 'inertia' shown by doctors in prescribing PPIs. Overall, the Court found that antihistamines did not exert a significant competitive pressure on PPIs, and that PPIs constituted their own relevant market.[111] The market definition was not organised around the IPR at issue, but around the therapeutic use of the pharmaceutical in question and its possible substitutes.[112] Only later, in analysing whether AstraZeneca was dominant in that market, did the Court refer to IPRs. It affirmed that IPRs had a role to play in determining whether a company holds a dominant position and that in this case, the strong patent protection that AstraZeneca enjoyed for its PPI Losec allowed it 'to exert significant pressure on its competitors'.[113] This has been held to mean that AstraZeneca's 'patent dominated the relevant market'.[114]

[104] ibid 430.
[105] ibid.
[106] ibid 442 f.
[107] ibid 423.
[108] Case T-321/05 *AstraZeneca v Commission* EU:T:2010:266, para 2.
[109] ibid para 23.
[110] ibid paras 28, 62.
[111] ibid paras 28–107.
[112] M Negrinotti, 'Abuse of Regulatory Procedures in the Intellectual Property Context: The AstraZeneca Case' (2008) 29 *European Competition Law Review* 446, 449.
[113] Case T-321/05 *AstraZeneca v Commission* EU:T:2010:266, 270, 272 (direct quote at para 272).
[114] M Maggiolino and ML Montagnani, 'AstraZeneca's Abuse of IPR-Related Procedures: A Hypothesis of Anti-trust Offence, Abuse of Rights, and IPR Misuse' (2011) 34 *World Competition* 245, 248.

Interestingly, the Court also acknowledged that strong IP protection did not necessarily mean that no competition whatsoever was possible on the relevant market.[115] The General Court's judgment was upheld on appeal, in particular as regards market definition.[116]

In a similar US lawsuit in the pharmaceutical industry, *American Sales v AstraZeneca*, a District Court judge dismissed a section 2 of the Sherman Act case for sham patent litigation in 2011 because the plaintiff had not properly pleaded a plausible relevant market. The Court found that while the plaintiff tried to rely on a market definition that only encompassed a patented PPI and its generic version, it was not shown that, based on product characteristics or actual buying patterns, there was a lack of interchangeability with other pharmaceutical products.[117] Indeed, the Court admonished that it was 'left to speculate whether a relevant product market could include any product used to treat frequent heartburn; a narrower market consisting of proton pump inhibitors; the single-product market proposed by plaintiff; or any number of formulations'.[118]

The *AstraZeneca* judgments both concerned the same product: a PPI to treat heartburn. They show that in the area of pharmaceuticals, the starting point for a market definition is of relevance and IPRs should not be too heavily relied on. The European case used therapeutic use as a starting point and subsequently accepted that PPIs constituted their own product market based on evidence regarding the therapeutic use, and AstraZeneca's patent for Losec was seen as central in this market. The US court, on the other hand, was asked to regard only AstraZeneca's patented pharmaceutical and its generic as the relevant antitrust market, and remained unconvinced in this respect as the market definition did not provide the required evidence.

In contrast to the conclusion that was reached in the *Geneva Pharmaceuticals* case discussed above, in *Ciprofloxacin* (2005) a US District Court found that the relevant market revolved around the active pharmaceutical ingredient ciprofloxacin. In that case, the price of branded ciprofloxacin fell considerably when the generic product entered the market. In addition, the generic was able to capture 50 per cent of the branded product's sales very soon after its market entry. All this indicated that the generic formed part of the relevant product market.[119] This shows that consumer response to products incorporating IPRs is essential in order to determine their interchangeability with other products for antitrust purposes. Other US courts have also been more wary than the *Geneva Pharmaceuticals* Court, with one District Court cautioning in an antitrust counter-claim by Sandoz to a patent infringement claim brought by Bayer that the relevant antitrust market

[115] Case T-321/05 *AstraZeneca v Commission* EU:T:2010:266, para 274.
[116] Case C-457/10P *AstraZeneca v Commission* EU:C:2012:770, paras 31–52.
[117] *American Sales v AstraZeneca and Others*, Case 1:10-cv-06062 (SDNY 14 April 2011) 6.
[118] ibid 7.
[119] *Ciprofloxacin Hydrochloride Antitrust Litigation*, 363 FSupp2d 514, 522 f (EDNY 2005); Hovenkamp et al (n 9) § 4.03[A][2].

should not be delineated along the lines of a pharmaceutical's active ingredient which was marketed under a successful brand ('unique market').[120] In another case, Judge Posner of the US Court of Appeals for the Seventh Circuit found in *Sheridan v Marathon Petroleum* (2007) that while Marathon certainly was a trademark, it did not constitute its own market of Marathon gasoline.[121] Similarly, in the area of copyright – and in contrast to the *PolyGram*, *Jackson Pollock* and *Andy Warhol* cases discussed above – a US District Court held that reggae sound recordings did not constitute their own antitrust product market and that, in any case, Bob Marley sound recordings were too narrow to constitute their own relevant product markets.[122] In *Compliance Marketing v Drugtest* (2010), a US District Court discussed the possibility of a relevant market consisting of a presumably copyrighted database with its entries and assessed whether *Kodak* could be applied to this case. However, it found that such a narrow market consisting of merely one brand only had to be defined where there was no interchangeability with other services – an aspect that the defendants had not addressed.[123]

These US cases show that it is rather difficult to distinguish cases in which the relevant antitrust market was essentially based on the IPR in question from other cases where the courts did not do so, for this analysis frequently hinges on the customer responses that can be proved to the court. It appears that some courts are willing to take customer response into account, while at other times the relevant market appears to be based on the judges' own intuitive response to the substitutability question.

In the European Commission case of *Qualcomm* (2018), this company was found to hold a dominant position on the market for LTE baseband chipsets. The Commission based its market definition on Qualcomm's high market shares on this market (exceeding 90 per cent), but also referred to the significant R&D that any potential competitor would need to invest in before entering this market. Qualcomm's IPRs were regarded as an entry barrier.[124]

The EU's General Court recently contributed to the development of market definition in pharmaceuticals in its *Servier* judgment (2018). In this case, the European Commission had alleged that the originator Servier had entered into pay-for-delay agreements as well as engaging in abusive conduct in order to keep generics manufacturers out of its lucrative market for perindopril, a cardiovascular medicine. The Commission's market definition was very narrow, only encompassing the original and generic versions of the molecule perindopril, while excluding

[120] *Bayer Schera Pharma v Sandoz*, Cases 08 Civ 03710 (PGG) and 08 Civ 08112 (PGG) (SDNY 29 March 2009).

[121] *Sheridan v Marathon Petroleum*, 530 F3d 590, 595 (7th Cir 2008).

[122] *Rock River Communications v Universal Music Group*, Case CV08-635 CAS (AJWx) (CD Cal 27 April 2011).

[123] *Compliance Marketing v Drugtest*, Case 09-cv-01241-JLK (D Colo 7 April 2010).

[124] *Qualcomm (Exclusivity Payments)* (Case AT.40220) Commission Decision of 24 January 2018 [2018] OJ C269/25; European Commission, 'Antitrust: Commission Fines Qualcomm €997 Million for Abuse of Dominant Market Position' (IP/18/421, 24 January 2018).

other ACE inhibitors of the same therapeutic class.[125] The General Court held that this market definition was erroneous, leading to an annulment of the Commission's abuse of dominance finding.[126] The Court drew particular attention to qualitative, non-price competitive factors in pharmaceuticals. Based on a wide range of evidence, it held that the therapeutic use of perindopril was the same as that of 15 other ACE inhibitors and that, contrary to the Commission's findings, doctors did not show inertia in switching patients on perindopril to other medicines. Servier's promotional efforts were also seen as a token of competition.[127] The *Servier* case is a reminder that market definition remains a central element under Article 102 TFEU.[128] A case can stand or fall with this analytical step. It also shows that the General Court does not welcome too narrow a view on therapeutic interchangeability. It now remains to be seen whether the CJEU agrees.

C. Analysis

There are stark differences between competition law and IP law in terms of market analysis. In IP law, and particularly in copyright and patent law, substitutability is not taken into consideration when courts analyse a market from an IP perspective. Instead, the market is from the outset delineated by the copyrighted work or by the patented invention (ie, the patent claims) as such.[129] By contrast, in antitrust law, IPRs may be used to delineate relevant antitrust product markets, but this is not necessarily the case. Instead, product markets may often be found to incorporate IPRs. As one author has noted: 'Patents … do not fit well into traditional notions of a product.'[130] Whether one regards the patent itself as the relevant antitrust product market (input market), the end product incorporating the patent (output market)[131] or a patent portfolio, none of these 'fully captures the dynamics of a

[125] *Perindopril (Servier)* (Case AT.39612) Commission Decision of 30 September 2016 [2016] OJ C393/7, para 2403. ACE inhibitors slow the activity of the ACE enzyme, which ultimately lowers blood pressure. Separate appeals to the *Perindopril* case were lodged, and the General Court decided them on 12 December 2018 (Cases T-677/14; T-679/14; T-680/14; T-682/14; T-684/14; T-701/14; T-705/14). These judgments are currently on appeal to the CJEU (Cases C-207/19 P; C-198/19 P; C-144/19 P; C-197/19 P; C-151/19 P; C-164/19 P; C-166/19 P). In particular, see Case T-691/14 *Servier v Commission* EU:T:2018:922; Case C-201/19 P *Servier v Commission* [2019] OJ C139/39 (appeal pending); Case C-176/19 P *Commission v Servier* [2019] OJ C139/37 (appeal pending).

[126] Case T-691/14 *Servier v Commission* EU:T:2018:922, paras 1591, 1622, 1633.

[127] ibid paras 1418–585.

[128] See also J Killick, J Jourdan and P Pêcheux, 'The *Servier* Judgment: The General Court Annuls the Commission's Market Definition But Confirms the Illegality of Certain Patent Settlement Agreements' (2019) 10 *Journal of European Competition Law & Practice* 25.

[129] AF Kingsbury, 'Market Definition in Intellectual Property Law: Should Intellectual Property Courts Use an Antitrust Approach to Market Definition?' (2004) 8 *Marquette Intellectual Property Law Review* 63, 91.

[130] Feldman (n 12) 13 § 44.

[131] See JDC Turner, *Intellectual Property and EU Competition Law*, 2nd edn (Oxford, Oxford University Press, 2015) § 3.04.

patent'.[132] This reasoning can be extended to other areas of IP. In fact, the 'notion of a product may become even more amorphous in the biotech world'.[133] In antitrust, one will require other, perhaps non-IP-specific criteria in order to delineate an antitrust market in which IPRs play a significant role.

Against the background of the case analysis in previous sections, in what follows we will identify some general trends in EU and US antitrust law as regards market definition and IPRs before embarking on a short comparative analysis of how various types of IPRs are assessed under antitrust market definition, covering copyright, patents and trademarks. This will lead to a more general view on product substitutability where IPRs are involved, including a discussion of the meaning of an IPR's success for market definition, issues of demand and supply substitutability as regards IPRs, and the question which initial product should be relied on when defining antitrust markets in which IPRs may be important.

In general, neither the European Commission and the EU Courts nor the FTC, the DoJ and US courts will simply narrowly define the relevant market based on any available IPR; instead, they will carry out a case-by-case analysis.[134] In the US, the patent law standard to the effect that a patent conveys monopoly power on the patent owner was analogously applied to antitrust law for a long time, but this assumption has now been overcome for good following *Independent Ink*. The short overview of EU landmark cases for market definition and IP has shown that in the EU, the mere possession of an IPR will only be held to convey market power under certain circumstances and, more precisely, where there are no readily available substitutes for the IPR. This, in turn, greatly depends on consumer perceptions and consumer behaviour, but also on the nature of the market in question.

A general distinction that could be seen in the many cases discussed above is that between relevant product markets proper that revolve around the IPR in question ('IP market') – particularly in licensing cases – and relevant product markets which centre on a particular product in which IPRs play a more or less determinative role. The former, of course, is a much narrower market, with possible consequences for the finding of market power.

For copyright, it was seen that particular copyrighted works were regarded as their own relevant product market in some licensing cases (eg, *Magill*, *IMS Health*, *PolyGram* and *Paramount Pictures*). In some instances, the works by a particular artist were also grouped together, thus constituting a market for Jackson Pollock paintings or for Andy Warhol art. This is not surprising, for the situation will often be such that consumers and customers at large do not regard one copyright holder's work – be it a musician's music, a choreographer's choreography, an author's

[132] Feldman (n 12) 13 §§ 45, 46 (direct quote); 14 §§ 50, 51; 15 § 54.

[133] ibid 18 § 67.

[134] But see Olivier Vrins, who argues that the Commission's narrow market definitions where IPRs are involved are 'prejudicial to intellectual property rightowners' interests'; O Vrins, 'Intellectual Property Licensing and Competition Law: Some News from the Front – The Role of Market Power and Double Jeopardy in the EC Commission's New Deal' (2001) 23 *European Intellectual Property Review* 576, 584.

bestseller book or a company's brick structure – as exchangeable with another, thus requiring the finding of narrow product markets for antitrust purposes. This issue can be taken a step further, of course, asking whether particular works by an artist – rather than an artist's entire oeuvre – should be regarded as an antitrust market. This is what occurred in the case of the Three Tenors. Successful copyrighted works are often seen as non-interchangeable – think Umberto Ecco's *Name of the Rose* or JK Rowling's *Harry Potter* series – while a customer looking for a good holiday read might be content with a broader array of copyrighted works by a range of authors that he or she will regard as largely interchangeable. Similarly, avid Metallica fans will not forego the purchase of an *S&M* special edition album if it is suddenly sold at a higher price, while parents looking for soothing lullabies for their baby might very well choose from a wider selection of children's albums. And a radio station that wants to license Wham's 'Last Christmas' will not be happy with the proposition of licensing a cheaper Christmas track by an unknown artist.

The criteria for awarding copyright for a work are such that a copyrighted work necessarily exhibits an individuality that sets it apart from all others.[135] In order to be awarded copyright protection, a work must fulfil certain standards – the standard of originality that is known under both EU[136] and US law.[137] To fulfil this standard of originality, a work must distinguish itself from previous works. However, simply because a work fulfils the copyright standard from an IP law perspective does not in itself mean that it is not regarded as interchangeable with other works by the relevant customer group for the purposes of antitrust law. Ultimately, the question for market definition is one of interchangeability, which may be limited for the very reason that a copyrighted work embodies a certain originality.

From a practical perspective, Amazon's suggestions in its 'Customers who bought this item also bought …' section might be of great interest for locating possible substitutes for a given copyrighted work, as these suggestions are based on actual customer behaviour and therefore provide a good indication of what consumers may regard as interchangeable with a certain copyrighted work. Of course, this is but a starting point, but one based on large amounts of actual consumer data. As a side note, this again points to the increasing importance of data and analytics already highlighted above (see section I.B.i in Chapter 4).

Turning to patents, a patent might be found to protect a successful invention so different from the closest available substitute that the patented invention constitutes its own antitrust market. As a patent is only granted for an invention that is new and contains an inventive step (EU) or is non-obvious (US), the criteria of patentability already ensure that there is limited substitutability between

[135] ibid 581.

[136] On originality in EU IP law, see E Rosati, *Originality in EU Copyright: Full Harmonization through Case Law* (Cheltenham, Edward Elgar, 2013).

[137] *Feist Publications v Rural Telephone Service Company*, 499 US 340 (1991).

patented inventions.[138] In this respect, there is a certain similarity between patented inventions and copyrighted creative works. However, just as in copyright, the mere fact that a patent is granted and thus, in the eyes of the patent agency, fulfils the criteria of novelty and inventiveness/non-obviousness does not necessarily mean that the invention is seen as non-interchangeable by customers. This, however, is the applicable test under the antitrust rules. As the US Court of Appeals for the Federal Circuit has aptly pointed out, 'aspects of an invention that may have led the PTO to issue a patent are not per se coterminous with the features of the patented product that may lead consumers to select that product over other similar ones'.[139] For competition law purposes, what is relevant is thus whether inventions are regarded as substitutable. Where the patent licensees view another invention or technology as substitutable because it performs the same function, albeit possibly with entirely different means, this invention or technology will form part of the relevant technology market. For pharmaceuticals that are subject to a patent, it has become customary to rely on therapeutic use in order to assess interchangeability; sometimes, a drug's brand will also need to be factored into the equation.

Trademarks are less frequently relied upon for market definition compared to other IPRs. The reason might lie within the very characteristics of this IPR. Trademarks do not protect the nature of a product, process or technology, but help the customer to identify the origin of a product. As such, there may be a number of trademarks relating to products that are readily interchangeable – a case in point are originator and generic pharmaceuticals. As could be seen in several cases, antitrust market definition has not yet 'properly taken account of the power of brands'.[140] If antitrust law wants to rely on customer perception and behaviour, then it may well need to increasingly factor trademarks into account when delineating a relevant market.

Where trademarks are at issue, the Nice classification can provide a useful first point of reference to identify the industry to which the trademark relates. This classification lays down in which goods or services the respective trademark receives protection.[141] Products with protection in the same or similar classes may turn out to be substitutes.[142] However, this first orientation based on the IP law-inherent classification system will then have to be narrowed down considerably. Products that are regarded as identical or similar for the purposes of trademark law may not be held to be interchangeable for antitrust law purposes.[143] A successful brand's product might constitute its own relevant product market

[138] Vrins (n 134) 581.

[139] *Delano Farms v California Table Grape Commission*, 655 F3d 1337, 1352 (Fed Cir 2011). See also Hovenkamp et al (n 9) § 4.02[E].

[140] DR Desai and SW Waller, 'Brands, Competition, and the Law' [2010] *Brigham Young University Law Review* 1425, 1468 (direct quote), 1472.

[141] See WIPO, 'Nice Classification', www.wipo.int/classifications/nice/en.

[142] Heide (n 10) 165.

[143] Vrins (n 134) 581.

where customers do not view virtually identical products as interchangeable with each other based on the brand. Going forward, consumers' actual buying patterns will need to be taken into account much more carefully where trademarks are used for branding products.[144] This insight can also be found in the European Commission's Vertical Guidelines, which state that '[b]randing tends to increase product differentiation and reduce substitutability of the product, leading to a reduced elasticity of demand and an increased possibility to raise price.'[145] Supply-side substitutability may also be reduced when trademarks are at play, thereby acting as barriers to (legal) market entry. These findings immediately link back to the question of market definition for branded goods. Especially in the area of character merchandising or personality trademarks, the resulting product differentiation leads to the fragmentation of markets under antitrust law.[146] If it is found that one branded product constitutes the relevant product market, then it will not come as a surprise when the product's manufacturer is deemed to have significant market power in that market, or indeed in the related aftermarket.[147] The *Geneva Pharmaceuticals* Court recognised that brand loyalty can come to bear on the market definition. Similarly, the Supreme Court in *Kodak* hinted at the possibility of delineating antitrust markets along the lines of trademarks, even if this does not occur often. *Consten and Grundig* was also based on such an assumption without explicitly spelling it out.

Having analysed market definition along the lines of different types of IPRs, the question remains as to what ties these cases together. First of all, antitrust law does not in itself distinguish cases by the type of IPR involved. Nevertheless, similar issues do tend to arise depending on the type of IPR involved, and some IPRs are more frequently found to delineate an antitrust market. Licensing cases invariably hinge on the IPR the licensing of which is being analysed, and this will be further explored below (see section IV). What all IPRs have in common is that their relevance for antitrust market definition depends on their success in inducing the kind of product differentiation that IP strives for, from an IP law point of view. Different IPRs have different techniques in order to increase differentiation to their benefit, with patents largely relying on innovation and technology, copyright on what works are seen as popular, and trademarks on marketing strategies.[148] For instance, supermarkets' own brands are often sold at a significantly lower price than the branded products. Even if the quality of those products may be the same (which, arguably, is not always the case), many consumers will continue to buy their trusted branded product. Two aspects are thus decisive for the role of IPRs in antitrust product market definition: the degree to which the IPR

[144] Lemley and McKenna (n 47) 2084, 2086.
[145] European Commission, Guidelines on Vertical Restraints [2010] OJ C130/1, para 104.
[146] Vrins (n 134) 580.
[147] Hovenkamp et al (n 9) § 12.03[B].
[148] WR Cornish, *Intellectual Property: Patents, Copyright, Trade Marks and Allied Rights* (London, Sweet & Maxwell, 1981) 22 (currently in its 9th edn).

manages to ensure product differentiation so as to make competition by imitation impossible or at least unlikely; and the degree to which the IP-bearing product is able to capture customers so as to make competition by substitution unlikely.[149] Whether the product differentiation is induced (through trademarks) or inherent in a work/invention (through copyright or patents) is not relevant; this conclusion applies with equal force in both jurisdictions under scrutiny.

The importance of demand-side substitutability and product differentiation can be observed in some of the cases discussed above: *AstraZeneca*, *Intel* and *Microsoft* were cases that were litigated, in various constellations, in both the EU and the US. In all these cases, highly innovative products were at issue and IPRs played a significant role. Yet, the markets were – rightly – not delineated along the lines of the IPRs in question, but based on the invention in a broader sense, ie, a pharmaceutical product's therapeutic use, the particular kind of CPU under investigation or the operating system at issue. In other cases, however, it was the perceived non-substitutability of the IP-protected subject matter that led to the finding of narrower IP markets; the European *Consten and Grundig* and *Nungesser* cases and the US cases of *Geneva Pharmaceuticals* and *Andy Warhol* are cases in point. Depending on the particular circumstances of the case, an IPR therefore may or may not be relied upon in order to delineate the relevant antitrust market. Based on the product differentiation that IPRs bring about, coupled with the very limited effect that small but significant non-transitory increases in price tend to have in IP-intensive product markets, antitrust markets delineated by the respective IPR may be the norm rather than the exception.[150]

Which test of substitutability is applied to an IPR case may influence the resulting market definition. For instance, the test centred around a product's physical characteristics, price and intended use, as found in the Commission's Market Definition Notice, is regarded as conducive to the finding of very narrow product markets where IPRs are involved.[151] In the US, there is a similar focus on product characteristics.[152] However, it should not be forgotten that it is precisely because of many IPRs that products' characteristics differ from each other – and indeed must be different in order to avoid IP infringement. The question, then, is how customers react to those differences in terms of demand substitutability. The *AstraZeneca* cases in the EU and US provide instructive examples of how market definition in IP-intensive industries may be approached.

In IP, consumer preferences in the downstream market, ie, at the product market level, also influence the market definition in the upstream market, ie, at the technology market level.[153] If consumers prefer to buy smartphones

[149] See Drexl, 'Economic Market Power' (n 9) 16, 18.
[150] Lemley and McKenna (n 47) 2059.
[151] Vrins (n 134) 580.
[152] Hovenkamp et al (n 9) § 4.03[B].
[153] Heinemann, 'Contestability' (n 12) 61.

that incorporate the LTE standard, this will have an influence on how the antitrust market revolving around the LTE standard is defined at the upstream level. And if radio listeners have a strong preference for a new hit song, this will influence the antitrust market definition of the licensing that is connected to this song.

As was already pointed out above, IPRs may also be relevant to market definition in terms of supply substitutability: As they grant certain exclusive rights, IPRs prevent the free copying of an IP-bearing product and as such can act as effective barriers to entry, thus reducing supply-side substitutability.[154] However, where black market goods and services are seen as substitutes by customers, they may indeed form part of the relevant antitrust market, IP protection or no IP protection.[155] Thereby, counterfeited goods, illegally imported goods, illegally copied music or illegally produced generics can all become part of the same relevant product market as their non-IP infringing, legally produced or legally imported counterparts. Similarly, an open-source software may very well restrict a proprietary software. These examples show that market definition in the presence of IPRs is in no way restricted to IP-protected goods or services, but can extend well beyond these.

For IPRs, it is particularly significant which initial product is chosen in order to further define the relevant product market.[156] Thereby, the perceived anti-competitive conduct may form the starting point. Where the relevant product market is delineated around a narrowly chosen initial 'product' (which is essentially defined by an IPR or a conglomeration of IPRs), a single product market – or, rather, IP market – will easily be found, and market power will more easily be established.[157] So which initial product should be relied on as the starting point for an antitrust case in which an IPR might play a potential role? In some cases, the IPR may just be one of several factors to consider (eg, the European *AstraZeneca* case). In other cases, however, the IPR will be at the centre of the anti-competitive conduct. This is the case for refusals to license IPRs (eg, *IMS Health*) or anti-competitive licensing agreements (eg, *PolyGram*), which all concern licensing or technology markets. In these markets, which will be discussed further below, the market definition will hinge on the IPR being licensed and might continue to do so even after the IPR has expired or been revoked.

[154] Hovenkamp et al (n 9) § 4.01[B]. Free copying might, of course, occur within the limits of the copyright legislation, for instance, for educational purposes.

[155] See also JM Sellers, 'The Black Market and Intellectual Property: A Potential Sherman Act Section Two Antitrust Defense?' (2004) 14 *Albany Law Journal of Science & Technology* 583, 607–31; OB Arewa, 'YouTube, UGC, and Digital Music: Competing Business and Cultural Models in the Internet Age' (2010) 104 *Northwestern University Law Review* 431, 449.

[156] SD Anderman and HSK Schmidt, *EU Competition Law and Intellectual Property Rights: The Regulation of Innovation*, 2nd edn (Oxford, Oxford University Press, 2011) 40 f.

[157] HKS Schmidt, 'Article 82's "Exceptional Circumstances" that Restrict Intellectual Property Rights' (2002) 23 *European Competition Law Review* 210, 211.

IV. Technology Markets and the Licensing of Intellectual Property Rights

Technology markets relate to IPRs that are being licensed and are an important way of conceptualising market definition in the presence of IPRs. When competition issues arise, the question poses itself as to whether the conduct or merger affects a technology market or a traditional product market. The competitive structure and market power could differ quite considerably between the (upstream) technology and the (downstream) product market.[158] The technology market approach needs to be distinguished from the innovation or R&D market approach discussed above (see section II in Chapter 5), as no innovation or technology has yet emerged from the latter.

A. Policy on Technology Markets

Technology markets are a well-known feature of US antitrust policy, although there appears to be a preference for actual product markets. The IP Guidelines state that in licensing cases in which the analysis can be conducted by relying on product markets rather than on technology or R&D markets, this will be done.[159] Where a technology is therefore incorporated into a product, this product will be the focus of the antitrust investigation, and it can be expected that the IPR will play a role in the characterisation of that market.

The US agencies will analyse the possibly anti-competitive effects of licensing agreements with respect to the relevant market(s) in which those effects occur. These can be traditional goods or services markets (together, product markets), but may also be technology markets or innovation/R&D markets. If the effects on a product market need to be assessed, then the agencies will define this market in accordance with the Horizontal Merger Guidelines 2010.[160] However, where products and know-how are being marketed separately from each other, the agencies might also need to make reference to a technology market that is distinct from the product market. This market is made up of the IPR that is being licensed and its close substitutes. In order to determine the close substitutes of a technology, the agencies might make use of the SSNIP test as set out in the Horizontal Merger Guidelines 2010. Where the data necessary for this analysis is not available, the agencies will look at which technologies buyers regard as substitutable with the licensed technology.[161] The agencies also recognise that the licensing of IPRs can promote innovation in various ways and pledge to take this into account.[162]

[158] Hovenkamp et al (n 9) § 4.03[D][1].
[159] US IP Guidelines 2017 (n 21) § 3.2.
[160] ibid.
[161] ibid § 3.2.2 (see also fn 35 of those Guidelines).
[162] ibid § 2.3.

Under the Competitor Collaboration Guidelines, the US agencies will usually analyse collaborations with reference to all relevant product and geographical markets affected.[163] They name three possible markets that will be looked at: product markets, technology markets and innovation markets.[164] A technology market will be delineated if IPRs and the products in which they are incorporated are sold separately. It encompasses the IPR that is being licensed and all IPRs or other technology regarded as substitutable or interchangeable with it.[165] Where product or technology markets are not able to account for (anti-)competitive effects on innovation, the agencies will resort to the definition of innovation/R&D markets as outlined in the IP Guidelines.[166]

The US policy on technology markets has had an important influence on how these markets are assessed under EU competition law, introducing a more economics-based approach in the latter.[167] Today, the European Commission holds that technology markets 'consist of the intellectual property that is licensed and its close substitutes', which can be defined with the help of the SSNIP test. Potential competition also has a special role to play in these markets.[168] However, there is no further indication of how the latter can be achieved. In its binding Technology Transfer Block Exemption Regulation (TTBER), the Commission distinguishes product and technology markets that, combined with the respective geographical market, can make up a relevant market for the purposes of the TTBER. There, the relevant technology market is defined as 'the market for the licensed technology rights and their substitutes'.[169] It is the licensee's point of view that is decisive when establishing substitutability, and the latter will be established based on the characteristics of the technology rights in question, the licensing fees and the intended use of the technology rights.[170] In order to calculate a licensor's market share on a technology market for the safe harbour of block exemption, one will need to rely on the sales of all products – by the licensor and its licensees – that incorporate the respective technology right.[171] This '"footprint" at the product level' is regarded as a good proxy for establishing the licensor's market position.[172] It includes in-house technologies that companies

[163] Federal Trade Commission and US Department of Justice, Antitrust Guidelines for Collaborations among Competitors (April 2000) (US CC Guidelines), § 3.32.

[164] ibid §§ 3.32(a)–(c).

[165] For technology markets, the CC Guidelines refer to the IP Guidelines for more detail; ibid § 3.32(b).

[166] ibid § 3.32(c). On innovation/R&D markets, see already section II in ch 5.

[167] SD Anderman, 'Substantial Convergence: The US Influence on the Development of the Regulatory Framework for IP Licensing in the EC' in Marsden (n 4) 335, 341.

[168] European Commission, Guidelines on the applicability of Article 101 of the Treaty on the Functioning of the European Union to horizontal co-operation agreements [2011] OJ C11/1, paras 116 f (direct quote at para 116).

[169] Commission Regulation (EU) 316/2014 on the application of Article 101(3) of the Treaty on the Functioning of the European Union to categories of technology transfer agreements [2014] OJ L93/17 (TTBER), art 1(1)(k).

[170] ibid art 1(1)(j)–(m).

[171] ibid art 8(d).

[172] European Commission, Guidelines on the application of Article 101 of the Treaty on the Functioning of the European Union to technology transfer agreements [2014] OJ C89/3, para 87.

rely on in order to produce products, even if they are not currently licensed out.[173] The footprint approach benefits newly introduced technologies, as they would not yet have left any product level footprint, resulting in a zero per cent market share.[174] This is an indication of how much competition analysis continues to rely on product markets, even where innovation and technology are at issue.

In its (non-binding) Technology Transfer Guidelines accompanying the TTBER, the Commission describes technology as an input; technology licensing can have an impact on competition in upstream (input) and downstream (output) markets. Consequently, it might become necessary to analyse competition effects on both of these distinct markets.[175] Outside the safe harbour of block exemption under Article 8(d) TTBER, the Commission will normally rely on the SSNIP test in order to ascertain which technologies licensees regard as interchangeable with the technology under scrutiny. Where this cannot easily be established, the Commission may alternatively establish the technology market 'on the basis of sales of products incorporating the licensed technology on downstream product markets'.[176] Both approaches may also be combined.[177]

B. Technology Markets in the Case Law

The use of technology markets, often next to more traditional goods markets, is well-established in the practice of the European Commission. The licensing case of *Nungesser*, as discussed above, was an early case that involved a technology market relating to the licensing of plant breeders' rights.[178] Another early case that involved technology markets was *IBM* (1984), in which the Commission closed an investigation based on IBM's commitments relating to the release of interface information for its System/370 CPU.[179] In 1994, the Commission had to rule on a joint venture between Shell and Montedison and concluded that one

[173] ibid para 88. This inclusion has been criticised as an additional complication in applying the market share threshold; see B Batchelor and T Jenkins, 'Commission Consults on Revisions to the Competition Rules on Technology Transfer Regime: Proposes Tightening of the Rules' (2013) 34 *European Competition Law Review* 348, 349.

[174] P Alexiadis and A Guerrero Pérez, 'European Commission Proposes Stricter EU Antitrust Rules on Technology Transfer' (2013) 35 *European Intellectual Property Review* 415, 417; Guidelines on the application of Article 101 of the Treaty on the Functioning of the European Union to technology transfer agreements (n 172) para 90.

[175] Guidelines on the application of Article 101 of the Treaty on the Functioning of the European Union to technology transfer agreements (n 172) para 20.

[176] ibid para 22.

[177] ibid para 25.

[178] Case 258/78 *Nungesser v Commission* EU:C:1982:211.

[179] *Undertaking Given by IBM* [1984] Bulletin of the European Communities 17/10, 96. See also C Madero Villarejo, 'Abuses of a Dominant Position in Information Technology Industries (IT)' in C-D Ehlermann and I Atanasiu (eds), *What is an Abuse of a Dominant Position?* (Oxford, Hart Publishing, 2006) 552 ff.

of the markets concerned was the technology market relating to polypropylene. It highlighted the significant role of IPRs and of licensing in this area of technology.[180] In another case from 1994, *Digital-Olivetti*, the Commission had to consider a concentration on the market for the know-how related to RISC technology, finding that this market was 'highly dynamic'.[181]

Cases revolving around copyright may also be seen as licensing markets. In the case of *Magill*, which was discussed above (see section II), the Court viewed the upstream market (copyrighted weekly television listings) and the downstream market (TV magazines including television listings) as separate relevant markets, with the upstream market constituting a licensing market.[182] It was the refusal to license a copyrighted work that was considered the abuse of the dominant position in this case. *IMS Health* was another case on a refusal to license a copyrighted work.[183]

In September 2016, the General Court ruled on pay-for-delay settlements in *Lundbeck*, a case in which the relevant product was found to be an antidepressant with the active pharmaceutical ingredient of citalopram.[184] Having unsuccessfully tried to extend patent protection for citalopram, Lundbeck entered into pay-for-delay agreements with four competitors.[185] Interestingly, market definition played virtually no role in this case, merely occupying a single paragraph in a Commission decision that is 464 pages long. There was no discussion of whether antidepressants with other active pharmaceutical ingredients could be regarded as interchangeable with citalopram; the Commission merely spoke of 'the product concerned'.[186] As this was a case under Article 101(1) TFEU, the question arises as to whether a more broadly defined market might have escaped antitrust scrutiny based on the *de minimis* rule or a lack of an effect on trade.[187]

[180] *Shell/Montecatini* (Case IV/M.269) Commission Decision 94/811/EC [1994] OJ L332/48, paras 28–32.

[181] *Olivetti-Digital* (Case IV/34.410) Commission Decision 94/771/EC [1994] OJ L309/24, paras 8, 10 (direct quote at para 10). RISC stands for reduced instruction set computer.

[182] Joined Cases C-241 and C-242/91 P *RTE and ITP v Commission (Magill)* EU:C:1995:98, para 24.

[183] See section III.A.

[184] Case T-472/13 *Lundbeck v Commission* EU:T:2016:449, para 15.

[185] ibid paras 16 ff, para 23; appeal pending in Case C-591/16 P *Lundbeck v Commission* [2017] OJ C30/25.

[186] *Lundbeck* (Case AT.39226) Commission Decision of 19 June 2013 [2015] OJ C80/13, para 2 ('the product concerned [is] the anti-depressant citalopram, whether in the form of an active pharmaceutical ingredient … or in the form of a medicinal product').

[187] Under the *De Minimis* Notice, agreements between competitors that do not exceed a market share of 10 per cent on any relevant market are not considered to appreciably affect competition unless they include a restriction by object; see European Commission, Notice on agreements of minor importance which do not appreciably restrict competition under Article 101(1) of the Treaty on the Functioning of the European Union [2014] OJ C291/1 (*De Minimis* Notice) paras 8(a), 13. However, one might argue that in this case, the object of the agreement was indeed to restrict competition, as was also held by the Commission and the General Court; *Lundbeck* (Case AT.39226) Commission Decision of 19 June 2013 [2015] OJ C80/13, paras 647 ff; Case T-472/13 *Lundbeck v Commission* EU:T:2016:449, paras 331 ff. Under the Effect on Trade Guidelines, of course, the agreements would have been caught due to the turnover of the companies involved, see European Commission, Guidelines on the effect on trade concept contained in Articles 81 and 82 of the Treaty [2004] OJ C101/81, para 52(b).

The FTC routinely uses the concept of technology markets when analysing possible anti-competitive outcomes from a merger, acquisition or joint venture. Only two such cases will be highlighted here. In *Montedison*, the FTC investigated the same proposed joint venture between Montedison and Shell, the world's two largest polypropylene producers, as the European Commission had shortly before it.[188] The FTC delineated several relevant markets, including the licensing of polypropylene technology, polypropylene technology itself, and the licensing, production and sale of specific polypropylene catalysts and catalyst technology.[189] The licensing markets are typical technology markets and are reminiscent of the market definition the European Commission carried out in its own *Montedison/Shell* investigation. The FTC found that there was a high concentration on the technology licensing markets concerned, leading to concerns for the competitiveness on the polypropylene licensing markets once the joint venture was put into place.[190] In order for the joint venture to be approved, Shell had to divest its polypropylene business.[191]

In another FTC investigation, *Dow Chemical/Union Carbide*, Dow Chemical wanted to acquire Union Carbide. While Dow Chemical was a science and technology company and had a diverse portfolio consisting of plastic, chemical and agricultural products, Union Carbide was a chemical and plastics producer.[192] One of the competition concerns voiced by the FTC related to the polyethylene market. In this market, both companies were found to be leaders, competing in the development of technology that led to IPRs and the subsequent licensing of those IPRs. The FTC found this aspect of the acquisition to infringe section 7 of the Clayton Act and section 5 of the FTCA, amongst other things because the merged entity would control the most widely licensed linear low-density polyethylene reactor process technology.[193] To counter this concern, Dow Chemical and Union Carbide had to divest and license some of their IPRs and assets that related to polyethylene to BP.[194] This particular technology market was only one amongst a number of markets concerned, but the detailed concerns related to this technology market show that the FTC takes competition in technology markets seriously.

US courts have also tackled technology markets and their relevance for the antitrust laws. The US *Microsoft* case discussed above, for instance, which regarded the licensing of operating systems as the relevant product market, concerned a technology market.[195] Despite the general recognition that IPRs do not bestow

[188] *Montedison*, 119 FTC 676, 678, 680 (1995).

[189] ibid 678 f.

[190] ibid 681, 683 f.

[191] ibid 694 in combination with 691 (§ I.Q), in particular (h) relating to IPRs, (o) relating to trademarks, and (p) relating to licences.

[192] *Dow Chemical Co and Union Carbide Corp*, 131 FTC 600, 601 (2001).

[193] ibid 603, 607–09.

[194] ibid 659 in combination with 622 (§ I.M.3) and 624 f (§ I.X).

[195] On this characterisation, see already Newberg (n 89) 113; *United States v Microsoft*, 253 F3d 34 (DC Cir 2001).

market power on their owner,[196] in the case of successful patents there is a sense that the relevant technology market needs to be construed around these, thus leading to very high market shares and perhaps the finding of market power. In the *Actavis* case before the Supreme Court, it was held that one could deduce market power from a successful patent.[197] And a US District Court found that despite the ruling in *Independent Ink*, a market share of 100 per cent in all technology patents for the flash memory technology demonstrated market dominance on the flash memory technology market that the plaintiff alleged.[198]

In another US District Court case, *Clean Conversion*, the relevant market was claimed to be the pressurised steam classification (PSC) conversion market, ie, the conversion of municipal solid waste into commercial materials such as plastics, glass or biomass.[199] Several patents related to this market, and the defendant was accused of buying up the patents related to PSC conversion so as to consolidate its portfolio. Subsequently, it had tried to end licence agreements with competitors with the aim of keeping the PSC conversion market to itself.[200] In this case, the relevant product market comprised several IPRs that were essential for its functioning. While this was not a technology market in the strict sense, it certainly came close to establishing one, particularly with regard to the importance that the licensing of the patents had.

C. The Influence of an Intellectual Property Right's Revocation on the Technology Market

A particular aspect of technology markets has preoccupied both US and EU courts: where an IPR forms the basis of a technology market, how should antitrust law deal with a licensing case when the IPR is either revoked or expires? Does the technology market cease to exist?

Such a question was at issue in *Genentech v Hoechst* (2016). The Court of Justice had to decide whether Article 101 TFEU precluded a party to a licensing agreement from enforcing it under specific circumstances. In the present case, the licensee was obliged to pay royalties for use of a human cytomegalovirus enhancer. This enhancer was the subject of a European patent that was revoked over six years after conclusion of the agreement, and two US patents that were only issued several years after conclusion of the agreement.[201] The Paris Cour d'Appel asked the Court of Justice whether it would run counter to Article 101 TFEU to enforce

[196] See the discussion above, section II.

[197] *FTC v Actavis*, 570 US ___ (2013), 133 SCt 2223, 2236 (2013).

[198] *PNY Technologies v SanDisk*, Case C-11-04689 YGR (ND Cal 20 April 2012) 11.

[199] *Clean Conversion Technologies v CleanTech Biofuels*, Case 12-cv-239-L (JMA) (SD Cal 20 August 2012) 1, 8.

[200] ibid 2 f.

[201] Case C-567/14 *Genentech v Hoechst & Sanofi-Aventis* EU:C:2016:526, para 3.

a licence agreement based on revoked patents if that licence agreement foresees that the licensee must pay royalties for using the rights attached to the licensed patent.[202] This case is interesting for the purposes of market definition because the relevant market at issue is a technology market which, by definition, is delineated by the licensed IPR (here the patent) and its available substitutes. Where the underlying IPR ceases to exist, it is questionable whether the once-defined relevant market continues to exist. Two scenarios are possible in such a case: (1) the agreement becomes void because the very basis of the transaction ceases to exist; or (2) the agreement needs to be terminated by the licensee. What should be borne in mind is that a concluded licence agreement may already incorporate the probability of the IPR not being valid (or infringed), meaning that solution (1) may not do justice to the bargain that was initially reached. It may be for this reason that the Court of Justice stated in *Genentech v Hoechst* – as it had already done in *Ottung* in 1989[203] – that if the licensee can terminate the licence agreement within a reasonable period of time, the licensor can request it to pay royalties for the whole duration of the agreement, even if the patent was revoked or not infringed, without incurring antitrust liability under Article 101(1) TFEU.[204] This analysis implicitly relies on the Court of Justice's assumption that in such a case, the technology market consisting of the (now-revoked) patent continues to form the relevant antitrust market which is the starting point for antitrust liability. Both the licence agreement and the technology market it relates to survive the revocation of the patent. The only reason for such a construction lies in the fact that the licensor and the licensee must have incorporated such considerations into their licensing negotiations. Retroactively negating the licence agreement or the technology market it (at least initially) related to would run counter to contractual principles. Thus, the technology market that previously related to the patent then relates to the know-how incorporated in the (revoked or expired) patent. The know-how does not cease to exist once the patent is revoked or expires; however, the market dynamics will change considerably following such an event.

A comparable case was decided in the US in 2015. In *Kimble v Marvel Entertainment*, Kimble had sold its patent on an action figure to Marvel for a lump sum and a three per cent royalty payment on all sales.[205] When the patent was close to expiring, Marvel sought declaratory judgment that it could then cease its royalty payments. Although no such clause was contained in the agreement, an earlier Supreme Court case had established a doctrine to that effect under patent law.[206]

[202] ibid para 19.

[203] Case 320/87 *Kai Ottung v Klee & Weilbach and Others* EU:C:1989:195, para 13. In that case, the patent was not revoked, but simply expired. Nevertheless, the Court of Justice found that it would not run counter to art 101(1) TFEU if the licensee had to pay royalties until it rightfully terminated the agreement.

[204] Case C 567/14 *Genentech v Hoechst & Sanofi-Aventis* EU:C:2016:526, para 43.

[205] *Kimble v Marvel Entertainment*, 576 US ___ (2015), 135 SCt 2401 (2015).

[206] *Brulotte v Thys*, 379 US 29, 32 (1964), holding that 'a patentee's use of a royalty agreement that projects beyond the expiration date of the patent is unlawful per se'. The Court cautioned that this was a patent rather than an antitrust case (ibid 38).

Kimble wanted to rely on an antitrust-type rule of reason analysis to assess the royalty agreement's lawfulness after the patent's expiry.[207] This case was decided within the realm of patent law and the Supreme Court noted that it might resolve the case differently within the realm of antitrust law, based on the competitive effects of royalties after a patent's expiry and a possible negative impact on technological innovation.[208] Implicit in this line of thought is the view that the antitrust relevant market in such a potential antitrust case might not cease to exist under US antitrust law either.

D. Analysis

The concept of technology markets as it is used in both the EU and the US can be seen as yet another type of input or upstream market. The product incorporating the licensed IPR would then constitute the downstream market. This might allow us to view licensed product downstream markets as proprietary aftermarkets,[209] entailing a market definition analysis similar to the one outlined below (see section I in Chapter 7). Importantly, technology markets in an antitrust sense should be inter-modal, ie, they should include different technologies that achieve similar outcomes.[210]

In order to find a technology market centred around a certain IPR, it is not necessary that the IP owner is willing to license the IPR;[211] instead, it may be enough for there to be a potential or hypothetical market with a discernible demand.[212] However, what is required is that the technology is marketable.

As the term 'technology' suggests, technology markets were conceptualised for technology-related IPR; the cases discussed above related to patents, but also to plant breeders' rights and interface information. However, any IPR that can be licensed may be at the centre of such a market. Perhaps, therefore, the term 'licensing market' would be more accurate. Particularly in relation to copyright licensing, similar issues pertaining to market definition arise.

Technology markets are defined in very much the same way under the US IP Guidelines and the EU's Technology Transfer Guidelines.[213] And both have been criticised for not giving the necessary attention to potential competition from

[207] *Kimble v Marvel Entertainment*, 576 US ___ (2015), 135 SCt 2401, 2411 (2015).

[208] ibid 2412 f.

[209] Heinemann, 'Contestability' (n 12) 60. See also MM Wirtz and M Holzhäuser, 'Die kartellrechtliche Zwangslizenz' (2004) 50 *Wettbewerb in Recht und Praxis* 683, 689.

[210] Newberg (n 89) 121.

[211] Anderman and Schmidt (n 156) 47.

[212] Similarly, see Case T 201/04 *Microsoft v Commission* EU:T:2007:289, para 335; Case C-418/01 *IMS Health v NDC* EU:C:2004:257, para 44.

[213] See section IV.A.

technologies that are being developed.[214] Both jurisdictions also display a certain preference for goods markets over technology markets. The question therefore poses itself as to whether a technology (or licensing) market can be delineated where one can or could also delineate a product market that centres around a product that incorporates the IPR in question. As highlighted above, the IP Guidelines give priority to product markets over technology markets.[215] And the European Commission may sometimes treat a technology (which may be covered by an IPR, such as a patent or know-how) as the product.[216] Cases have shown that both types of markets are sometimes relied upon in parallel. Only where the licensee's views as to the interchangeability of technologies does not yield useful results will the EU go a step further and analyse the licensor's market position through the products in which the technology is incorporated, thereby relying more on product market definition. The concept of the technology market, being an upstream market, allows antitrust authorities to apply traditional antitrust doctrine to this dynamic market. However, the antitrust analysis that is later based on this market will need to be particularly sensitive to innovation competition (see section III in Chapter 5).

V. Standard-Essential Patents and Product Market Definition: Just Another Technology Market?

Standard-essential patents (SEPs) are IPRs that relate to a technology standard,[217] for instance, to the USB standard for connecting electronic devices or to the Wi-Fi standard used in wireless internet connections. They are the typical case of a technology market, and yet the fact that they relate to a technical standard rather than a 'normal' invention begs the question of whether they represent a special case amongst technology markets.

A. Policy Regarding Market Definition in Standard-Essential Patents

Only a limited amount of guidance is available on market definition in the realm of SEPs. In January 2013, the DoJ and the US Patent & Trademark Office jointly

[214] M Glader, *Innovation Markets and Competition Analysis: EU Competition Law and US Antitrust Law* (Cheltenham, Edward Elgar, 2006) 202.

[215] US IP Guidelines 2017 (n 21) § 3.2.

[216] SD Anderman and J Kallaugher, *Technology Transfer and the New EU Competition Rules: Intellectual Property Licensing after Modernisation* (Oxford, Oxford University Press, 2006) § 6.55.

[217] C Graham and J Morton, 'Latest EU Developments in Standards, Patents and FRAND Licensing' (2014) 36 *European Intellectual Property Review* 700, 700.

issued a Policy Statement on remedies for SEPs subject to voluntary F/RAND commitments.[218] This Statement refers to the market power a SEP may confer on the patent owner,[219] thereby implying that a single patent could constitute the (narrowly defined) relevant market.

In its standardisation communication from 1992, the European Commission underlined that whether or not a SEP owner was found to be in a dominant position 'depends heavily on the definition of the relevant product market. Obviously, the narrower the relevant product market is the greater the likelihood of dominance being established'.[220] Referring to previous cases in a non-SEP environment, the Commission concluded that in the realm of standardisation, one might also find that companies are dominant on the market for their own products, which must be understood to include their IPRs.[221]

For standardisation agreements, the European Commission identifies four distinct markets that could be affected, one of which is the technology market that the IPRs (mostly patents) incorporated in the standard form part of. The other markets are: the product market that the standard relates to; the market for standard-setting as such; and possibly also the market for testing/certification.[222] The Commission has underlined that while the holder of an IPR that is essential to a technical standard may derive substantial market power from its IPR, 'there is no presumption that holding or exercising IPR essential to a standard equates to the possession or exercise of market power'.[223]

B. Cases Delineating Markets in the Area of Standard-Essential Patents

In an early case on a patent ambush in SEPs, the FTC reached a consent agreement with Dell in 1996. Dell had participated in adopting a new standard for a computer bus design and during the standard-setting process had repeatedly confirmed that this new standard did not infringe on its IPRs. However, following the successful implementation of the standard, Dell informed several users of the standard that they were violating its exclusive rights.[224] Challenged by the FTC, Dell agreed to

[218] US Department of Justice and US Patent & Trademark Office, 'Policy Statement on Remedies for Standards-Essential Patents Subject to Voluntary F/RAND Commitments' (8 January 2013). FRAND stands for fair, reasonable and non-discriminatory.

[219] ibid 4, 6.

[220] European Commission, Intellectual Property Rights and Standardization: Communication from the Commission' (27 October 1992) COM(1992) 445 final, para 5.1.7.

[221] ibid.

[222] Guidelines on the applicability of Article 101 of the Treaty on the Functioning of the European Union to horizontal co-operation agreements (n 168) para 261.

[223] ibid para 269.

[224] *Dell Computer Corporation*, 121 FTC 616, 617 (1996).

cease and desist from enforcing its patents relating to the bus standard.[225] The FTC noted that it had reason to believe that at the point in time when the 'bus standard had become widely accepted, the standard effectively conferred market power upon Dell as the patent holder'.[226] The relevant market in this case was therefore Dell's SEP itself, which had become non-substitutable through the process of successful standardisation.

Referring to the special nature of SEPs, a US District Court held in 2008 that the landmark case of *Independent Ink*, which had stated that there was no longer to be a presumption of market power for patents in tying cases, 'does not address holders of *essential* patents, that is, patents that are incorporated into a standard'.[227] In *Broadcom v Qualcomm* (2007), the US Court of Appeals for the Third Circuit held that '[i]t is the incorporation of a patent into a standard – not the mere issuance of a patent – that makes the scope of the relevant market congruent with that of the patent'.[228] The relevant product market in the case at hand was Qualcomm's IP-protected WCDMA technology, which was essential for the UMTS standard. There were said to be no substitutes for that technology, so a user of the UMTS standard necessarily needed access to Qualcomm's technology.[229] Other US District Courts have since relied on these findings.[230]

Early in 2017, the FTC launched an antitrust claim under section 5 of the FTCA against Qualcomm in relation to its SEPs, alleging that the chip producer engaged in anti-competitive licensing practices, refused to license SEPs to competitors and entered into exclusive agreements with Apple in order to boost its market position.[231] The FTC held that Qualcomm was 'a dominant supplier of baseband processors and a licensor of patents that Qualcomm has declared essential to widely adopted cellular standards'.[232] Qualcomm was a leading manufacturer of these chips ('baseband processors') on a global scale.[233] In essence, therefore, the relevant markets at issue were technology markets with SEPs at their heart, as well as product markets that relied on proprietary technology, and the alleged anti-competitive harm stemmed from Qualcomm's licensing practices.

In the European *Qualcomm* case (2018), the European Commission fined Qualcomm for foreclosing competitors from the market for LTE baseband chips. The Commission highlighted that the relevant market was characterised by high barriers to entry, which resulted from the need to invest in considerable R&D for the development of such chipsets and the fact that Qualcomm owned several

[225] ibid 620. Dell was not precluded from enforcing its patents against infringers that were not implementing the standard; ibid 625.

[226] ibid 624 fn 2.

[227] *Research in Motion v Motorola*, 644 FSupp2d 788, 793 (ND Tex 2008) (emphasis in original).

[228] *Broadcom v Qualcomm*, 501 F3d 297, 315 (3d Cir 2007).

[229] ibid.

[230] See, eg, *Apple v Samsung*, Case 11-CV-01846 (ND Cal 14 May 2012) 7.

[231] *FTC v Qualcomm* (Complaint for Equitable Relief, ND Cal 17 January 2017).

[232] ibid para 2.

[233] ibid paras 20, 31.

IPRs in that market.[234] It appears that in this case, the Commission focused on the chipsets incorporating patented technology rather than on the licensing of the technology on which the FTC had relied, thus in effect delineating a product market with an important IPR impact.

In *Rambus* (2006), the FTC brought a monopolisation claim against that technology licensing company, based on Rambus' patent ambush. The FTC relied on four relevant markets in this case, all of which were technology markets consisting of Rambus' patent rights related to the DRAM chip JEDEC standard. These were the latency technology, the burst length technology, the data acceleration technology and the clock synchronisation technology.[235] Rambus explicitly did not contest this market definition before the Court of Appeals.[236] The same patent ambush was the subject of a separate European investigation, which led to a commitment decision (2009) in which the European Commission focused on the 'technology market for DRAM interface technology'.[237] This market was characterised by important standardisation, with 96 per cent of all DRAM chips complying with the JEDEC standard.[238] The patents owned by Rambus read on this standard.

The *N-Data* consent decree (2008) of the FTC is another case in point. At issue was an IP-protected technology that had read on the highly successful IEEE Ethernet standard since the mid-1990s. N-Data bought these patents in 2001 and started renegotiating the licensing terms for its SEPs. The FTC held this to be anticompetitive under section 5 of the FTCA. The complaint was resolved through a consent agreement.[239]

In its merger decision on *Google/Motorola Mobility* (2012), the European Commission held that 'there is by definition no alternative or substitute for each [standard-essential] patent. Therefore, each SEP constitutes a separate relevant market'.[240] This led the Commission to conclude that, in the case at issue, '[t]he relevant market … is thus the (at least) EEA-wide market for the licensing of each of the relevant SEPs that Google will acquire from Motorola Mobility following the transaction.'[241] This very narrow market definition necessarily impacts the assessment of the SEP holder's market power. However, a word of caution is in order. The Commission's statement should not be understood as a rule; there may not be

[234] *Qualcomm (Exclusivity Payments)* (Case COMP/40.220) Commission Decision of 24 January 2018 [2018] OJ C269/25; European Commission, 'Antitrust: Commission Sends Two Statements of Objections on Exclusivity Payments and Predatory Pricing to Qualcomm' (IP/15/6271, 8 December 2015).

[235] *Rambus*, 142 FTC 98, 139 ff (2006). The *Rambus* decision was later overturned; see *Rambus*, 143 FTC 85 (2007).

[236] *Rambus v FTC*, 522 F3d 456 (DC Cir 2008).

[237] *Rambus* (Case COMP/38.636) Commission Decision of 9 December 2009 [2010] OJ C30/17, para 16.

[238] ibid paras 18 f.

[239] *N-Data*, 146 FTC ___ (22 September 2008).

[240] *Google/Motorola Mobility* (Case COMP/M.6381) Commission Decision of 13 February 2012 [2012] OJ C75/1, para 54.

[241] ibid para 61.

an alternative for the SEP, but there may be an alternative standard that is viable. If this standard is regarded as substitutable by market participants, then this will curtail the leveraging power of the single SEP owners. This ultimately leads to the question of whether in the standardisation context there needs to be a two-stage market definition, proceeding from the standards market to the technology market(s).

On the same *Google/Motorola Mobility* merger, the FTC issued a consent order in July 2013. It defined the relevant market as 'the technology covered by any Google-owned SEP and all substitutes for that technology'.[242] It also held that Google held monopoly power over that technology from the point in time onwards when that technology was incorporated into the relevant industry standard.[243] In essence, it found the relevant market to be delineated by the SEP because inclusion in the industry standard had 'eliminated viable technology alternatives for implementers'.[244]

Slightly before the *Google/Motorola Mobility* merger, the FTC was confronted with the market power that SEPs might confer. In its consent order allowing the acquisition of SPX Service Solutions by Robert Bosch (2013), the relevant product market was found to be the manufacture and sale of air conditioning recycling, recovery and recharge devices used for the repair of motor vehicle air conditioning systems.[245] SPX held a number of SEPs that related to the relevant market. These were FRAND-encumbered, but SPX nevertheless sought injunctions for them.[246] The FTC held this to constitute an unfair method of competition.[247] Nowhere did the FTC explicitly hold that SPX held market power over a market that was delineated by the individual SEPs, but its antitrust analysis was obviously based on this assumption or else it would not have seen a need for the remedy.

In two decisions adopted in 2014, *Samsung* and *Motorola*, the European Commission repeated its principal stance that there is no presumption that the ownership of SEPs amounts to market power.[248] Nevertheless, it found that the SEP owners held dominant positions in both cases. In the commitment decision of *Samsung*, the Commission held that there were several relevant technology markets at issue, each comprising the technology that was protected by one of Samsung's SEPs and that was essential for the specifications of the UMTS standard.[249] For companies in the market, there was no substitute for the UMTS standard.[250]

[242] *Motorola Mobility & Google*, 156 FTC 147, 152 (2013).
[243] ibid.
[244] ibid.
[245] *Robert Bosch GmbH*, 155 FTC 713, 716 (2013).
[246] ibid 718 f.
[247] ibid 719, 739–41 (2013); FTCA, § 5.
[248] *Samsung – SEPs* (Case AT.39939) Commission Decision of 29 April 2014 [2014] OJ C350/8, para 46; *Motorola – SEPs* (Case AT.39985) Commission Decision of 29 April 2014 [2014] OJ C344/6, para 223.
[249] *Samsung – SEPs* (Case AT.39939) Commission Decision of 29 April 2014 [2014] OJ C350/8, para 41.
[250] ibid paras 42, 44.

Although it is estimated that approximately 10,000 patents are essential for the UMTS standard,[251] the Commission found that Samsung was dominant on these very narrow markets consisting of single SEPs as it is impossible to implement the UMTS standard without licensing (or infringing) Samsung's UMTS SEPs.[252]

In *Motorola*, which was an infringement decision, the Commission analysed the issue of the relevant market in even more detail. It distinguished an input and an output market and subsequently focused on the input, ie, the technology market. It defined the latter as the market for the licensing of the technology that was protected by Motorola's Cudak GPRS SEP, which was essential for the GPRS standard.[253] The Commission analysed whether companies that were active in the manufacturing of mobile devices could turn to any other standards in mobile telecommunications as substitutes, but found that no such substitutes existed.[254] For this reason, the market had to be defined so narrowly.

In *Huawei v ZTE*, the first ever case on SEPs to come before the Court of Justice, Advocate General Wathelet opined in 2014 that 'the fact that an undertaking owns a SEP does not necessarily mean that it holds a dominant position within the meaning of Article 102 TFEU'.[255] The technology market to which the SEP belongs is therefore not necessarily confined to the technology incorporated in the single SEP. This is in line with the general policy statements of the European Commission. Nevertheless, the Regional Court of Düsseldorf as the referring court had assumed that a dominant position was held, and this was not further questioned by the Court of Justice.[256]

C. Analysis

In the context of standardisation, the question of whether a SEP constitutes its own technology market under competition law largely depends on the success of the standard.[257] Where a technical standard is so successful that market participants cannot avail themselves of its use and cannot viably opt for alternative standards, each single SEP that reads on the standard will be found to constitute its own technology market. This is because in such a case, each single SEP can act as a gatekeeper for access to the standard and, as such, a narrow market definition is warranted. This inevitably leads to the finding that each single SEP owner has a

[251] R Bekkers, R Bongard and A Nuvolari, 'Essential Patents in Industry Standards: The Case of UMTS' (DRUID Summer Conference, Copenhagen, June 2009) 7.

[252] *Samsung – SEPs* (Case AT.39939) Commission Decision of 29 April 2014 [2014] OJ C350/8, paras 42, 45 f.

[253] *Motorola – SEPs* (Case AT.39985) Commission Decision of 29 April 2014 [2014] OJ C344/6, paras 184, 186.

[254] ibid paras 190 ff, especially 193, 210.

[255] Case C-170/13 *Huawei v ZTE* EU:C:2014:2391, Opinion of AG Wathelet, para 57.

[256] Case C-170/13 *Huawei v ZTE* EU:C:2015:477, para 28.

[257] MA Lindsay, 'Safeguarding the Standard: Standards Organizations, Patent Hold-up, and Other Forms of Capture' (2012) 57 *Antitrust Bulletin* 17, 18.

market share of 100 per cent on each of these very narrowly construed technology markets,[258] with all the antitrust consequences that such a finding entails.

Where a standard is less successful, and depending on available alternative technologies, the relevant antitrust market may or may not be delineated by the SEP in question.[259] Where standards are found to be competing with each other, and no one standard has a prominent, must-have position, this may indicate that no single standard – and no single patent reading on one of the standards – constitutes its own relevant technology market.[260] Importantly, in a next step, the individual SEP owner's degree of market power will then also depend on the success of the standard. This, on the other hand, might be determined by the strength of the network effects present on the market.[261]

Whether or not a SEP is FRAND-encumbered – ie, whether or not a SEP owner has agreed to license its SEP on fair, reasonable and non-discriminatory terms – can have a decisive influence on the ensuing antitrust analysis of a given set of facts. However, it does not have any bearing on the antitrust market definition as such.

In the various SEP cases analysed in the previous section, the IPR in question was repeatedly found to constitute the relevant market, sometimes more and sometimes less explicitly. The conclusions on market definition reached in these cases are consistent when one takes into account that the standards that the SEPs in question related to were highly successful ones that companies could not circumvent. However, as a general matter and despite their somewhat special nature, market power should never be assumed for SEPs, or indeed for any IPRs. Instead, a careful analysis is warranted, with market definition as its starting point.[262] In that respect, some courts' approach to market definition in SEPs appears too restrictive.

VI. Conclusion

There is a friction between the exclusivity granted by IPRs and the competition law notion of monopoly, which is also palpable when delineating relevant antitrust markets. This alludes to the more general discussion on the rivalry or complementarity of IP laws and competition law. Both the US and the EU have

[258] See, eg, *Samsung – SEPs* (Case AT.39939) Commission Decision of 29 April 2014 [2014] OJ C350/8, para 46; *Motorola – SEPs* (Case AT.39985) Commission Decision of 29 April 2014 [2014] OJ C344/6, para 225.

[259] TA Väisänen, *Enforcement of FRAND Commitments under Article 102 TFEU: The Nature of FRAND Defence in Patent Litigation* (Baden-Baden, Nomos, 2011) 34.

[260] Similarly, see U Petrovčič, *Competition Law and Standard Essential Patents: A Transatlantic Perspective* (Alphen aan den Rijn, Kluwer Law International, 2014) 77.

[261] D Telyas, *The Interface between Competition Law, Patents and Technical Standards* (Alphen aan den Rijn, Kluwer Law International, 2014) 46.

[262] G Bruzzone and M Boccaccio, 'Standards under EU Competition Law: The Open Issues' in G Caggiano, G Muscolo and M Tavassi (eds), *Competition Law and Intellectual Property: A European Perspective* (Alphen aan den Rijn, Kluwer Law International, 2012) 89.

moved away from the presumption that the ownership of an IPR is synonymous with market power. This also means that an IPR is no longer seen as necessarily constituting its own antitrust market. Instead, IPRs are seen as a factor – albeit an important one – in antitrust market definition.

Some EU and US cases continue to delineate the relevant market along the lines of an IPR. This is primarily the case for successful IPRs that have led to customers not regarding other products as interchangeable with the IP-protected one, such as in the case of a strong brand image or works by famous artists. Concerning the role of IPRs in cases in which the relevant market was not defined along the lines of IPRs, it is clear that IPRs can constitute an important factor in the analysis – such as an entry barrier, eg, in the case of interoperability information not being shared. However, the focus of these market delineations is not on IPRs, but on other product markets (eg, work group servers, a certain CPU, active pharmaceutical ingredients), with the IPR being a factor in that analysis. Technology markets, then, are a useful way to conceptualise licensing markets, and this aspect of market definition is routinely handled by the antitrust authorities. Standard-essential patents do not represent a special case for market definition, as they can be delineated like any other technology market.

As of now, it appears that competition authorities and courts are more easily satisfied that a successful patent delineates the relevant antitrust product market, while customer behaviour is not yet regarded as equally decisive when it comes to copyrighted works or trademarked signs. If market delineation continues to rest on the criterion of substitutability, then customer response and the influence of successful marketing need to be taken into account more seriously. In some instances concerning generic and originator pharmaceuticals, this has already been put into practice.

7

Innovative Aftermarkets

An aftermarket or secondary market consists of goods or services that are complements to a long-lasting primary product and that are typically bought after acquisition of the primary product. There are many types of constellations involving aftermarkets. The definition can apply to consumables (eg, oil for an oil tank or replacement filters for a water filter) as well as to durables (eg, a charger for a smartphone or spare parts for a car).[1] Aftermarkets are a common feature in innovative market environments.[2] This may lead to competition law-related questions when innovators keep these lucrative aftermarkets to themselves. Regularly, this is achieved through intellectual property (IP) protection, leading to what is called a proprietary aftermarket. Further means are contractual obligations with exclusivity clauses, technical barriers or marketing strategies that allow the company on the primary market to apportion a large part of the secondary market. In the present context, the market definition stage of the antitrust analysis of such secondary markets in innovative market environments is of interest. The success of many antitrust claims involving aftermarket considerations – eg, where competitors want access to an aftermarket through a licence – requires the finding of an aftermarket that is separate from the primary market.

I. Policy Documents on Delineating Aftermarkets

The delineation of aftermarkets for antitrust law purposes has only received attention from European antitrust authorities. This may be due to the stricter substantive treatment that tying and excessive pricing receive in Europe, whereas in the US, tying is today viewed more leniently and exploitative abuses are not prohibited.

The European Commission's Market Definition Notice (1997) stresses that the general principles on market definition need to be diligently applied where primary and secondary markets are concerned, especially in the realm of Article 102 TFEU. While the general method of market delineation remains the same,

[1] GT Gundlach, 'Aftermarkets, Systems, and Antitrust: A Primer' (2007) 52 *Antitrust Bulletin* 17, 18; J Temple Lang, 'Practical Aspects of Aftermarkets in European Competition Law' (2011) 7 *Competition Policy International* 199, 200.

[2] See also JP Bauer, 'Antitrust Implications of Aftermarkets' (2007) 52 *Antitrust Bulletin* 31, 35.

ie, substitutability patterns need to be studied, 'constraints on substitution imposed by conditions in the connected markets' need to be considered.[3] Depending on the particular market circumstances, the Commission will find narrower or broader markets. Where compatibility issues between the primary and secondary products are relevant, this might lead to the finding of a narrow secondary market. If, on the other hand, secondary products are freely interchangeable or if primary products are such that consumers quickly react to relative price increases, this might lead to the finding of a broader relevant market.[4]

In its 2004 publication on market definition, the UK's Office of Fair Trading (OFT, now the Competition and Markets Authority or CMA) suggested that where primary and secondary products are involved, there are three ways of delineating the relevant market: as a system market, as multiple markets or as dual markets. A system market encompasses the primary and the secondary product, thus forming one comprehensive market. In multiple markets, there is one market for all the primary products and separate markets for the secondary products linked to the individual primary products. Finally, in dual markets, there is one market for the primary product and a separate market for the secondary product, with all secondary products generally compatible with all primary products.[5] The OFT believes that a system market is appropriate where consumers engage in lifecycle pricing. If customers are locked in once they have bought the primary product, then multiple markets should be defined.[6] Only multiple markets will lead to antitrust concerns in connection with aftermarkets.[7] This distinction between system markets and multiple markets will frequently be relied upon in the following. As a first indicator, pricing practices on a particular market may allow an insight into the existence of a system market or of multiple markets.[8] However, this may also simply indicate that two goods are being bundled.

In its 2005 Discussion Paper, the European Commission emphasised that in order to delineate a secondary market, one must focus on the customers that have already acquired the primary product.[9] Normally, the Commission expects the secondary market to consist of one brand's products. There are two exceptions to this: (1) if the customer of the primary product can switch to another producer's secondary products, then no brand-specific secondary market will be found – here, a broader market delineation is required; and (2) cases in which the customer can

[3] European Commission, Notice on the definition of relevant market for the purposes of Community competition law [1997] OJ C372/5 (EU Market Definition Notice 1997) para 56.

[4] ibid.

[5] Office of Fair Trading, 'Market Definition' (OFT 403, December 2004) § 6.2.

[6] ibid §§ 6.3 f. For a detailed analysis of the arguments leading to the finding (or not) of a system market, based on a comparative legal analysis, see RS Gohari, *Verweigerung von Geschäftsbeziehungen – Kartellrechtliche Analyse nach Schweizer, EU- und US-Recht* (Bern, Stämpfli, 2017) 33–51.

[7] Temple Lang, 'Aftermarkets' (n 1) 201.

[8] ibid 205 f.

[9] European Commission, 'DG Competition Discussion Paper on the Application of Article 82 of the Treaty to Exclusionary Abuses' (December 2005) para 247.

switch to another primary product altogether and in this way dodge the overpriced secondary market. This is only possible where switching costs are not prohibitively high. Where no standalone aftermarket can be established, the competition law assessment must be carried out within a more comprehensive system market.[10] The Discussion Paper appeared to adopt the view that one could establish dominance in a secondary market even if the primary market was competitive.

The Commission's 2010 Vertical Guidelines provide further insight into that competition authority's views on market definition in aftermarkets. When discussing which market to base the 30 per cent market share threshold on in the presence of aftermarkets, the Commission states that this 'may be the original equipment market including the spare parts or a separate original equipment market and after-market depending on the circumstances of the case'.[11] As relevant factors, the Commission cites the competition restrictions at issue, the lifetime of the equipment and the magnitude of the repair or replacement costs. The Commission will in particular look at whether a substantial number of customers take lifecycle costs into account when buying a primary product, as this may indicate that there is a system market encompassing the primary and secondary markets.[12] The latter depends on a number of factors, such as the customers' sophistication, the expected life of the equipment or switching costs.[13]

The OECD addressed the issue of market definition in aftermarkets in a 2012 Policy Roundtable. It was stressed that there are two different kinds of aftermarkets: those that are open and those that are closed. In an open aftermarket, the secondary market does not underlie any innate constraints related to the primary market. The classic example would be a diesel car that can tank diesel at any petrol station in the country, not only at petrol stations run by the specific car manufacturer. Under such circumstances, the supplier of the primary product (here the car) usually has no market power over the secondary market (here the diesel). These two markets therefore need to be regarded as separate or – in the terminology relied upon by the OFT – dual markets.[14] They will usually not attract antitrust scrutiny for aftermarket-related competition issues. In a closed aftermarket, on the other hand, the secondary market underlies constraints related to the primary market. As mentioned above, these constraints can stem from IP protection or from contractual obligations. An example is a spare part protected by design rights that only the IP owner can (legally) produce. In such a case, the system market approach allows us to analyse the relationship between the primary and

[10] ibid paras 248 f. Contrary to the Commission's Guidance Paper of 2009, the Discussion Paper does not have enforcement status. The former no longer refers to aftermarkets; see European Commission, Guidance on the Commission's enforcement priorities in applying Article 82 of the Treaty to abusive exclusionary conduct by dominant undertakings [2009] OJ C45/7 (Guidance Paper).

[11] European Commission, Guidelines on Vertical Restraints [2010] OJ C130/1, para 91.

[12] ibid.

[13] Temple Lang, 'Aftermarkets' (n 1) 203.

[14] U Schwalbe and F Maier-Rigaud, 'Background Note' in OECD (ed), *Policy Roundtable: Market Definition* (2012) DAF/COMP(2012)19, 46.

the secondary market, and how competition on the primary market may influence competition or market power on the secondary market. The following section is devoted to this question.

II. The Relationship between Primary and Secondary Markets

An issue that frequently surfaces where aftermarkets are concerned is the question whether a company needs to have market power in the primary market in order to have market power in a narrowly defined secondary market or – to put it another way – whether lively competition on the primary market may constrain anti-competitive conduct on the secondary market. While at first glance this question seems to relate to market power more than to market definition, it is also of interest for the latter. If it is assumed that a competitive primary market constrains the secondary market(s), then one might need to delineate a system market rather than multiple antitrust markets. This also has repercussions on the substantive antitrust analysis of that secondary market, the scope of which would equally need to be broadened.

A. Primary and Secondary Markets in the Case Law

The US Supreme Court's *Kodak* judgment of 1992 is the leading US case on aftermarkets and their relationship with the primary market.[15] In this landmark case, the Supreme Court had to decide whether, 'as a matter of law', 'a defendant's lack of market power in the primary equipment market precludes … the possibility of market power in derivative aftermarkets'.[16] What was at issue in this case were Kodak's photocopiers and micrographic equipment, relating to which Kodak also offered servicing and replacement parts. Eighteen independent service organisations (ISOs) sued Kodak because the latter had started adopting policies that resulted in replacement parts only being sold to customers who either used Kodak's own servicing or repaired their own machines. This severely limited the availability of replacement parts to ISOs, thus making it harder for them to compete with Kodak's servicing activities. The ISOs argued that Kodak had violated section 1 of the Sherman Act by tying its servicing to the sale of replacement parts and section 2 of the Sherman Act by monopolising and attempting to monopolise the

[15] For an early US case on the relationship between the primary and secondary markets, see *Jefferson Parish*, where the Supreme Court ruled that competition on the primary market (for hospitals, the tying market) meant that there would be no competitive concerns on the secondary market (anesthesiology services, the tied market): *Jefferson Parish Hospital District v Hyde*, 466 US 2 (1984).

[16] *Eastman Kodak v Image Technical Services*, 504 US 451, 455 (1992).

servicing of Kodak equipment. Kodak's market share in the servicing of Kodak equipment was about 80–95 per cent. Kodak urged the Supreme Court to adopt a legal rule that should say that 'equipment competition precludes any finding of monopoly power in derivative aftermarkets'.[17] However, the Supreme Court stated that no such legal rule could be adopted. Instead, one had to determine whether customers adopted lifecycle pricing and whether this constrained market power in the aftermarket(s).[18] This assessment had to be carried out based on 'actual market realities'.[19]

The *Kodak* Court found that it was perfectly possible to have a competitive primary market, yet secondary markets in which a company has substantial market power.[20] This separate regard for aftermarkets was reaffirmed in *Virtual*.[21] From the point of view of behavioural antitrust, this judgment relies on the insight that consumers often do not act rationally (ie, adopt lifecycle pricing), even if they were in a position to do so.[22] Factors to which the *Kodak* Court pointed in order to make this assessment were information and switching costs, lifecycle pricing, and how these affect market power in the aftermarket. It carefully set out how customers often do not engage in lifecycle pricing even where it is theoretically possible, due to the information costs (including time) that are required for such an exercise.[23] In the Court's opinion, 'there is a question of fact whether information costs and switching costs foil the simple assumption that the equipment and service markets act as pure complements to one another'.[24] These criteria are quite closely aligned with what became known as the *EFIM* criteria in Europe (see below).

In *Kodak*, the tie-in between replacement parts and servicing represented a policy change which Kodak implemented. Some commentators held that it was only Kodak's policy change after customers had purchased the equipment that gave rise to the anti-competitive tie-in.[25] If Kodak had made its tie-in known before the purchase, customers could have engaged in lifecycle pricing,[26] calling for a system market rather than multiple markets to be delineated. However, *Kodak* does not

[17] ibid 466.

[18] ibid 473–77.

[19] ibid 466.

[20] ibid 471.

[21] *Virtual Maintenance v Prime Computer*, 957 F2d 1318 (6th Cir 1992); *Virtual Maintenance v Prime Computer*, 506 US 910 (1992); *Virtual Maintenance v Prime Computer*, 995 F2d 1324 (6th Cir 1993).

[22] ME Stucke, 'Behavioral Antitrust and Monopolization' (2012) 8 *Journal of Competition Law & Economics* 545, 567; A Tor, 'The Market, the Firm, and Behavioral Antitrust' in E Zamîr and D Teichman (eds), *The Oxford Handbook of Behavioral Economics and the Law* (Oxford, Oxford University Press, 2014) 551.

[23] *Eastman Kodak v Image Technical Services*, 504 US 451, 473 (1992).

[24] ibid 477.

[25] In this vein, see AH Silberman, 'The Myths of Franchise "Market Power"' (1996) 65 *Antitrust Law Journal* 181, 184; B Klein, 'Market Power in Franchise Cases in the Wake of *Kodak*: Applying Post-contract Hold-Up Analysis to Vertical Relationships' (1999) 67 *Antitrust Law Journal* 283, 299 ff. See also *Digital Equipment v Uniq Digital Technologies*, 73 F3d 756, 763 (7th Cir 1996).

[26] *Digital Equipment v Uniq Digital Technologies*, 73 F3d 756, 763 (7th Cir 1996).

need to be interpreted as necessitating a post-contractual policy change in order to delineate multiple markets rather than a system market.[27] The 'actual market realities'[28] that *Kodak* stresses appear to favour an approach more akin to the multiple markets approach, thus including findings from behavioural economics and acknowledging that parties are not as rational as they perhaps could or should be.

A long line of cases following *Kodak* applied and interpreted this landmark judgment, regularly concluding that a brand-specific aftermarket could not be established where the seller of the primary product had always relied on a policy of tying its aftermarket sales to its primary products, and customers were therefore encouraged to engage in lifecycle pricing. This, in the view of many Courts of Appeals, meant that the secondary market was sufficiently restricted by competition in the primary market.[29]

In the EU, the influence of the primary market's competitiveness on the aftermarket has been litigated on several occasions. Both in *Pelikan/Kyocera* and *Info-Lab/Ricoh*, the European Commission addressed the legal question of whether an undertaking could be dominant on the secondary market if it was not dominant on the primary market.[30] It established four criteria that, if fulfilled in their entirety, showed that functioning competition on the primary market had a disciplining effect on the secondary market:

> [A] customer (i) *can* make an informed choice including lifecycle-pricing … he (ii) is *likely* to make such choice accordingly, and that, in case of an apparent policy of exploitation being pursued in one specific aftermarket, a (iii) *sufficient number* of customers would adapt their purchasing behaviour at the level of the primary market (iv) within a *reasonable time*.[31]

In these two cases, the Commission initially delineated very narrow secondary markets, but then found that the incumbent was not dominant on these markets based on the disciplining effect of the competitive primary market, thus essentially finding for a system market rather than multiple markets.

In *EFIM*, the EU Courts approved of this decisional practice.[32] In that case, the European Federation of Ink and Ink Cartridge Manufacturers (EFIM) challenged the aftermarket practices of Hewlett Packard, Lexmark, Epson and Canon regarding ink cartridges. EFIM wanted access to this aftermarket and requested

[27] WS Grimes, 'Market Definition in Franchise Antitrust Claims: Relational Market Power and the Franchisor's Conflict of Interest' (1999) 67 *Antitrust Law Journal* 243, 258.

[28] *Eastman Kodak v Image Technical Services*, 504 US 451, 466 (1992).

[29] On cases in this regard, see *PSI Repair Services v Honeywell*, 104 F3d 811 (6th Cir 1997); *Queen City Pizza v Domino's Pizza*, 124 F3d 430 (3d Cir 1997); *Alcatel USA v DGI Technologies*, 166 F3d 772 (5th Cir 1999); *SMS Systems Maintenance Services v Digital Equipment Corporation*, 188 F3d 11 (1st Cir 1999); *DSM Desotech v 3D Systems*, 749 F3d 1332 (Fed Cir 2014).

[30] *Pelikan/Kyocera* (Case IV/34.330) Commission Decision of 22 September 1995, para 58; *Info-Lab/Ricoh* (Case IV/E-2/36.431) Commission Decision of 7 January 1999, para 35. See also Case C-56/12 P *EFIM v Commission* EU:C:2013:575, para 13.

[31] *Pelikan/Kyocera* (Case IV/34.330) Commission Decision of 22 September 1995, para 61 (emphasis in original).

[32] Case T-296/09 *EFIM v Commission* EU:T:2011:693; Case C 56/12 P *EFIM v Commission* EU:C:2013:575.

information on the IPRs relating to these markets or licensing of these IPRs.[33] Relying on *Pelikan/Kyocera* and *Info-Lab/Ricoh*, amongst others, the Commission underlined that there was a possibility that the markets for ink cartridges compatible with the printer of a certain brand each represented their own product markets. However, where certain criteria are fulfilled – such as in the present case – one can assume a disciplining effect of the primary market on the secondary market.[34] Upon appeal, the Court of Justice approved the Commission's established decisional practice on these four criteria.[35]

Importantly, a merely theoretical possibility of switching to another primary product is decidedly not enough in the EU.[36] Rather, as the General Court explained in *Swiss Watchmakers*, if the Commission wants to define a system market consisting of the primary products (luxury watches) and their aftermarkets (repair and maintenance services, spare parts), then it has to demonstrate 'that a sufficient number of consumers would switch to other primary products if there were a moderate price increase for the products or services on the after markets and thus render such an increase unprofitable'.[37] This is again in line with the insight of behavioural economics,[38] as it then becomes decisive whether customers actually act on the information they are given, as the second *EFIM* criterion requires.

B. Analysis

The US and EU approaches to aftermarkets and their relationship with the primary markets are not quite aligned, although there certainly appears to be some convergence. In particular, there is growing awareness of the importance of actual consumer behaviour in aftermarkets, as opposed to assumed rational behaviour. With the *Kodak* judgment pre-dating the *EFIM* judgment by over 20 years, it is hardly surprising that the former is less explicit in its behavioural approach. An additional hurdle to convergence lies in the lower US courts, which have consistently interpreted various parts of *Kodak* in a less behavioural way.

Behavioural economics demonstrates that while customers might be aware of the higher price of certain products on the aftermarket, the cheaper price of the primary product may nevertheless be able to lure them into buying it. Few customers will actually engage in lifecycle costing.[39] The CJEU's *EFIM* criteria seem to recognise this, as they place considerable importance on consumers'

[33] *EFIM* (Case COMP/C-3/39.391) Commission Decision of 20 May 2009, paras 1, 3.
[34] ibid para 25.
[35] Case C 56/12 P *EFIM v Commission* EU:C:2013:575, paras 12, 37–39.
[36] Case T-427/08 *CEAHR v Commission (Swiss Watchmakers)* EU:T:2010:517, para 102.
[37] ibid para 105, referring to Case T-30/89 *Hilti v Commission* EU:T:1991:70, para 75.
[38] See already above, text accompanying n 22.
[39] Tor (n 22) 551 f.

actual behaviour rather than mere theoretical possibilities. This was reaffirmed by the General Court in *Swiss Watchmakers*.[40] The *EFIM* criteria provide an elaborate legal test in order to ascertain the primary market's constraining effect on the secondary market and are in this respect more detailed than the *Kodak* judgment.

One might argue that, as with any dynamic market, where there is high innovation in the primary market, competition authorities do not need to be overly concerned with market power in the aftermarket, even if that aftermarket is protected by IPRs. Once a company's market power in the primary market is eroded based on new innovations, its market power in the proprietary aftermarket will automatically vanish – even if there will be a time lag because locked-in consumers will need to continue buying consumables or spare parts for the primary product until the latter has concluded its lifecycle. However, innovation and its timing are highly uncertain, as discussed above (see section II.A in Chapter 3). Therefore, competition law may be significantly under-enforced if aftermarkets for innovative products were not subjected to antitrust scrutiny. On the other hand, the prospect of reaping substantial profits from the lucrative aftermarkets of an innovation may constitute an (additional) incentive for engaging in innovative activities in the first place. However, even if this were the case, it is out of the question to reserve non-proprietary aftermarkets to an innovator as this would result in a considerable loss of competition, to the detriment of consumers.

In economics, several papers have argued that irrespective of the competitiveness of the primary market, companies will tend to use any market power they have over their aftermarkets in order to raise prices above the competitive level.[41] Together with the findings from behavioural economics already alluded to throughout, this gives a solid backing to the multiple markets approach. While this does not preclude the finding of a system market, it does entail that a closer analysis of the factors relevant in each jurisdiction needs to be undertaken rather than rejecting such an approach offhand.

III. Proprietary Primary Markets and Proprietary Aftermarkets

Proprietary aftermarkets are secondary markets the access to which is protected by an IPR. While Chapter 6 considered whether IPRs can be construed as relevant antitrust markets in their own right and the extent to which they determine the delineation of a relevant market in which IPRs play a decisive role, the question in

[40] On these cases, see section II.A above.

[41] S Borenstein, JK MacKie-Mason and JS Netz, 'Exercising Market Power in Proprietary Aftermarkets' (2000) 9 *Journal of Economics & Management Strategy* 157, 184; NW Hawker, 'Automotive Aftermarkets: A Case Study in Systems Competition' (2011) 56 *Antitrust Bulletin* 57, 74 (containing further references).

the context of proprietary aftermarkets is to what extent these constitute separate antitrust markets. In particular, the question is whether they need to be construed as a system market together with the respective primary market. Another question is whether the fact that an aftermarket is protected by an IPR – rather than, for example, by contractual obligations – is of any relevance. A further scenario to be considered is the case in which IPRs play an important role in a primary market, while the aftermarkets are not IP-protected. Under such circumstances, a system market approach to market definition would attract heightened antitrust scrutiny for companies.

A. Proprietary Aftermarkets in the Case Law

The EU's Court of Justice traditionally relies on narrow market definitions for proprietary aftermarkets, regularly delineating them as company or brand markets. In *Hugin* (1979), the relevant product market was delineated as the market for spare parts for Hugin cash registers that were required by independent undertakings.[42] The reason for this narrow market definition was that the Commission found the replacement parts for this particular brand of cash registers not to be interchangeable with replacement parts for any other cash registers, as well as not being economically reproducible.[43] The Court of Justice agreed.[44] The Commission's reasoning was that Hugin controlled the supply of its spare parts via IPRs.[45] Hugin held a monopoly position on the market for new replacement parts and a dominant position on the overall market for Hugin replacement parts.[46] The Court did not make this finding dependent on whether or not Hugin was dominant on the primary market, ie, the market for cash registers.[47]

In *Volvo v Veng* (1988), the English High Court of Justice sent a preliminary reference to the Court of Justice. The case revolved around a refusal to license by the owner of a registered design and its possible liability under Article 102 TFEU.[48] The referring court asked whether a car manufacturer's ownership of registered designs relating to replacement parts conferred a dominant position in replacement parts on that manufacturer.[49] While the Court of Justice did not answer this question directly,[50] its analysis in regard to the abuse itself indicates that it did not find it far-fetched that a car manufacturer could derive a dominant position on

[42] Case 22/78 *Hugin v Commission* EU:C:1979:138, para 8.
[43] ibid 1874.
[44] ibid para 7.
[45] See P Andrews, 'Aftermarket Power in the Computer Services Market: The Digital Undertaking' (1998) 19 *European Competition Law Review* 176, 177.
[46] Case 22/78 *Hugin v Commission* EU:C:1979:138, paras 9 f.
[47] Temple Lang, 'Aftermarkets' (n 1) 207.
[48] Case 238/87 *Volvo v Veng* EU:C:1988:477, para 1.
[49] ibid para 4.
[50] ibid para 10.

the market for replacement parts from IPRs over the design of those replacement parts. Similarly, in the *Maxicar v Renault* preliminary ruling of that same year, industrial property protection for the design of spare parts for cars was at issue.[51] Here, the Court of Justice ruled that the exercise of those design rights might contravene Article 102 TFEU if a company was dominant, but did not further explore the issue of market definition.[52]

Spare parts for cars, and the repair and maintenance services related to them, have also been discussed in the US. It has been claimed that car manufacturers, by relying on IPRs, have successfully kept third-party competitors out of these aftermarkets.[53] However, third parties have tried to gain access to these markets through means other than antitrust law suits. For instance, in *Aro v Convertible Top* (1961), an independent repairer managed to counter a patent infringement claim for a patent relating to the construction of a convertible top, with the US Supreme Court finding that Aro could replace the cloth top, which formed part of the convertible top, without infringing the patent.[54]

The US case of *Eastman Kodak* (1992) concerned a tying case in which one aftermarket (replacement parts for Kodak's photocopiers and similar machines) was to some extent proprietary,[55] while the other (servicing of Kodak equipment) was not.[56] The primary market, Kodak's photocopiers and similar machines, was also proprietary. This landmark case has already been discussed above (see section II.A).

The Supreme Court's *Kodak* judgment was reaffirmed in the case of *Virtual v Prime* (1992).[57] Prime, a manufacturer of minicomputers, relied on an IP-protected architecture. The aftermarkets concerned were hardware maintenance on the one hand and software support including software upgrades on the other hand, with the software upgrades again being IP-protected. Virtual wanted to compete in the hardware maintenance market, but Prime's exclusivity contracts prevented it from doing so. Prime's customers could only obtain software support including

[51] Case 53/87 *CICRA & Maxicar v Renault* EU:C:1988:472, para 4.

[52] ibid para 18(ii).

[53] Hawker (n 41) 58.

[54] *Aro v Convertible Top*, 365 US 336 (1961). Right to repair legislation was enacted in Massachusetts in 2012, upon which the car industry reached an agreement with independent repairers on making these rules binding in all 50 US states; see G Nelson, 'Automakers Agree to "Right to Repair" Deal' *Automotive News* (25 January 2014), www.autonews.com/article/20140125/RETAIL05/301279936/automakers-agree-to-right-to-repair-deal. In the EU, the Commission has tried to make access to original equipment manufacturer spare parts easier for independent service organisations by including a respective hardcore restriction in the Motor Vehicle Block Exemption Regulation; Commission Regulation (EU) 461/2010 on the application of Article 101(3) of the Treaty on the Functioning of the European Union to categories of vertical agreements and concerted practices in the motor vehicle sector [2010] OJ L129/52, art 5.

[55] Indeed, only a minority of Kodak's replacement parts was patented; see Bauer (n 2) 50.

[56] *Eastman Kodak v Image Technical Services*, 504 US 451 (1992).

[57] *Virtual Maintenance v Prime Computer*, 506 US 910 (1992). For the original judgment, see *Virtual Maintenance v Prime Computer*, 957 F2d 1318 (6th Cir 1992).

upgrades if they agreed to also source their hardware maintenance from Prime.[58] By tying the IP-protected secondary product to the non-proprietary secondary product, Prime extended its market power to hardware maintenance. Based on the Supreme Court's reasoning in *Kodak*, the Court of Appeals for the Sixth Circuit reversed its first verdict. In its amended opinion (1993), it agreed with Virtual that one could delineate a proprietary tying market for 'the sale of software revisions and support of software necessary to do business with [major client] Ford Motor Company', adding that this market definition not only included Ford, but also encompassed all Ford suppliers which were affected by Ford's initial adoption of Prime's product.[59]

The European Commission's commitment decision in *Digital* (1997)[60] is reminiscent of *Virtual v Prime*, for the antitrust investigation revolved around the servicing aftermarkets for proprietary Digital computer systems.[61] Like in *Virtual*, third-party maintenance companies had brought the complaint.[62] The two aftermarkets at issue were again hardware maintenance on the one hand and software support, which included proprietary software updates, on the other. Both of these aftermarkets only related to computers of the Digital brand. These were thus single-brand secondary markets. Through contractual tying and its pricing strategy, Digital had managed to tie together software support and updates as the tying product with hardware maintenance as the tied product.[63] The narrow market delineation was also based on the fact that Digital had successfully implemented its contractual tying.[64] The complainants wanted access to the tied market, and the investigation concluded with Digital's commitment to offer third parties access to the hardware maintenance market for Digital computer systems.[65] A case against Digital was also litigated in the US, although in that case it was a warranty for the primary product that, an ISO alleged, extended Digital's market power into the aftermarket for services in an anti-competitive way.[66]

In *Hilti*, the undertaking concerned wanted to rely on a system market for powder-actuated fastening systems, which would include its nail guns, cartridge strips and nails. However, the European Commission found three separate

[58] For the facts of the case, see *Virtual Maintenance v Prime Computer*, 995 F2d 1324 (6th Cir 1993).

[59] ibid 1328.

[60] European Commission, 'The European Commission Accepts an Undertaking from Digital Concerning its Supply and Pricing Practices in the Field of Computer Maintenance Services' (IP/97/868, 10 October 1997); V Pickering and M Dolmans, 'The 1997 Digital Undertaking' (1998) 19 *European Competition Law Review* 108, 112–15.

[61] On further competition law issues investigated in this case, see Pickering and Dolmans (n 60) 108 f.

[62] Andrews (n 45) 176; Pickering and Dolmans (n 60) 109. The American *Virtual* case was privately litigated, while the European *Digital* case triggered public enforcement action by the Commission.

[63] Andrews (n 45) 179; Pickering and Dolmans (n 60) 109.

[64] See European Commission, 'Digital Undertaking' (n 60); European Commission, 'XXVIIth Report on Competition Policy 1997' (Luxembourg, 1998) para 69.

[65] European Commission, 'Digital Undertaking' (n 60).

[66] *SMS Systems Maintenance Services v Digital Equipment Corporation*, 188 F3d 11 (1st Cir 1999).

relevant markets, namely the market for nail guns, the market for Hilti-compatible cartridge strips and the market for Hilti-compatible nails. Although these markets were undeniably connected, they were subject to different conditions of supply and demand. Hilti held that several patents related to these markets. Despite this, some manufacturers had managed to produce nails for use in Hilti nail guns.[67] On the (primary) market for nail guns, Hilti had a market share of about 55 per cent, which suggested dominance under the European standard, while its market shares for Hilti-compatible accessories was higher (about 70–80 per cent for nails).[68] The Court attached importance to the fact that these were proprietary aftermarkets, underlining that Hilti's patent and claimed copyright protection on its cartridge strips strengthened its dominant position. On the aftermarket's relationship with the primary market, it held that Hilti's patents reading on its DX 450 nail gun, which formed part of the primary market, additionally reinforced its dominant position in the aftermarkets.[69]

The European *Microsoft* case (2007) again showed that the Commission was sensitive to a dominant undertaking's capturing of its aftermarket with the help of its highly successful, proprietary primary market.[70] One of the aspects litigated in this case was the inclusion of the Windows Media Player – a proprietary software – in the highly successful Windows operating software (OS).[71] In the *Microsoft (Tying)* commitment decision of 2009, it was the inclusion of Internet Explorer in the Windows OS that led to competitive concerns by the European Commission.[72] In a similar US case, *Digital v Uniq* (1996), the US Court of Appeals for the Seventh Circuit found that the inclusion of an operating system in every Digital computer did not infringe the antitrust laws.[73]

B. Non-proprietary Aftermarkets Related to Proprietary Primary Markets

Early US case law only provides limited guidance on market power in non-proprietary aftermarkets related to proprietary primary markets, due to the 2006

[67] Nevertheless, supply-side substitutability is notably absent from the market definition in *Hilti*; see TE Kauper, 'The Problem of Market Definition under EC Competition Law' (1997) 20 *Fordham International Law Journal* 1682, 1713.

[68] Case T-30/89 *Hilti v Commission* EU:T:1991:70, paras 14, 18, 66, 89; affirmed in Case C-53/92 P *Hilti v Commission* EU:C:1994:77.

[69] Case T-30/89 *Hilti v Commission* EU:T:1991:70, para 93.

[70] SD Anderman, 'Innovation, IPRs and EU Competition Law' in I Lianos and D Geradin (eds), *Handbook on European Competition Law* (Cheltenham, Edward Elgar, 2013) 577.

[71] Case T-201/04 *Microsoft v Commission* EU:T:2007:289, paras 43–45.

[72] *Microsoft (Tying)* (Case COMP/39.530) Commission Decision of 16 December 2009 [2010] OJ C36/7, para 32. In 2013, Microsoft was fined for not complying with the commitments it had offered; see *Microsoft (Tying)* (Case COMP/39.530) Commission Decision of 6 March 2013 [2013] OJ C120/15.

[73] *Digital Equipment v Uniq Digital Technologies*, 73 F3d 756 (7th Cir 1996).

case of *Independent Ink* discussed below.[74] Many of these early cases arose in the context of tying, where the market power related to a proprietary primary market was extended to a non-proprietary aftermarket. In *Morton Salt* (1942), a company held a patent in machines for depositing salt tablets and used exclusivity agreements in order to increase its sales of (unpatented) salt tablets.[75] The US Supreme Court ruled that this extension of the company's patent monopoly into an unpatented area of business was not within the scope of protection granted by the IP laws.[76] Without deciding on a possible violation of the Clayton Act, the Court did not grant the applicant an injunction as it was relying on its patent in a way that contradicted public policy.[77] This case was therefore resolved by exclusively relying on the IP laws. Five years later, in *International Salt* (1947), the Supreme Court restated this reasoning and held that the tying of an unpatented market to a patented market constituted a per se violation of section 1 of the Sherman Act and section 3 of the Clayton Act.[78] While there was no discussion of the relevant market in this early case, the presumption that there could be market power in the patented tying market which could be extended to the non-proprietary aftermarket shows that the primary market and the aftermarket were, in fact, regarded as two separate relevant antitrust markets. The non-proprietary nature of the aftermarket was not discussed.

The European Commission has repeatedly narrowly delineated aftermarkets consisting of consumables that related to a proprietary primary market along the lines of a company's product. In *Pelikan/Kyocera* (1995), the Commission concluded that the supply of toners and other consumables for printers of one specific brand had to be delineated as their own product markets, as they were not interchangeable with those of other brands. However, Kyocera was found not to have a dominant position on the secondary market, as effective competition on the primary market – based on the four *EFIM* criteria – had a disciplining effect on the secondary market.[79] In the end, therefore, the Commission implicitly concluded that there was a system market rather than two separate markets.[80]

In *Tetra Pak II* (1996), aftermarkets were again narrowly construed in the EU, this time in relation to proprietary carton machinery and non-proprietary cartons.[81] In another Commission decision, *Info-Lab/Ricoh* (1999), Info-Lab alleged that the relevant product market should be defined as the market for empty toner cartridges compatible with Ricoh copying machines that are meant to be

[74] For additional early case law, see *Standard Oil v United States*, 337 US 293 (1949); *Continental TV v GTE Sylvania*, 433 US 36 (1977); *Jefferson Parish Hospital District v Hyde*, 466 US 2 (1984).

[75] *Morton Salt v Suppiger*, 314 US 488, 489 f (1942).

[76] ibid 491 f.

[77] ibid 492.

[78] *International Salt v United States*, 332 US 392, 395 f (1947).

[79] *Pelikan/Kyocera* (Case IV/34.330) Commission Decision of 22 September 1995, paras 54, 61.

[80] Temple Lang, 'Aftermarkets' (n 1) 202.

[81] Case T-83/91 *Tetra Pak v Commission* EU:T:1994:246, paras 43 ff, upheld on appeal in Case C-333/94 P *Tetra Pak v Commission (Tetra Pak II)* EU:C:1996:436.

filled with toner and sold to final customers.[82] The Commission did not accept this market definition, as Ricoh – the holder of design rights – was the only company able to produce the toner cartridges at issue. Ricoh did not license these rights, nor did it sell empty cartridges to any third party. Instead, the Commission found that the relevant product market was the market for Ricoh-compatible (filled) toner cartridges. It held that the market for photocopier machines was not relevant in this case and did not have to be precisely delineated. In any case, Ricoh did not have a dominant position on that market. Thereupon, the Commission applied the four criteria from *Pelikan/Kyocera* in order to establish whether the secondary market was disciplined by the primary market. As this was found to be the case, the Commission held that Ricoh had no dominant position on the secondary market, despite it being the only manufacturer on that narrowly defined market.[83] By rejecting a multiple market approach in this case, the Commission assumed that a system market was at issue, just as in *Pelikan/Kyocera*.

Replacement ink for proprietary toner cartridges has also been reviewed under US antitrust law. *Illinois Tool Works v Independent Ink* (2006) was a significant case for US antitrust law as it established that a patent does not as such bestow market power on the patentee; rather, market power also had to be proved in relation to IP-protected markets. The case revolved around Illinois Tool Works' printing systems, which consisted of patented and unpatented parts: a patented print head and ink container, and unpatented ink. Through exclusivity agreements, the company tied its unpatented ink to its patented machinery. Independent Ink wanted access to this tied market.[84] In deciding the case, the Supreme Court noted that its 'strong disapproval of tying arrangements has substantially diminished' over the years.[85] It overturned its previous case law, which had equated the possession of a patent with the possession of market power, and stated that the parties should be allowed to bring forward a proper market definition as well as evidence relating to market power before the District Court.[86]

In the *Systems Maintenance Services v Digital* case (1999), the US Court of Appeals for the First Circuit had to decide whether Digital had monopolised the services aftermarket for Digital computers 'by integrating three-year warranties with sales of computer systems'.[87] Systems Maintenance Services was an ISO specialising in servicing Digital computers. It maintained that the relevant antitrust market was the aftermarket for servicing Digital's mid-range servers. The Court, however, underlined the particular nature of aftermarkets cases and detailed how aftermarket reputation might affect the primary market.[88]

[82] *Info-Lab/Ricoh* (Case IV/E-2/36.431) Commission Decision of 7 January 1999, para 13.
[83] ibid paras 14, 19–22, 31, 36 ff, 47.
[84] *Illinois Tool Works and Others v Independent Ink*, 547 US 28, 28, 31, 43 (2006).
[85] ibid 28.
[86] ibid 46.
[87] *SMS Systems Maintenance Services v Digital Equipment Corporation*, 188 F3d 11, 13 (1st Cir 1999).
[88] ibid 14, 16 f.

It asserted that 'the aftermarket is the relevant market for antitrust analysis only if the evidence supports an inference of monopoly power in the aftermarket that competition in the primary market appears unable to check'.[89] As Digital's customers knew of the warranty, there was no policy change, thus distinguishing this case from *Kodak*.[90] The Eighth Circuit reached a similar conclusion in *Marts v Xerox* (1996), a warranty case involving Xerox's warranty that was conditional upon the use of original Xerox cartridges.[91]

In *Alcatel USA v DGI Technologies* (1999), a jury had found DSC (later Alcatel) to be 'liable under § 2 of the Sherman Act for monopolization of the expansion and enhancement market for DSC-manufactured switches'.[92] Upon appeal, the US Court of Appeals for the Fifth Circuit found that as customers were aware of the cost of the aftermarket good at the time of buying the primary product and many of them engaged in lifecycle pricing, the aftermarket did not constitute a separate relevant antitrust market.[93] The Court thereby agreed with the Sixth Circuit, which had held in *PSI Repair Services v Honeywell* (1997) that 'an antitrust plaintiff cannot succeed on a *Kodak*-type theory when the defendant has not changed its policy after locking-in some of its customers, and the defendant has been otherwise forthcoming about its pricing structure and service policies'.[94] The Court also pointed out that there were alternatives available for the expansion cards that DGI relied upon as the relevant aftermarket.[95]

In the *Swiss Watchmakers* case in 2010, the General Court was confronted with the Commission's view that the repair and maintenance services market for luxury watches and the spare parts market for luxury watches were not separate markets, but had to be seen as forming a system market together with the primary market, ie, luxury watches.[96] However, the Court suggested that, based on the available evidence, the Commission might just as well have found that 'separate brand-specific markets for spare parts existed as a function of their substitutability'.[97] The presence of important manufacturers specialising in the production of spare parts – rather than complete watches – pointed to the existence of a separate relevant market for spare parts. On the repair and maintenance services concerned, the Court found that the available evidence did not allow the Commission to

[89] ibid 17.
[90] ibid 19.
[91] *Marts v Xerox*, 7 F3d 1109, 1112 (8th Cir 1996).
[92] *Alcatel USA v DGI Technologies*, 166 F3d 772, 780 (5th Cir 1999).
[93] ibid 783.
[94] *PSI Repair Services v Honeywell*, 104 F3d 811, 820 (6th Cir 1997); *PSI Repair Services v Honeywell*, 520 US 1265 (1997) (certiorari denied); explicitly relied on in *Alcatel USA v DGI Technologies*, 166 F3d 772, 783 (5th Cir 1999).
[95] Telephone switching systems were the primary market in this case. The licensing agreement for the operating software for these systems had tied the software to the equipment. DGI wanted to enter the market for cards that expand a telephone network's handling capacity. The main aspects of this case were IP-related, and it may indeed be argued that the aftermarket was also proprietary; *Alcatel USA v DGI Technologies*, 166 F3d 772, 777 ff (5th Cir 1999).
[96] Case T-427/08 *CEAHR v Commission (Swiss Watchmakers)* EU:T:2010:517, paras 53 f.
[97] ibid para 109.

conclude that the market for repair and maintenance services for luxury watches did not constitute its own relevant market. In particular, it was not shown how a price increase in any aftermarket would affect demand in the primary market. The Commission decision of 2008 was thus annulled.[98] It can safely be assumed that the watches constituting the primary markets were protected by design rights, making this a proprietary primary market.

Following the annulment of the first *Swiss Watchmakers* decision in December 2010, the Commission had to re-open the proceedings. It again rejected the complaint brought before it in July 2014. It reanalysed the relevant market in the light of the General Court's judgment, finding that there was a primary market for prestige watches, and aftermarkets for repair and maintenance services and for the supply of spare parts.[99] Each of these aftermarkets, according to the Commission, was 'associated, in principle, with a particular watch brand'.[100]

In 2011, the Commission accepted commitments from IBM to counter its concerns regarding the market for IBM mainframe computers and the aftermarkets for IBM inputs and IBM maintenance services for IBM mainframes.[101] The concerns related to a possible abuse of a dominant position on that market, in the form of unreasonable terms vis-à-vis ISOs. Based on the commitments, ISOs would receive technical information to enable them to carry out maintenance on IBM mainframe systems and be promptly delivered spare parts at the same price as self-maintainers.

In *DSM Desotech v 3D Systems* (2014), Desotech was a manufacturer of resins used in stereolithography (SL), a specific form of 3D printing, while 3D Systems made and sold SL printing machines as well as resins. Desotech sued 3D Systems for five counts of antitrust infringement under the Sherman and Clayton Acts and state antitrust law, alleging that the technological lock that 3D Systems had installed on its machines prevented customers from using resins not approved by 3D Systems.[102] This technological lock consisted of wireless technology that automatically switched the machine off if non-approved resin was detected. The US Court of Appeals for the Federal Circuit held that Desotech had not succeeded in showing that SL machines were their own antitrust market, as SL printing was substitutable with other rapid-prototyping technologies.[103] The second product

[98] ibid paras 107 f, 116, 119, 178.

[99] *Watch Repair* (Case AT.39097) Commission Decision of 29 July 2014, paras 80 ff ('prestige watches' are such that it is worth repairing or maintaining them; see para 81).

[100] ibid para 96.

[101] As the Commission discussed: 'Whether the aftermarkets for IBM's inputs and maintenance services for mainframes constitute separate product markets or are part of a single market for "systems", mainly depends on the likely reaction of customers to moderate price increases in the aftermarkets.' See *IBM Maintenance Services* (Case COMP/C-3/39.692) Commission Decision of 13 December 2011 [2012] OJ C18/6, para 21 (footnote in original omitted).

[102] *DSM Desotech v 3D Systems*, 749 F3d 1332, 1336 f (Fed Cir 2014).

[103] ibid 1337, 1344 f. Desotech had particularly relied on four of the *Brown Shoe* indicia (ibid 1342), but did not succeed in providing a convincing hypothetical monopolist test as its sample was limited to four out of 268 customers.

market considered was the aftermarket, SL resin, and the Court discussed this with reference to the *Kodak* judgment. As SL machines are very costly, Desotech relied on a lock-in theory 'to show that SL resin constitutes an independent market in which anticompetitive conduct cannot be regulated by competition in the primary market'.[104] However, as Desotech did not manage to prove that a substantial number of customers were actually locked in, it did not succeed. The Court held that while in *Kodak*, customers had only learned about the lock-in after purchasing the expensive equipment, in *DSM Desotech v 3D Systems* they were aware of this at the time of purchase.[105]

At present, Keurig Green Mountain is subject to a consolidated antitrust lawsuit in the US, facing claims that its ambition to limit compatibility of coffee portion packs to its own K-Cups for its Keurig coffee maker 2.0 violates US antitrust laws.[106] It is accused of using technical tying to keep the non-proprietary aftermarket, consisting of coffee portion packs, to itself.

C. Analysis

The wealth of case law on proprietary aftermarkets on the one hand and non-proprietary aftermarkets related to proprietary primary markets on the other hand, as well as the policy documents discussed above, allow for some insights into how secondary markets are delineated for the purposes of EU and US competition law. When the aftermarket is proprietary, neither the EU Courts nor the European Commission are reticent to delineate secondary markets along the lines of a company's product, despite companies' repeated insistence that they operate in a system market rather than in separate primary and secondary markets (multiple markets). *Hugin*, *Hilti* and *Digital* are all cases in point. Where only a certain secondary product is compatible with a chosen primary product, even a company with a very low market share on the primary market will regularly be seen as dominant on the secondary market,[107] for in the EU, the company's market position on the primary market is frequently not taken into account.

In the US, the Supreme Court has shown a willingness to delineate single-brand aftermarkets in its landmark *Kodak* judgment, as well as in *Virtual.* Both of these cases concerned non-proprietary aftermarkets that were tied to a proprietary aftermarket. However, Federal Courts of Appeals since *Kodak* have regularly distinguished cases before them from *Kodak*, finding that lifecycle pricing was

[104] ibid 1346.

[105] ibid 1346 f.

[106] M Leonard, 'Keurig Must Face Most of Sprawling Antitrust Suit over K-Cups' *Bloomberg Law* (24 April 2019), https://news.bloomberglaw.com/mergers-and-antitrust/keurig-must-face-most-of-sprawling-antitrust-suit-over-k-cups.

[107] A Jones and BE Sufrin, *EU Competition Law: Text, Cases, and Materials*, 6th edn (Oxford, Oxford University Press, 2016) 303.

theoretically able to exert competitive pressure on the single-brand aftermarket and that the latter therefore did not constitute a relevant market in its own right. A long line of case law taking such an explicit stance can be traced, including *Marts v Xerox*, *PSI Repair Services v Honeywell*, *Alcatel USA v DGI Technologies*, *Systems Maintenance Services v Digital* and *DSM Desotech v 3D Systems*. Herbert Hovenkamp observed that '[m]ost lower courts have bent over backwards to construe *Kodak* as narrowly as possible.'[108] Rather than finding multiple markets, and thus viewing the respective aftermarkets as their own antitrust relevant markets, all these cases came to the conclusion that the customers allegedly locked in by the aftermarkets in question could have engaged in lifecycle pricing as the companies had not changed their aftermarket policies after the purchase of the primary product. The courts therefore held a system market approach to be applicable. This change of policy, as was already pointed out above in the discussion on the relationship between primary and secondary markets (see section II.A), seems to have become the lynchpin of market definition in US aftermarket cases, at least in Federal Courts of Appeals. Whether the Supreme Court approves of this interpretation of its *Kodak* judgment remains to be seen. Some judgments that followed *Kodak*, such as *Trinko*, indicate that it might.[109]

Whether products on a secondary market are only compatible with one particular product on the primary market because of successfully implemented contractual clauses, IPRs, technical tying or any other reason is irrelevant for market definition. What needs to be established is the presence or absence of competitive constraints on the secondary market, no matter how those are held at bay. Therefore, proprietary aftermarkets can be analysed much in the same way as any other aftermarket. While this might appear counter-intuitive, it is simply a matter of looking at the actual market context, something that the *Kodak* Court strongly encouraged. However, in the ensuing competition law analysis, the fact that an IPR protects an aftermarket may nevertheless have ramifications. Where the legislator has allocated a certain aftermarket to an IP owner, such as the sales of spare parts through the award of design rights or the secondary use of a copyright through the protection of translations or broadcasting, antitrust law might be reticent to interfere.[110] While some view IPRs as a legitimate way to keep an aftermarket to oneself, others do not want to analyse refusals to deal in aftermarket cases any differently in the presence of IPRs than in their absence.[111] A middle ground can be found if one recognises that IPRs may provide an objective justification for

[108] HJ Hovenkamp, 'Post-Chicago Antitrust: A Review and Critique' [2001] *Columbia Business Law Review* 257, 286.

[109] *Verizon Communications v Trinko*, 540 US 398, 407 (2004); WE Kovacic, 'The Intellectual DNA of Modern US Competition Law for Dominant Firm Conduct: The Chicago/Harvard Double Helix' [2007] *Columbia Business Law Review* 1, 19.

[110] A Heinemann, 'The Contestability of IP-Protected Markets' in J Drexl (ed), *Research Handbook on Intellectual Property and Competition Law* (Cheltenham, Edward Elgar, 2008) 66.

[111] Bauer (n 2) 51.

keeping a proprietary aftermarket to oneself – unless this practice is found to be an abusive refusal to license, which is for competition law to determine. What is undisputed is that IP protection in one aftermarket may not be used as a lever to extend that 'monopoly' (in an IP sense) to other aftermarkets. This was recognised early on in *Morton Salt*, where this argument was entirely based on patent law. More recently, this was confirmed in a number of cases, such as *Digital* (EU) and *Virtual* (US). Also, undertakings with considerable market power have a special competition law responsibility under Article 102 TFEU[112] or might be faced with claims of monopolisation under section 2 of the Sherman Act, no matter what the basis of their market power.

While the US Supreme Court no longer views tying with outright suspicion,[113] this is not true for the EU. Here, market definition in aftermarkets appears closely related to the competition concerns involved. The narrow product markets in aftermarkets that are delineated by the EU Courts and the Commission inevitably lead to a situation in which, 'unless a consumable itself is protected by the IPR, it is viewed as a separate product and cannot be bundled with another product which is protected by an IPR'.[114]

In the case law on proprietary aftermarkets and proprietary primary markets discussed above, there was a notable shift in focus. While cases such as *Hugin* and *Digital* in the EU and *Kodak* and *Virtual* in the US concerned a proprietary and a non-proprietary aftermarket that were tied together, other cases such as *Pelikan/Kyocera* and *EFIM* in the EU and *Independent Ink* and *3D Systems* in the US related to proprietary primary markets to which non-proprietary aftermarkets (in most of these cases ink) were tied. In fact, all of the cases discussed above that followed *Kodak* and *Virtual* concerned non-proprietary aftermarkets and their relationship with a proprietary primary market: cartridges and Xerox copiers with warranties in *Marts v Xerox*, expansion cards and switches in *Alcatel USA v DGI Technologies*, maintenance services and Digital computer systems with warranties in *Systems Maintenance Services v Digital*, and resin and SL machines in *DSM Desotech v 3D Systems*. While *Kodak* had concerned proprietary replacement parts and maintenance services (for Kodak machines) and *Virtual* related to proprietary software updates and maintenance services (for Digital computers), and thus aftermarket/aftermarket tying, the focus of these follow-on cases shifted to primary market/aftermarket tying. Likewise, *Digital* (EU) concerned proprietary software updates and maintenance services (for Digital computers), while *Pelikan/Kyocera*, *Info-Lab/Ricoh* and *EFIM* all concerned some form of ink tied to proprietary copiers or printers.

This shift makes a difference because in the first set of cases, the antitrust analysis concentrates on the two aftermarkets concerned, on the relationship between

[112] See on this Guidance Paper (n 10) para 1.
[113] *Illinois Tool Works and Others v Independent Ink*, 547 US 28, 28 (2006).
[114] SD Anderman and HSK Schmidt, *EU Competition Law and Intellectual Property Rights: The Regulation of Innovation*, 2nd edn (Oxford, Oxford University Press, 2011) 43.

the proprietary and non-proprietary aftermarket. The company's success in the primary (usually also proprietary) market is frequently taken into consideration in these cases and is used in order to ascertain whether there is a system market rather than multiple markets at play. However, it is the proprietary nature of one of the aftermarkets and how it is used to extend the 'monopoly' rights granted by an IPR that appears to be the tipping point. Underlying this, perhaps there is an assumption that if a company is able to engage in this kind of tying, there must be market power in the aftermarket – and where there is market power, there must be a market. In the second set of cases, the relationship between the primary (proprietary) and the secondary (non-proprietary) market is at the heart of the competitive concern, and it is here that the company's success in the primary market and the customers' ability and actual willingness to engage in lifecycle pricing particularly come into play. This might be a different way of accounting for the outcomes of the cases that followed on from *Kodak* and *Virtual* in the US. As was held above in relation to tying, market definition in aftermarkets appears to be closely related to the particular competition concerns associated with the market behaviour in question.

IV. The Special Case of Aftermarkets in Franchises

Franchise agreements can represent an efficient way to distribute products or services. They often broach similar issues as conventional aftermarkets and can involve several forms of IPRs, such as trademarks to help convey their brand image, but also protected know-how in order to pass on their business-specific knowledge to franchisees. In such cases, the franchise – or indeed the trademark and know-how intricately associated with it – is usually conceptualised as the primary product.[115]

A. Franchises in the Case Law

In the US, an early case on franchising was *Principe v McDonald's* (1980), where the Court of Appeals for the Fourth Circuit had to decide 'whether a fast food franchisor that requires its licensees to operate their franchises in premises leased from the franchisor is guilty of an illegal tying arrangement in violation of § 1 of the Sherman Act'.[116] The Court agreed with McDonald's that the lease of the premises, the licence and the note regarding a security deposit were not separate markets, but were all components of the franchise agreement. As no two distinct markets could be established, there could be no anti-competitive tie-in.[117]

[115] Bauer (n 2) 35.
[116] *Principe v McDonald's*, 631 F2d 303, 304 (4th Cir 1980).
[117] ibid 308.

In its first and only judgment on franchise agreements to date, the European Court of Justice was asked to rule on a franchise for wedding attire. It found that distribution franchises such as the one at issue in *Pronuptia de Paris* (1986) relied on the transmission of know-how from the franchisor to the franchisee, as well as on the upkeep of the network's identity and reputation.[118] The latter particularly involved the franchisor's trademark. In order for such a franchising system to work, some restrictions as to the sourcing of supplies were necessary and permitted under EU competition law.[119] Ultimately, therefore, the aftermarkets of the franchising agreement itself – ie, the markets for trademarked goods and services, and for products involving the franchisor's know-how – were closely linked to the primary market, in which the franchise's trademark and know-how played an important role. Based on the business model involved, the franchisor was allowed to keep these aftermarkets to itself where this was necessary in order for the system to work.

In 1997, the US Court of Appeals for the Third Circuit issued a judgment in *Queen City Pizza v Domino's Pizza*. In this post-*Kodak* case, 11 of Domino's franchisees alleged that Domino's Pizza had market power in 'the $500 million aftermarket for sales of supplies to Domino's franchisees',[120] which the Court took to signify the market for 'Domino's-approved ingredients and supplies used by Domino's Pizza franchisees'.[121] The franchise agreement at issue required the franchisee to buy certain approved ingredients and supplies from the franchisor, Domino's. The Court noted the importance of product uniformity and consistency for the success of a franchise. It also found that general rules on market definition led to the conclusion that the relevant market was pizza dough, tomato sauce, paper cups and other ingredients and supplies in general, and was not restricted to Domino's pizza dough, tomato sauce, paper cups etc. The plaintiffs could not rely on *Kodak* or *Virtual*, as the franchisees knew of the contractual restrictions they had entered into.[122]

B. Analysis

In the area of franchising, the franchisor's grip on the aftermarket serves important functions, such as quality control and protecting the franchise's goodwill.[123]

[118] Case 161/84 *Pronuptia de Paris* EU:C:1986:41, paras 15–17.

[119] ibid paras 18–22. That same year, the Commission issued a decision applying the franchising principles set out by the Court to Pronuptia's standard franchise agreement. Its decision contains a section on the relevant product market, clearly highlighting the difficulties in overall assessing the market(s) a franchise system is active on and the multitude of products (most of them bearing the Pronuptia trademark, but some of them not) involved in such a system; see *Pronuptia* (Case IV/30937) Commission Decision 87/17/EEC [1987] OJ L13/39, paras 7–9.

[120] *Queen City Pizza v Domino's Pizza*, 124 F3d 430, 434 (3d Cir 1997).

[121] ibid 435.

[122] ibid 433, 438 ff.

[123] Bauer (n 2) 40.

It is for these reasons that the cases briefly discussed above sanctioned restrictions within franchise systems that would often be deemed anti-competitive without this particular distribution background.

Market definition in the area of franchising faces the difficulty that the products at issue are, from a theoretical point of view, certainly interchangeable with others, as the Court underlined in *Domino's Pizza*. However, due to the contractual limitations stemming from the franchise agreement, for the franchisee they are not. In this respect, franchise aftermarkets are no different than conventional aftermarkets – they may be tied to a primary market through contractual exclusivity, IPRs, technical tying or any combination of these factors. Therefore, it can be useful to apply the same kind of antitrust market definition analysis to these markets.

Kodak proposed 'a more refined basis for explaining established decisions holding that a franchisor can exercise market power over its franchisee'.[124] However, lower courts have been wary of this new approach, frequently rejecting a narrow market definition that only included the franchisor's supplies.[125] As in the post-*Kodak* cases discussed above (see section III.B), the Court in *Domino's Pizza* relied on the policy change element in Kodak in order to ascertain that franchisees – like customers locked into any other type of aftermarket – cannot rely on tying claims if they could have carried out lifecycle costing before entering into the franchise agreement. This view can also be found in the literature.[126] However, just as in more conventional aftermarkets, this misses the point that franchisees, like regular customers, are not as rational as many economists assume.

In the EU, it would seem that the Court of Justice has given franchise agreements quite some leeway in its *Pronuptia* preliminary ruling some 30 years ago, but no new cases have come before it since its landmark *EFIM* ruling. Therefore, the applicability of the four *EFIM* criteria to franchising aftermarkets has not been tested in court. However, it does seem fathomable that the Court would also apply these criteria to franchise aftermarkets. Slightly tweaking the EU's *EFIM* criteria, the questions would then be as follows. Can franchisees engage in lifecycle pricing for this franchise? Do they? Would they switch franchisor in the event of a price rise in the aftermarkets? And would they do so in a timely manner?

Another possible entry point for competition law are rules on relational market power, which are contained in a number of national competition laws.[127] These provisions foresee that companies in superior bargaining positions are deemed to be dominant vis-à-vis their customers or suppliers, particularly if the latter mainly depend on this business relationship.[128] The assumption must be that the relevant

[124] Grimes (n 27) 244.

[125] ibid 245; JL McDavid and RM Steuer, 'The Revival of Franchise Antitrust Claims' (1999) 67 *Antitrust Law Journal* 209.

[126] Klein (n 25) 284, 286.

[127] Grimes (n 27) 257.

[128] See, eg, Austrian Competition Act 2005 (Kartellgesetz 2005), Federal Law Gazette I 61/2005, as amended, § 4(3); Japanese Antimonopoly Act 1947, Act 54/1947, as amended, art 2(5); South Korean Monopoly Regulation and Fair Trade Act, Law No 3320 of 1980, as amended, art 23(1)(4).

markets in these cases are delineated along the line of the business relationship. Franchise agreements, it would seem, can be likened to such superior bargaining positions.

V. Conclusion

Innovative aftermarkets represent a challenge for market delineation, particularly when IPRs are used to seal them off. Depending on the market conditions, either a system market or multiple markets can be delineated. Concerning the relationship between primary and secondary markets, case law in the US and the EU is not congruent when it comes to the primary market's ability to keep the secondary market competitive. The EU places more emphasis on consumers actually switching primary products in the case of exploitation on the secondary market. Only when switching actually occurs will a system market be delineated. In the US, and particularly in lower courts, a stricter view prevails that focuses on the mere possibility of switching to another primary product. Only then will multiple markets be delineated.

For proprietary aftermarkets, both the EU and the US frequently delineate these as brand-specific, although US Federal Courts of Appeals like to deviate from this. However, it is not the proprietary nature of the aftermarket that is decisive for this narrow market delineation – technical or contractual tying are equally sufficient.

The delineation of aftermarkets can serve as a useful blueprint for the delineation of multi-sided markets (see sections II.B and II.C.vii in Chapter 10). In that scenario, antitrust authorities and courts need to ask whether the mere possibility to switch platforms ('primary products') can act as a competitive constraint on the secondary markets, thus leading to the finding of one comprehensive system market instead of multiple antitrust markets.

8

Platform Markets

I. Market Definition in Platforms

Two- or multi-sided markets, frequently also referred to as platforms, represent a business model which relies on capturing indirect network effects between different customer groups that the latter could not realise without the intermediation of the platform.[1] While direct network effects relate to the consumer benefits that directly increase with the number of owners of a certain good or participants in a certain service, indirect network effects relate to 'improved opportunities to trade with the other side of a market'[2] or, indeed, the value that different user sides obtain 'from interacting with users of the opposite type over a common platform'.[3] Indirect network effects are known for their positive feedback loop.[4] These cross-platform externalities have the consequence that an increased number of market participants on one market side translates into an increased number of market participants on the other market side.[5] Depending on the platform market at issue, these indirect network effects will be bidirectional or unidirectional.[6] The focus in platform markets is on the role of the intermediary, ie, the platform.[7] Increasingly, this is also the case in competition law.

Despite some pioneering research into the theory of two-sided markets,[8] economic literature in this area still appears to be 'in a state of flux'[9] and various

[1] See VHSE Robertson, 'Delineating Digital Markets under EU Competition Law: Challenging or Futile?' (2017) 12 *Competition Law Review* 131, 136 ff.

[2] J Farrell and P Klemperer, 'Coordination and Lock-in: Competition with Switching Costs and Network Effects' in M Armstrong and RH Porter (eds), *Handbook of Industrial Organization*, vol 3 (Amsterdam, North Holland, 2007) 1974. For a discussion of network effects in innovative market environments, see ch 3, section II.C.

[3] J Wright, 'One-Sided Logic in Two-Sided Markets' (2004) 3 *Review of Network Economics* 44, 44.

[4] J Pohlmeier, *Netzwerkeffekte und Kartellrecht* (Baden-Baden, Nomos, 2004) 29, 32 ff.

[5] D Auer and N Petit, 'Two-Sided Markets and the Challenge of Turning Economic Theory into Antitrust Policy' (2015) 60 *Antitrust Bulletin* 426, 429.

[6] See also G Luchetta, 'Is the Google Platform a Two-Sided Market?' (2014) 10 *Journal of Competition Law & Economics* 185, 199.

[7] M Rysman, 'The Economics of Two-Sided Markets' (2009) 23 *Journal of Economic Perspectives* 125, 126.

[8] See B Caillaud and B Jullien, 'Chicken & Egg: Competition among Intermediation Service Providers' (2003) 34 *RAND Journal of Economics* 309; DS Evans, 'The Antitrust Economics of Multi-sided Platform Markets' (2003) 20 *Yale Journal on Regulation* 325; J-C Rochet and J Tirole, 'Platform Competition in Two-Sided Markets' (2003) 1 *Journal of the European Economic Association* 990; M Armstrong, 'Competition in Two-Sided Markets' (2006) 37 *RAND Journal of Economics* 668.

[9] Auer and Petit (n 5) 428.

definitions of what constitutes a multi-sided market exist. The multi-sided nature of a platform may often have an important impact on the antitrust analysis,[10] but not always.[11] This greatly complicates the question of whether and how to incorporate multi-sided market theory into competition law. Nevertheless, it would seem that the main elements of the theory of two-sided markets are advanced enough in order for competition law to incorporate them into its analytical framework,[12] a process that should be led by lawyers rather than economists.[13]

The theory of multi-sided markets as developed by economists allows antitrust an insight into the unique competitive structure of these markets.[14] This must also have repercussions on how antitrust law delineates markets. There have been various attempts at framing multi-sided markets from a competition law perspective – eg, by framing one market side as the input market and the other as the output market, thus likening them to a typical vertical relationship,[15] by focusing on their data-gathering activities[16] or by understanding such scenarios as a tying and a tied market.[17] In the end, however, multi-sided market theory allows for the capturing of the particular substitution patterns and interdependencies between the different market sides that provide the necessary link so that we can properly understand and assess platform markets in competition law.

Platforms are present in a great number of innovative markets, but are not limited to these. For instance, credit card systems with their market sides of the issuing and the acquiring party can be framed as platform markets. It is within these markets that many antitrust cases on multi-sided platforms arose both in the EU and in the US (see sections III.B.i and III.B.iv below).[18] Nevertheless, the most prominent platforms have certainly emerged in the digital marketplace,[19] as this arena gives them manifold opportunities to innovate and succeed commercially. Examples of such platforms include online search engines such as Baidu, Bing or Google, with their market sides of search users, advertisers and content providers (websites), and social networking sites such as Facebook or LinkedIn, with their market sides of social networkers and advertisers. However, there are a great number of other innovative platforms as well, such as video or music

[10] Rysman (n 7) 138; F Thépot, 'Market Power in Online Search and Social Networking: A Matter of Two-Sided Markets' (2013) 36 *World Competition* 195, 197.

[11] DS Evans, 'Background Note' in OECD (ed), *Policy Roundtable: Two-Sided Markets* (2009) DAF/COMP(2009)20, 25.

[12] A Lamadrid de Pablo, 'The Double Duality of Two-Sided Markets' (2015) 64 *Competition Law Journal* 5, 6.

[13] ibid.

[14] Auer and Petit (n 5) 457.

[15] Luchetta (n 6) 201.

[16] I Graef, 'Market Definition and Market Power in Data: The Case of Online Platforms' (2015) 38 *World Competition* 473.

[17] Oxera course on 'Using Economics in Competition Cases' (Oxford, 7–8 November 2016).

[18] Not all credit card systems rely on platform markets; see *United States v American Express*, 838 F3d 179, Addendum, figures 1 through 4 (2d Cir 2016).

[19] T Hoppner, 'Defining Markets for Multi-Sided Platforms: The Case of Search Engines' (2015) 38 *World Competition* 349, 349.

streaming services (eg, Amazon Prime, Facebook Live, Netflix, Spotify and YouTube), online video-gaming sites (eg, IGN and Xbox Live), online shopping platforms (eg, Amazon Marketplace and eBay), platforms of the sharing economy (eg, Airbnb, Lyft and Uber) etc. Common online retailers, on the other hand, which engage in typical vertical relationships with their suppliers and their customers rather than capturing the indirect network effects between those user groups, do not constitute platform markets within the meaning of this chapter.

When defining a platform market in an innovative setting for antitrust purposes, several issues need to be addressed. To what degree can conventional antitrust market definition be applied? How does one take into account the market side which receives services for free? How are direct and indirect network effects present in a platform market best captured? Which market(s) should antitrust base its analysis on in a platform market? Emerging case law in the EU and the US is already dealing with these relatively new issues in the context of innovative markets, but they have yet to develop a coherent approach.

II. Free Services in Innovative Platform Markets

In digital platforms, it is frequently the case that one user group receives the services provided for 'free', while another – usually the advertising side – pays a price. This may lead to 'dramatic asymmetry' in terms of how much platforms charge their market sides.[20] Examples of digital platforms that operate in this way include advertisement-based services such as social networks (Facebook, LinkedIn), e-mailing services (Gmail, Outlook), comparison shopping engines (Kelkoo, Twenga) or video streaming services (YouTube). However, a free market side might also be encountered in more traditional 'offline' markets, for instance, in free print newspapers that are advertisement-based. The question arises whether the 'free' market side in these scenarios can be called a commercial activity, thus subjecting it to antitrust scrutiny, or whether antitrust should back out of scrutinising these markets. A particular feature of digital platforms is that they are data-driven. The data that users provide by using the platform and agreeing to the platform's tracking practices, as well as users' willingness to be exposed to targeted advertising, frequently constitutes the very reason why the user side is free.[21]

[20] DS Evans and R Schmalensee, 'The Antitrust Analysis of Multisided Platform Businesses' in RD Blair and DD Sokol (eds), *The Oxford Handbook of International Antitrust Economics*, vol 1 (Oxford, Oxford University Press, 2015) 408.

[21] See A Ezrachi and VHSE Robertson, 'Competition, Market Power and Third-Party Tracking' (2019) 42 *World Competition* 5.

A. Cases on Free Market Sides

A number of cases from both the EU and the US have openly or implicitly addressed free market sides and the application of competition law to them. Some scepticism towards the aptness of competition law to address these markets can be discerned in these cases, along with a preference for relying on revenue-generating markets where possible.

In 2007, a US District Court discussed the relevant markets relating to Google's search business in the *KinderStart* case. The Court held that in order for KinderStart's claim that Google had attempted to monopolise to be upheld, the former had to define a relevant market. KinderStart alleged that two markets were concerned: the search market and the search ad market. However, the Court found that KinderStart's claims as to internet search as the relevant market could not succeed because the plaintiff had relied on 'no authority indicating that antitrust law concerns itself with competition in the provision of free services'.[22] Two years later, China's leading internet search engine, Baidu, similarly defended itself from an abuse of dominance claim by relying on the argument that its free online search services could not constitute an antitrust market.[23] However, the free nature of many online markets did not deter the Chinese Supreme People's Court in *Qihoo 360 v Tencent* (2014) to rule that these could form relevant antitrust markets.[24]

In a merger inquiry decided in 2011, the DoJ assessed Google's proposed acquisition of ITA, which provided the market's leading independent airfare pricing and shopping (P&S) system software to online travel websites.[25] In its complaint, the DoJ regarded comparative flight search services as one of the relevant product markets,[26] but did not mention the fact that all the providers it cited (ie, Bing Travel, Expedia, Kayak, Orbitz and Travelocity) generally provide these for free to users. By implication, it thus regarded such a market as a relevant antitrust market that could come under antitrust scrutiny.

A US class action about rate parity clauses agreed upon amongst online hotel booking portals was dismissed in a District Court in 2014. The market alleged by the plaintiffs was the US market for the 'direct online sale of hotel room

[22] *KinderStart.com v Google*, Case C 06-2057 JF (RS) (ND Cal 16 March 2007) § III.1.a.i.

[23] Beijing No 1 Intermediate People's Court, *Baidu v Renren* (18 December 2009). See AH Zhang, 'Using a Sledgehammer to Crack a Nut: Why China's Antimonopoly Law was Inappropriate for *Renren v Baidu*' (2011) 7 *Competition Policy International* 277, 283; Z Li, 'New Developments in Civil Antitrust Litigation in China' (2012) 2 *Competition Policy International Antitrust Chronicle* 1, 2 fn 3.

[24] The Court held that the SSNIP cannot be properly applied here because of these markets' particular dynamics; see Chinese Supreme People's Court, *Qihoo 360 v Tencent* (16 October 2014); DS Evans and VY Zhang, '*Qihoo 360 v Tencent*: First Antitrust Decision by the Supreme Court' *Competition Policy International Asia Column* (21 October 2014), www.competitionpolicyinternational.com/assets/Uploads/AsiaOctober214.pdf.

[25] *United States v Google/ITA Software*, Case 1:11-cv-00688 (Complaint, DDC 8 April 2011) para 1.

[26] ibid para 19.

reservations'.[27] The fact that this was a platform market with two market sides and that one of these market sides (ie, the hotel bookers) generally received services for free was not mentioned in that case.

Germany has seen a number of antitrust cases in which free user sides were disregarded for the purposes of market definition, most notably in several cases relating to hotel portal services such as HRS or Booking.com. In January 2015, the Higher Regional Court Düsseldorf affirmed the Bundeskartellamt's finding that where one market side (hotel bookers) receives services for free, while the other market side (hotels) pays for participating in the platform, only the remunerated part of the service is considered when defining the relevant market.[28] While the Bundeskartellamt had found that this was a two-sided market with network effects, it did not further incorporate this characterisation into its market definition.[29] The Court asserted that while the market was delineated based on the remunerated service, the 'free' user side could be considered in the context of market conditions, thus ultimately influencing the antitrust assessment.[30]

In its infringement decision against Booking.com of December 2015, the German Bundeskartellamt stuck to its market definition of hotel portal services, again excluding the free user side from market definition.[31] Its market definition in that case corresponded to the market definitions that a number of other national competition authorities had relied on in commitment decisions concerning the same infringing party, Booking.com.[32] All of these decisions held that Booking.com was active on the market for online hotel reservation services.[33]

In its recent *Facebook* decision (2019), the Bundeskartellamt found that Facebook relied on its terms of use in order to abuse its dominant position on the market for social networks under German competition law.[34] The authority considered the market for social networks to be the relevant product market. This case therefore relies on the 'free' user side of the platform at issue. Arguably, this market definition might be at odds with the same authority's market definition

[27] *Online Travel Company (OTC) Hotel Booking Antitrust Litigation*, 997 FSupp2d 526, 529, 533 (ND Tex 2014). At another point, the Court referred to this market as the 'online hotel bookings market'; ibid 534. The Court also mentioned the then-ongoing online hotel booking investigations in the UK and in Switzerland, which the plaintiffs had tried to rely on as an additional factor; ibid 540.

[28] Higher Regional Court Düsseldorf, Case VI – Kart 1/14 (V) *HRS* (9 January 2015) para 43; Bundeskartellamt, Case B 9-66/10 *HRS* (20 December 2013) para 71.

[29] Bundeskartellamt, Case B 9-66/10 *HRS* (20 December 2013) para 81; see also Higher Regional Court Düsseldorf, Case VI – Kart 1/14 (V) *HRS* (9 January 2015) para 43.

[30] Higher Regional Court Düsseldorf, Case VI – Kart 1/14 (V) *HRS* (9 January 2015) para 43.

[31] Bundeskartellamt, Case B 9-121/13 *Booking.com* (22 December 2015) paras 129 f; decision annulled in Higher Regional Court Düsseldorf, Case VI – Kart 2/16 (V) *Booking.com* (4 June 2019).

[32] *Booking.com* (n 31) paras 146 f; Autorità Garante della Concorrenza e del Mercato, Case I779 *Booking.com* (21 April 2015) para 9; Autorité de la concurrence, Case 15-D-06 *Booking.com* (21 April 2015) para 9; Konkurrensverket, Case 596/2013 *Booking.com* (15 April 2015) para 15.

[33] The UK's (then) Office of Fair Trading had also launched an investigation into online hotel booking portals, but eventually the (now) Competition and Markets Authority closed this investigation; see M Colangelo, 'Parity Clauses and Competition Law in Digital Marketplaces: The Case of Online Hotel Booking' (2017) 8 *Journal of European Competition Law & Practice* 3, 8.

[34] Bundeskartellamt, Case B 6-22/16 *Facebook* (6 February 2019).

in its hotel portal cases, in which it repeatedly disregarded the free user side for market definition purposes. However, this change in approach is well in line with an amendment to the German Cartel Act that came into force in June 2017, which explicitly foresees that even where a service is offered without remuneration, a relevant antitrust market may be found.[35]

In the European Commission's case on *Google Shopping* (2017), general online search was regarded as one of the relevant markets. The fact that this was a free service was highlighted by the Commission, which gave three reasons why it nevertheless considered this to constitute a relevant market for antitrust purposes: first of all, while users did not pay a fee for using online search, they did pay in kind by providing Google with their data when entering a search query; second, the free user side formed part of the business model, which consisted of a two-sided platform; and, third, general search services competed on parameters other than price.[36] Particularly the second reasoning behind this market definition strongly related to the multi-sided nature of this platform market, without mentioning two-sided market theory. In *Google Android* (2018), the Commission considered that Google was dominant in a number of markets, namely general online search, licensable smart mobile operating systems and app stores for the Android mobile operating system. Again, online search is the free market side in Google's search platform.[37]

B. On the Incorporation of Free Market Sides into the Analysis

Whether or not free services should be made the object of antitrust scrutiny is far from controversial. Neither the European nor the US courts and authorities provide a clear answer in this respect.[38] However, if one understands a multi-sided platform to constitute an antitrust market – rather than an array of antitrust markets – then it becomes clear that 'free' services within digital platform markets are not identical to free services in many other scenarios. This is especially so because of the personal data and attention that users offer up in return for the

[35] German Restriction of Competition Act (Gesetz gegen Wettbewerbsbeschränkungen), Federal Law Gazette I 2114/2005, as amended, § 18(2a).

[36] *Google Search (Shopping)* (Case AT.39740) Commission Decision of 27 June 2017 [2018] OJ C9/11, paras 158–60. The European Commission has also developed a proposal to address the anti-competitive behaviour of certain digital intermediaries; see European Commission, 'Proposal for a Regulation on Promoting Fairness and Transparency for Business Users of Online Intermediation Services' (26 April 2018) COM(2018) 238 final.

[37] *Google Android* (Case AT.40099) Commission Decision of 18 July 2018; European Commission, 'Antitrust: Commission Fines Google €4.34 Billion for Illegal Practices Regarding Android Mobile Devices to Strengthen Dominance of Google's Search Engine' (IP/18/4581, 18 July 2018).

[38] For the EU, see also M Sousa Ferro, '"Ceci n'est pas un marché": Gratuity and Competition Law' (2015) 1) *Concurrences* 1, paras 78–82.

free service. In addition, the pricing decision of the platform owner includes all market sides. It is a platform owner's conscious decision to charge no fees on one market side, but instead gather data from it and then monetise it, eg, by selling advertisements to the other market side. Overall, therefore, the platform with all its market sides constitutes a commercial activity to which competition law can and should apply.

It is often possible to characterise free market sides that are subsidised through advertising as non-transactional markets, as the two market sides involved do not directly enter into a transaction – as compared to transactional markets such as booking portals, in which a direct transaction between the two sides occurs. Based on this distinction, one may be able to delineate two separate yet interrelated antitrust markets for non-transactional markets, but one all-encompassing antitrust market for transactional markets.[39] This distinction is based on the insight that in the case of a non-transactional two-sided market, one does not necessarily need to take both market sides into account in order to analyse an antitrust case.[40] This approach is consistent with many developments in the case law, for instance the 1953 case of *Times-Picayune*, in which the US Supreme Court held in relation to newspapers (ie, non-transactional two-sided markets) that only one of the two markets at issue was concerned.[41] However, the difficulty in defining two separate – albeit interrelated – antitrust markets for non-transactional markets precisely lies in the fact that the free market side by itself cannot be properly understood as a commercial activity outside the framework of the multi-sided platform. In order to depict the market at stake more accurately, one could frame the non-transactional market side as one platform market with various market sides, one of which receives the services for free precisely because of the advertisements that the other market side places with the platform provider.

One also finds the view that non-transactional market sides should not form part of antitrust market definition, but may be taken into account in the antitrust assessment of a situation.[42] This approach might be missing out on the fact that a multi-sided platform allocates prices to all its user sides. It consciously chooses to charge a zero price to one user side (eg, social networkers) and compensates this through a higher price charged to the other user side (eg, advertisers).[43] If the delineation of an antitrust market is limited to the paying user side, then we risk losing sight of the network effects that are so vital in multi-sided markets. While the outcome might not be any different if free market sides can enter the antitrust analysis at a later stage, it would be short-sighted to exclude the free product side

[39] L Filistrucchi et al, 'Market Definition in Two-Sided Markets: Theory and Practice' (2014) 10 *Journal of Competition Law & Economics* 293, 302, 322.

[40] ibid 322.

[41] *Times-Picayune Publishing Co v United States*, 345 US 594 (1953).

[42] Sousa Ferro, 'Gratuity and Competition Law' (n 38) 10–12, 13; Luchetta (n 6) 207. Opposing the latter, see Hoppner (n 19) 356 fn 31.

[43] On pricing considerations in two-sided platforms, see C Kehder, *Konzepte und Methoden der Marktabgrenzung und ihre Anwendung auf zweiseitige Märkte* (Baden-Baden, Nomos, 2013) 63 ff.

from a multi-sided platform market against the background of the insights that multi-sided market theory has already provided.

In online search and social networking, several possible markets can be discerned: the market for users (ie, internet searchers and social networkers), the market for advertising, the market for content and the market for data.[44] As many digital platforms do not currently commercialise the data they gather through their search functionality or through their social network vis-à-vis third parties, data as such does not form part of the platform market unless one goes a step further and includes potential markets.

Instructive examples of how closely free services are related to paid-for services can be found in Spotify Premium, YouTube Premium and other apps in which advertisements can be 'bought out' for a fee, as discussed above.[45] These freemium models demonstrate how a business can encompass both 'free' services subsidised by advertisements and paid-for services that are ad-free.

Both European and US courts and authorities have been somewhat reticent in accepting the fact that 'free' services in platform markets are not equivalent to the non-existence of a commercial activity – and thus the inapplicability of antitrust law.[46] Rather, these free services are but an element in the business model of platforms,[47] allowing the platform provider to capture indirect network effects that may be one- or bi-directional. Where possible, free markets should not constitute a standalone market. When conceptualising a platform market as separate relevant markets, the free market side may wrongly be considered benign in terms of competition law, and the inherent link between the various market sides of the platform may be underestimated. At the stage of market definition, one will need to acknowledge that the antitrust market cannot be limited to the functionality that one user side values or to the advertising that the advertising side places. Rather, it is necessary to obtain a broader picture of how the market works and how the market sides – free or not – are related to each other. Only then can a meaningful antitrust law analysis be conducted. Especially for market definition's characterisation function, such a comprehensive picture is vital. Importantly, it also needs to take the role of the platform as such into account.

[44] SW Waller, 'Antitrust and Social Networking' (2012) 90 *North Carolina Law Review* 1771, 1780–86; DA Crane, 'Search Neutrality and Referral Dominance' (2012) 8 *Journal of Competition Law & Economics* 459, 463–66; JD Ratliff and DL Rubinfeld, 'Is There a Market for Organic Search Engine Results and Can Their Manipulation Give Rise to Antitrust Liability?' (2014) 10 *Journal of Competition Law & Economics* 517, 534–38; T Körber, 'Internet Search Engines and Competition Law' (2014) 9 *Journal of Intellectual Property Law & Practice* 517, 517.

[45] See ch 4, section II.C.

[46] Similarly, see MS Gal and DL Rubinfeld, 'The Hidden Costs of Free Goods: Implications for Antitrust Enforcement' (2016) 80 *Antitrust Law Journal* 521, 549.

[47] For further business models involving 'free' goods or services, see DS Evans, 'The Antitrust Economics of Free' (2011) 7 *Competition Policy International* 71, 71; ch 4, section II.C.

III. Delineating Multi-sided Markets

In order to succeed in a multi-sided market, the platform must perform well on all market sides.[48] This creates a link between the different sides of a platform that makes it fruitless to exclusively consider one market side as an isolated relevant antitrust market, without also taking the other into account.[49] While this insight might seem clear enough in theory, there are practical challenges associated with it, as the cases discussed below demonstrate. In addition, each market side can compete with another platform's market side, or with a company that is not based on a multi-sided business model.[50] This further complicates the competitive assessment.

Where innovative platform markets are analysed without acknowledging their multi-sidedness, the analysis will easily be flawed. The 'free' services on one user side may not be paid the attention they require (see section II), the interconnectedness of the various market sides may be lost from sight and the functioning of competition within such markets may be poorly understood. In the following, we will trace case law that has evolved in courts and before competition authorities in order to see how these enforcers have ignored, acknowledged or tried to incorporate multi-sided market theory in delineating antitrust markets. This allows us to appreciate the difficulties of applying multi-sided market theory in practice and sheds light on ways to overcome these difficulties.

A. Cases Ignoring Multi-sidedness

In a number of cases which arguably constitute prime examples of multi-sided platforms, the courts or competition authorities ignored this economic reality. In doing so, they possibly missed an important opportunity to understand these markets – an understanding which constitutes the very basis of antitrust enforcement.

In 2007, the FTC investigated the *Google/DoubleClick* merger under US antitrust law, in parallel with a European Commission investigation of the same merger. The FTC had to delineate a number of relevant markets relating to internet search engines and online advertising. The multi-sided nature of online search engines was never even mentioned in the FTC statement released at the closing of the investigation.[51] The European Commission's decision, on the other hand, was

[48] Armstrong (n 8) 669.

[49] In this vein, see also Monopolkommission, 'Wettbewerbspolitik: Herausforderung digitale Märkte' (Sondergutachten 68, 2015) para 58.

[50] Evans, 'Background Note' (n 11) 28.

[51] Federal Trade Commission, 'Statement Concerning *Google/DoubleClick*, Case 071-0170' (20 December 2007).

issued a few weeks later and did highlight the two-sidedness of the ad networks at issue, albeit with little or no consequences.[52]

In another US case from 2007, *LiveUniverse v MySpace*, a US District Court discussed the relevant antitrust market in a case involving social networking. While the Court made some important observations regarding substitutability as the basis for delineating the relevant antitrust market and concluded that online dating sites were not interchangeable with social networking sites, it did not refer to the two-sidedness of the markets in question and thus left out an important element when considering this case.[53]

E-books would seem like an archetypal multi-sided market, with the e-book reader as the platform, readers on the one market side and publishers or perhaps even individual authors as content providers on the other.[54] Apple, for instance, behaves exactly 'as a typical two-sided platform would, acting as a conduit for others' content, which the content owner controls'.[55] Yet, neither the US nor the European *E-Books* cases relied on multi-sided markets in their analysis, despite these cases being quite recent.[56] Perhaps the authorities and courts did not want to complicate their analysis in this way? Another interesting point to note in this respect is that a majority of cases concerning innovation revolve around mergers or abuse of dominance/monopolisation; however, the *E-Books* cases concerned anti-competitive agreements. Where companies enter into contractual relationships in order to organise the platform environment, then these contracts may well be subjected to antitrust scrutiny.

Following a number of mergers in which a relevant market's multi-sided nature was at least mentioned, the European Commission adopted a different approach in its *Facebook/WhatsApp* merger decision of October 2014. It found that three relevant markets were concerned: (1) consumer communication services – which were analysed in the narrowest market of consumer communication applications for smartphones; (2) social networking services – the boundaries of which were left open regarding consumer communication applications; and (3) online advertising services – again leaving open whether segments of that market constituted their own relevant markets.[57] The Commission devoted some of its analysis in the context of consumer communication services to network effects, but ultimately

[52] *Google/DoubleClick* (Case COMP/M.4731) Commission Decision of 11 March 2008 [2008] OJ C184/10, paras 20, 290.

[53] *LiveUniverse v MySpace*, Case CV 06-6994 AHM (RZx) (CD Cal 4 June 2007).

[54] Auer and Petit (n 5) 455.

[55] GA Manne and W Rinehart, 'The US E-Books Case against Apple: The Procompetitive Story' (2012) 3 *Concurrences* 18, 19.

[56] For the EU, see *E-Books* (Case AT.39847) Commission Decision of 25 July 2013; *E-Book MFNs and Related Matters (Amazon)* (Case AT.40153) Commission Decision of 4 May 2017 [2017] OJ C264/7. For the US, see *United States v Apple and Others*, Case 12 Civ 2826 (DLC) (SDNY 10 July 2013), affirmed by *United States v Apple and Others*, 791 F3d 290 (2d Cir 2015).

[57] *Facebook/WhatsApp* (Case COMP/M.7217) Commission Decision of 3 October 2014 [2014] OJ C417/4, paras 34, 62, 79.

found that they were unlikely to be strengthened after the transaction.[58] Throughout its merger decision, the Commission did not mention the two-sidedness of the markets it was investigating, nor did it address the possibility of conceptualising a broader platform market encompassing social networking and online advertising.[59]

In Europe, national competition authorities and courts, in applying EU competition law, have also encountered numerous market definition challenges when delineating innovative platforms. However, in several of these cases, the authorities and courts outright ignored the multi-sidedness of the market in question. A French court saga saw Google's mapping service at the competition law forefront. In *Bottin Cartographes v Google*, Google was accused of abusing its dominant position on the market for mapping web applications, in the form of predatory pricing. After a favourable first judgment for Bottin by the Tribunal de Commerce de Paris,[60] the Paris Cour d'Appel asked the Autorité de la concurrence for an expert opinion in this case.[61] In its opinion, the Autorité held that the industry concerned was that of inserting maps displaying customer data on companies' websites.[62] However, despite the Court's express insistence upon examining the relevant market,[63] the authority held that the question of precise market definition did not need to be resolved in that case, and neither did the issue of whether or not Google was dominant in any relevant market. Instead, the authority focused on the costs analysis provided by Google, which showed that the latter had not engaged in predatory pricing.[64] In November 2015, the Paris Cour d'Appel dismissed Bottin's allegations of predatory pricing. It relied on the authority's opinion, which had found that in a case such as the present, the freemium business model was not aimed at driving out competitors, but at increasing Google's user base on another market – and this was regarded as legitimate by the Court.[65] The fact that the authority did not rely on two-sided market theory in its expert opinion at a time when it had already appeared in judgments by the EU Courts and in Commission and national competition decisions is somewhat problematic. The authority's unwillingness to even engage in a discussion of the relevant market may hint at a lack of confidence in this sensitive area of antitrust. Therefore, although the case reflects many insights from two-sided market theory, it is classified amongst the more 'ignorant' cases in relation to market definition.

[58] ibid paras 127 ff, 140. In 2017, the Commission fined Facebook, alleging that the latter had provided it with misleading information during these merger proceedings; *Facebook/WhatsApp* (Case M.8228) Commission Decision of 18 May 2017 [2017] OJ C286/6.

[59] Robertson, 'Delineating Digital Markets' (n 1) 138.

[60] Tribunal de Commerce de Paris, Case 2009061231 *Bottin Cartographes v Google* (31 January 2012).

[61] Cour d'Appel de Paris, Case 12/02931 *Google v Bottin Cartographes* (20 November 2013).

[62] Autorité de la concurrence, Case 14-A-18 *Bottin Cartographes v Google* (16 December 2014) para 18.

[63] Cour d'Appel de Paris, Case 12/02931 *Google v Bottin Cartographes* (20 November 2013) 10.

[64] Autorité de la concurrence, Case 14-A-18 *Bottin Cartographes v Google* (16 December 2014) para 24.

[65] Cour d'Appel de Paris, Case 12/02931 *Google v Evermaps (Previously Bottin Cartographes)* (25 November 2015) 11.

B. Acknowledging a Platform Market's Multi-sidedness

Going beyond the stage of ignoring a market's multi-sidedness, a vast number of cases since the early 2000s have started to acknowledge a platform market's multi-sidedness. However, this acknowledgement was often no more than that; consequences for antitrust market delineation did not arise.

i. EU Courts

In *Microsoft* (2004), an early case that pre-dated the landmark papers on two-sided markets,[66] the European Commission acknowledged the different market sides of the relevant markets without engaging with multi-sided market theory. It initially found three product markets: the client PC operating systems, work group server operating systems and streaming media players.[67] Discussing PC operating systems, the Commission analysed both the consumer side and the software developer side, but for the latter focused on whether software developers could move into producing operating systems rather than simply applications.[68] This discussion highlights that there is both a consumer and a developer side to operating systems. The operating system that was at the heart of the *Microsoft* cases is a typical two-sided platform, serving end users on the one side and application developers on the other.[69] Referring to streaming media players, the Commission also acknowledged their importance for consumers as well as for software developers and content providers.[70] Thereby, it essentially described a platform market.

Two Court of Justice judgments dealt with two-sided markets, albeit in the context of credit or payment cards rather than more dynamic markets: *MasterCard* and *Groupement des cartes bancaires (CB)*. Both cases concerned appeals to General Court judgments that had affirmed Commission decisions finding an infringement of Article 101(1) TFEU.[71] In *MasterCard*, the appellants alleged that 'the Commission failed to take the two-sided nature of the market into account'.[72] The General Court recognised that there were a number of elements that connected the two sides of the credit card system under scrutiny. On the one hand, there was the issuing side, ie, that part of the market in which banks issued

[66] On these, see n 8; Auer and Petit (n 5) 454 fn 184.

[67] *Microsoft* (Case COMP/C-3/37.792) Commission Decision 2007/53/EC [2007] OJ L32/23, paras 324 ff, paras 343 ff, 402 ff.

[68] ibid paras 340 f.

[69] DS Evans and R Schmalensee, 'Markets with Two-Sided Platforms' in ABA Section of Antitrust Law (ed), *Issues in Competition Law and Policy* (Chicago, ABA Publishing, 2008) 682.

[70] *Microsoft* (Case COMP/C-3/37.792) Commission Decision 2007/53/EC [2007] OJ L32/23, para 415. These market definitions were upheld in Case T-201/04 *Microsoft v Commission* EU:T:2007:289.

[71] Case T-111/08 *MasterCard v Commission* EU:T:2012:260; Case T-491/07 *Groupement des cartes bancaires v Commission* EU:T:2012:633.

[72] Case T-111/08 *MasterCard v Commission* EU:T:2012:260, para 168.

credit cards to customers, and on the other hand, there was the acquiring side on which banks entered into a contract with merchants who would then accept credit cards as means of payment. The General Court acknowledged that indirect network effects were at work 'since the extent of merchants' acceptance of cards and the number of cards in circulation each affects the other'.[73] It nevertheless found that the Commission was allowed to conclude that two separate markets were at issue in the present case. On appeal to the Court of Justice, market definition was no longer at issue.[74] In one instance, the Court of Justice referred to the 'two-sided system' of the MasterCard scheme,[75] thus acknowledging the Commission's terminology.

In *CB*, the Commission had defined the relevant market as the 'market for the issue of payment cards in France'[76] and thus concentrated on one of the platform's market sides. The General Court rightly found this two-sided market system to lead to interconnections and indirect network effects.[77] On appeal, the General Court's judgment was set aside and referred back to it.[78]

While these cases show that two-sided markets are slowly beginning to find their way into the Court of Justice's market definition, their analysis requires further sophistication and clarification, especially regarding the question of how the interconnectedness of two-sided markets shall be taken into account when defining the market(s). The various sides of a two- or multi-sided market cannot and would not exist without the other side, and this dependency of (at least) one side of the market on the other will need to be reflected more carefully in market definition.

It is not only the EU Courts but also national courts that have acknowledged a market's multi-sidedness. In *Streetmap v Google* (2016), the English High Court dealt with Google's mapping services. Streetmap had accused Google of abusing its dominant position in two relevant product markets: general search engines and online search advertising.[79] The Court discussed the business model of both general search engines and online maps, finding that both services are offered for free to users and thus are 'classic examples of what economists refer to as multi-sided markets'.[80] It emphasised that while online maps and general search engines were separate relevant markets, they were also interconnected and, for this reason, could not – and did not need to – be precisely delineated.[81]

[73] ibid para 176.
[74] Case C-382/12 P *MasterCard v Commission* EU:C:2014:2201, paras 149, 159, 175.
[75] ibid para 237.
[76] Case C-67/13 P *Groupement des cartes bancaires v Commission* EU:C:2014:2204, para 8.
[77] Case T-491/07 *Groupement des cartes bancaires v Commission* EU:T:2012:633, para 104.
[78] Case C-67/13 P *Groupement des cartes bancaires v Commission* EU:C:2014:2204, para 99.
[79] High Court of Justice (Chancery Division), Case HC-2013-000090 *Streetmap v Google* (12 February 2016) paras 4, 35.
[80] ibid para 8.
[81] ibid para 15.

ii. European Commission

The Commission has acknowledged that two-sided markets 'generally affect each step of standard antitrust analysis, [including] product market definition'.[82] Over the past 15 years or so, it has dealt with a string of merger cases in innovative platform markets in which it at least mentioned two-sided market theory. What is common to all these cases is the Commission's close characterisation of the market and the market dynamics at issue, its lack of actually relying on two-sided market theory when delineating the relevant product market and its reluctance to commit itself to one market definition. Instead, it regularly found that no competitive concerns arose under the narrowest possible market definition, thus ultimately leaving the point of market definition open in a great many merger cases.

The *Google/DoubleClick* merger decision of March 2008 revolved around three markets, the first of which was online advertising.[83] The Commission found online advertising to constitute a two-sided market exhibiting network effects when characterising the market.[84] Second, the Commission regarded intermediation in online advertising to constitute a relevant market. It held that for this case, it did not have to analyse possible further sub-segmentation of this market. Advertising networks were explicitly held to constitute a two-sided market for websites and advertisers. The provision of online display ad serving technology was regarded as the third relevant market, with a possibility of sub-segmenting this market depending on whether advertisers or websites (publishers) were served.[85] Two-sided market theory was thus mentioned throughout, but the various market sides of a platform were not connected in the market analysis. In all three relevant markets, the Commission did not settle on a definitive market definition.

The *Microsoft/Yahoo! Search Business* merger decision of February 2010 gave the Commission a chance to further develop its stance on two-sided markets. It found that internet search engines constituted two-sided platforms, but did not explicitly apply two-sided market logic to its actual market delineation. Instead, it analysed four different markets: online advertising, intermediation in online advertising, distribution agreements on entry points and internet search.[86] In this case, it devoted considerably more space to two-sided platforms when discussing market characteristics and pledged that it would 'examine the position of search advertising platforms [as to] their ability to generate search traffic as well as their ability to

[82] European Commission in OECD (ed), *Policy Roundtable: Two-Sided Markets* (2009) DAF/COMP(2009)20, 158.

[83] *Google/DoubleClick* (Case COMP/M.4731) Commission Decision of 11 March 2008 [2008] OJ C184/10, para 56 (not settling the issue of whether search and non-search advertising space belong to the same market).

[84] ibid para 290; see also para 8, which refers to the market sides as 'main players'.

[85] ibid paras 20, 73, 81.

[86] *Microsoft/Yahoo! Search Business* (case COMP/M.5727) Commission Decision of 18 February 2010, paras 30, 33, 47, 61–86, 100.

sell that search traffic to advertisers'.[87] It then analysed both the user side and the advertiser side.[88]

The Commission's *Google Shopping* decision only mentions the multi-sidedness of online search platforms in a single paragraph[89] – relating to the incorporation of free user sides – while not using this knowledge in order to build a proper understanding of the market(s) at issue. In *Google Android*, general search services and search advertising are characterised as the two sides of a general search platform. The Android app store is seen as a two-sided market in which indirect network effects create a barrier to entry.[90]

In the credit and payment cards cases discussed above (see section III.B.i), the Commission also underlined its understanding of two-sided markets and the network effects at play.[91]

iii. US Courts

Decades before the theory of two-sided market appeared in economists' papers, the US Supreme Court stated in *Times-Picayune* (1953) that 'every newspaper is a dual trader in separate though interdependent markets; it sells the paper's news and advertising content to its readers; in effect, that readership is, in turn, sold to the buyers of advertising space'.[92] The Court concluded that the case at issue only concerned one of these markets. This, one would argue today, is a classic characterisation of a two-sided market in an offline setting. There has thus long been an awareness of two-sided market issues in the US, even if most US courts and antitrust authorities continue to analyse the multiple sides of platform markets separately.

The *Microsoft* judgment (2001) asked whether Microsoft had unlawfully maintained its monopoly position on the market for operating systems for Intel-compatible PCs.[93] The market at issue was defined as 'the licensing of all Intel-compatible PC operating systems worldwide'.[94] The Court of Appeals highlighted that operating systems 'function as platforms for software applications'.[95] Some criticised the DoJ's case against Microsoft as having been based on a contradictory theory that did not adequately take the positive externalities of the

[87] ibid para 100.

[88] ibid paras 101–08.

[89] *Google Search (Shopping)* (Case AT.39740) Commission Decision of 27 June 2017 [2018] OJ C9/11, para 159.

[90] *Google Android* (Case AT.40099) Commission Decision of 18 July 2018, paras 328, 638.

[91] *Visa International (Multilateral Interchange Fee)* (Case COMP/29.373) Commission Decision 2002/914/EC [2002] OJ L318/17, paras 82 f; *MasterCard* (Cases COMP/34.579, COMP/36.518, COMP/38.580) Commission Decision of 19 December 2007 [2009] OJ C264/8, paras 257 ff.

[92] *Times-Picayune Publishing Co v United States*, 345 US 594, 611 (1953).

[93] *United States v Microsoft*, 56 F3d 1448 (DC Cir 1995). For a detailed account of the succession of Microsoft cases in the US, see M Glader, *Innovation Markets and Competition Analysis: EU Competition Law and US Antitrust Law* (Cheltenham, Edward Elgar, 2006) 170 fn 259.

[94] *United States v Microsoft*, 253 F3d 34 (DC Cir 2001).

[95] ibid.

operating system into account.[96] While *Microsoft* may not have clearly incorporated any multi-sided market theory,[97] the market characterisation tentatively acknowledged the presence of a platform.

The *First Data* merger (2004) revolved around networks for debit transactions authorised through a personal identification number (PIN). First Data wanted to acquire Concord, and these companies owned the third-biggest and the biggest PIN debit networks in the US, respectively.[98] The complainants were concerned about a reduction of competition in the market for PIN debit network services.[99] The DoJ described the relevant product market as PIN debit network services, ie, the provision of 'the telecommunications and payments infrastructure that connects a network's participating financial institutions with merchant locations throughout the United States'.[100] This market was characterised as a two-sided one, with the market sides of financial institutions and merchants.[101] The DoJ also argued that the hypothetical monopolist test could be applied to two-sided markets, and in this case to the merchant side.[102] Eventually, the merger was allowed to go ahead, on the condition that First Data divested its PIN debit network.[103]

In the *KinderStart* case of 2007 (see section II.A above), the plaintiff had not alleged that Google's search engine generated revenue from other sources; essentially, the Court therefore stated that the plaintiff had missed the opportunity to rely on two-sided market theory when defining the relevant product market. It held that instead of relying on the free internet search market, KinderStart could have argued that search and search ads constituted one relevant market under antitrust law. However, it immediately stated that 'such a combined market … would suffer from [a] lack of breadth'.[104]

As in the EU, US courts have been confronted with two-sided markets in the credit card industry. Antitrust litigation in this area has produced a string of cases. For instance, in *Visa* (2003), the US Court of Appeals for the Second Circuit agreed with the District Court that two 'interrelated, but separate, product markets' existed, namely the market for general purpose cards (charge cards and credit cards) and the network services provided for these cards.[105] This general

[96] GL Priest, 'Networks and Antitrust Analysis' in ABA Section of Antitrust Law (ed), *Issues in Competition Law and Policy* (Chicago, ABA Publishing, 2008) 656.

[97] Auer and Petit (n 5) 452.

[98] *United States v First Data Corporation*, Case 1:03CV02169 (Competitive Impact Statement, DDC 23 January 2004) § I.

[99] *United States v First Data Corporation*, Case 1:03CV02169 (Verified Complaint, DDC 20 October 2003) para 5.

[100] *United States v First Data Corporation*, Case 1:03CV02169 (Competitive Impact Statement, DDC 23 January 2004) § II.B.

[101] *United States v First Data Corporation*, Case 1:03CV02169 (Plaintiff's Pretrial Brief, DDC 10 December 2003) § III.D.

[102] ibid §§ V.B, V.C.

[103] *United States v First Data Corporation*, Case 1:03CV02169 (DDC 25 May 2004).

[104] *KinderStart.com v Google*, Case C 06-2057 JF (RS) (ND Cal 16 March 2007) § III.1.a.i.

[105] *United States v Visa and MasterCard*, 344 F3d 229, 238 (direct quote), 239 (2d Cir 2003).

purpose market theory was again posited as one of two possible market definitions in the *Payment Card Interchange Fee* case (2008). Under it, MasterCard would be one of several competitors. The other possible market definition was the single brand market theory, in which MasterCard would 'by definition [be] the sole supplier'.[106] More recently, the same District Court got a chance to weigh in on two-sided markets in *American Express* (2015).[107] Siding with the plaintiffs' economic expert, the Court held the relevant product market to be 'the market for general purpose credit and charge card network services', refusing to include debit network services in the relevant market.[108] The Court also held that it would go too far to find for one transaction-based relevant market that would consist of the network services market and the card issuance market. Instead, it thought that it was 'both necessary and appropriate to define separate product markets that reflect the competitive realities', while at the same time acknowledging the two-sidedness of that market.[109] The Court found no authority for defining a relevant antitrust market consisting of the whole multi-sided platform.[110] Nevertheless, it underlined that it must take the two-sidedness of the credit card industry into account – not only when defining the relevant antitrust market, but also in its later antitrust analysis.[111] While references to two-sided market theory abound in *American Express*, the judgment has been criticised for not relying on the theory 'in the resolution of the case's thorniest issues'.[112] This judgment was subsequently reversed by the Court of Appeals for the Second Circuit (discussed below in section III.C), where the Court of Appeals more fully implemented two-sided market theory into its market definition. It has now been decided by the Supreme Court.

iv. US Agencies

The two US antitrust agencies have acknowledged a market's multi-sidedness on occasion. In her dissenting statement regarding the US *Google/Double Click* merger (2007), Commissioner Pamela Jones Harbour discussed network effects in online advertising and alluded to the intrinsic connection between Google's search engine and search advertising, thus implicitly analysing the case from the point of view of multi-sided markets.[113] She departed from the majority's opinion, showing that awareness of multi-sided market issues has started to reach the FTC.

[106] *Payment Card Interchange Fee and Merchant Discount Antitrust Litigation*, 562 FSupp2d 392, 399 f (EDNY 2008).
[107] *United States v American Express*, Case 10-CV-4496 (NGG) (RER) (EDNY 19 February 2015).
[108] ibid § III.
[109] ibid § III.A.1.
[110] ibid (citing *Times-Picayune* as authority to the contrary).
[111] ibid.
[112] Auer and Petit (n 5) 445.
[113] P Jones Harbour, 'Dissenting Statement, in the Matter of Google/DoubleClick' (20 December 2007) 5 f.

In 2010, the DoJ's Antitrust Division closed an investigation into an agreement between Yahoo and Microsoft that related to internet search and paid search advertising. Rather than posing a threat to competition, the DoJ found that this agreement was likely to increase competition by combining the search advertising technology of Microsoft and Yahoo, thus enabling them to better face competition from Google. While no antitrust market definition was carried out as such in the DoJ's statement at the closing of the case, the market at issue was characterised as the 'search and paid search advertising industry'.[114] This implies an understanding that a two-sided market is at play, without any explicit adoption of two-sided market theory.

C. The Incorporation of Multi-sidedness into the Antitrust Framework

To date, full incorporation of the economics of multi-sided markets into competition law cases is the rare exception rather than the norm. The first such case only occurred in 2016. In September of that year, the US Court of Appeals for the Second Circuit reversed the District Court's *American Express* judgment discussed above.[115] The Court of Appeals discussed the nature of two-sided markets at length, citing a number of respected economists, and focused on the interdependencies within a platform market.[116] After having explored the credit card industry, the Court returned to the District Court's market definition, which centred on network services only and warned that the latter court's 'definition of the relevant market in this case [was] fatal to its conclusion that Amex violated § 1' of the Sherman Act.[117] Highlighting that market definition must reflect the 'commercial realities' of a case, it found that the District Court had been wrong to exclude the market for cardholders from its market definition.[118] It also emphasised that when applying a SSNIP to a two-sided market, one 'must consider the feedback effects inherent on the platform' when prices rise on one market side.[119] The fact that the District Court refused to delineate the relevant product market as the 'entire multi-sided platform' constituted an error, according to the Court of Appeals.[120]

[114] US Department of Justice, 'Statement of the Department of Justice Antitrust Division on its Decision to Close its Investigation of the Internet Search and Paid Search Advertising Agreement between Microsoft Corporation and Yahoo! Inc.: Investigation Shows that Agreement Not Likely to Reduce Competition' (18 February 2010).

[115] *United States v American Express*, 838 F3d 179 (2d Cir 2016).

[116] ibid 185 f.

[117] ibid 196.

[118] ibid 197. The Court then went on to distinguish the *American Express* case from its earlier *Visa* case.

[119] ibid 200.

[120] ibid.

The *American Express* case was decided by the US Supreme Court in June 2018. In a tight 5:4 majority, the Court relied on a wealth of economic literature and characterised the market at issue as a two-sided transaction platform, with American Express' two market sides consisting of cardholders and merchants. In this type of two-sided market, indirect networks effects are important, and the platform must carefully make pricing decisions based on demand on both market sides. Only other two-sided transactional platforms, which equally have cardholders and merchants on board, can compete with such a platform that is in the business of credit card transactions between its market sides. The Supreme Court concluded that the platform market for credit card transactions had to be assessed as one single relevant antitrust market.[121] The *American Express* approach to delineating multi-sided markets only applies to transaction markets, meaning that many types of digital or big tech platforms will not fall under this newly minted rule.[122] In fact, the Court acknowledged that '[n]ontransaction platforms, by contrast, often do compete with companies that do not operate on both sides of their platform.'[123]

The dissenting opinion held that case law contradicted the majority's conclusion that one should treat credit card transactions as one single market.[124] In the words of Justice Breyer, 'our precedent provides no support for the majority's special approach to defining markets involving "two-sided transaction platforms"'.[125] Merchant- and shopper-related card services are not substitutes, but merely complements. Therefore, they should not be included in the same relevant market for the purposes of a rule of reason analysis under section 1 of the Sherman Act. As the anti-competitive conduct that American Express was accused of only related to the merchant side, he held this to be the market that the majority should have considered.[126] However, the minority opinion also highlighted that the case did not need to rely on market definition in the first place, as direct evidence had already suggested that American Express' steering provisions had anti-competitive effects. It interpreted the economic literature as meaning that a flexible approach to two-sided markets is necessary, depending on whether or not the behaviour in question revolves around the network effects.[127] That was arguably not the case in *American Express*.

To date, there do not appear to be any European cases that have as fully considered two-sided market theory in their antitrust analysis as the US Court of Appeals

[121] *Ohio v American Express*, 585 US ___ (2018), 138 SCt 2274, 2286 f (2018).

[122] T Wu, 'The *American Express* Opinion, the Rule of Reason, and Tech Platforms' (2019) 7 *Journal of Antitrust Enforcement* 117, 118, 123–26. Remarking on the static nature of the transaction/non-transaction distinction that may hinder a dynamic analysis, see D Mandrescu, 'Applying (EU) Competition Law to Online Platforms: Reflections on the Definition of the Relevant Market(s)' (2018) 41 *World Competition* 453, 461.

[123] *Ohio v American Express*, 585 US __ (2018), 138 SCt 2274, 2287 fn 9 (2018).

[124] ibid 2294 f (citing *Times-Picayune*).

[125] ibid 2300.

[126] ibid 2295 f.

[127] ibid 2296, 2300 f.

for the Second Circuit and the US Supreme Court did in *American Express*. The three *Google* cases in Europe would have been ideal candidates for applying insights on multi-sided markets in innovative environments. In these cases, the European Commission investigated anti-competitive behaviour in platform markets under Article 102 TFEU – and thus for the first time left the realm of merger control or anti-competitive agreements. The *Google* cases also differ from *American Express* in two important respects: they concern non-transaction multi-sided markets and they relate to a digital market rather than a traditional market. However, as was seen above, the Commission's market definition in *Google Shopping* was more traditional than anticipated. The first decision, which was published in December 2017, saw no issue in the fact that one customer group receives services for free.[128] It also acknowledged that '[g]eneral search services and online search advertising constitute the two sides of a general search engine platform'.[129] Nevertheless, it then focused on general online search while eclipsing the revenue-generating side of search advertising and the possibility of the overall platform to constitute the relevant market. The Commission proceeded in a similar fashion when delineating the market for comparison shopping services.[130] This market definition lays the legal basis for the case and is 'perhaps too simple'.[131] In its *Google Android* decision, the Commission referred to seven 'products concerned', but ultimately relied on four relevant product markets: the market for the licensing of smart mobile operating systems, the market for Android app stores, the market for the provision of general search services and the market for mobile web browsers that are not specific to a particular operating system.[132] In that decision, the Commission may have delineated the market so as to fit the case it wanted to make, excluding competing search functionalities from the market or overly narrowly viewing the market as pre-installed apps on the Android operating system.[133]

As a side note, it is interesting to remark that in early 2013, the FTC decided to close its investigation into antitrust concerns relating to Google's online search results which treated its own products more favourably than those of its competitors, holding that this search bias should be understood as 'innovations that improved Google's product and the experience of its users'.[134] In its press release, the FTC did not comment on the relevant market at issue in this investigation.

[128] *Google Search (Shopping)* (Case AT.39740) Commission Decision of 27 June 2017 [2018] OJ C9/11, paras 158–60.

[129] ibid para 159.

[130] ibid paras 191 ff.

[131] C Bergqvist, 'Google and the Search for a Theory of Harm' (2018) 39 *European Competition Law Review* 149, 149.

[132] *Google Android* (Case AT.40099) Commission Decision of 18 July 2018, paras 73 (direct quote), 217. As the decision was only published when this study went to press, a detailed analysis will follow at a later stage.

[133] On this concern, see already GA Manne, 'The EU's Android Antitrust Complaints Are Contrived' *Wired* (20 November 2016), www.wired.com/2016/11/eus-android-antitrust-complaints-contrived.

[134] Federal Trade Commission, 'Google Agrees to Change its Business Practices to Resolve FTC Competition Concerns in the Markets for Devices like Smart Phones, Games and Tablets, and in Online Search' (3 January 2013).

However, it is likely that it only considered online search as the relevant market, rather than the non-transaction platform market comprising online search and search advertising. The FTC investigation had been preceded by a hearing on the power of Google before the US Senate's Subcommittee on Antitrust, Competition Policy and Consumer Rights (2011), in which the question whether Google's online search favoured its own operations was debated.[135] In mid-2016, it seemed that the FTC was re-evaluating its previous position and possibly re-opening investigations into Google's search bias.[136] The political climate in the US more generally seems to gear up to subjecting big tech platforms to antitrust scrutiny.[137]

D. Analysis

i. Stages of Development Regarding Multi-sidedness in the Case Law

In the case law discussed above, it was seen that antitrust cases can be found at one of three stages of development.[138] At the first stage, courts and authorities essentially ignore – knowingly or inadvertently – the fact that the markets they are scrutinising are two- or multi-sided markets. At this level, we find a 2015 French case, but also a 2014 European Commission decision and an FTC investigation from 2007. In these cases, neither the characterisation of the industry nor the market definition itself referred to the possibility that online maps, online advertisements or electronic communication services could form part of a multi-sided innovative platform market. Given the recent nature of some of these cases, this is rather surprising.

At the second stage, courts and authorities acknowledge the multi-sidedness of the markets at issue, but do not yet apply that insight to their actual antitrust market definition. It is noteworthy how EU and US courts and competition authorities repeatedly defined separate product markets for each market side rather than connecting them based on network effects. In the case of the European Commission, this takes the form of referring to two-sided market theory in the characterisation of the market, but without actually incorporating this knowledge

[135] Subcommittee on Antitrust, Competition Policy and Consumer Rights, 'The Power of Google: Serving Consumers or Threatening Competition?' (J–112–43, Washington, DC, 21 September 2011).

[136] N Scola, 'Sources: Feds Taking Second Look at Google Search' *Politico* (11 May 2016), www.politico.com/story/2016/05/federal-trade-commission-google-search-questions-223078.

[137] See, eg, B Brody, 'U.S. Antitrust Chief Blasts Google and Amazon, Citing Historic Breakups' *Bloomberg* (11 June 2019), www.latimes.com/business/la-fi-tn-antitrust-google-amazon-technology-20190611-story.html.

[138] In a similar attempt at making sense of the case law, Auer and Petit distinguished between cases from the EU and US in which two-sided market theory was (somewhat) taken into account, and cases in which it was not explicitly taken into account; see Auer and Petit (n 5) 440–57. Auer and Petit's reading of the case law somewhat differs from the one proposed here, perhaps because we here exclusively focus on market definition.

into the market delineation itself. While the Court of Justice acknowledged the two-sided nature of the markets for credit cards and banking cards in two 2014 judgments, two US District Courts did the same in their judgments relating to PIN debit network services (2004) and credit cards (2015). Indeed, the US Supreme Court already referred to something akin to two-sided market theory as early as 1953. Interestingly, these cases did not relate to archetypes of innovation, but to more conventional markets (credit cards, debit card networks, newspapers), and in each of these cases, the courts concluded that it was sufficient to scrutinise one market side for the purposes of the case at hand. An FTC dissenting statement, a District Court case and a DoJ investigation serve as evidence that there is growing awareness of indirect network effects and two-sided market theory in the US, especially in relation to more innovative markets.

At a third stage, courts and authorities attempt to incorporate two-sided market theory into the antitrust law framework, also at the stage of market definition. Multi-sided market theory has an important impact on market definition at stage three. To date, only the US Supreme Court in *American Express* (2018) has relied on two-sided market theory in order to entirely inform its market definition. This was heavily criticised by the minority opinion, which held that the 5:4 majority erred in the way it applied economic theory to market definition. The *American Express* majority set out that the relevant market had to consist of the platform market as such, comprising both the merchant side and the cardholder side. In Europe, there was a certain expectation that the ongoing *Google* cases might allow the Commission to develop a coherent approach to the multi-sided markets at issue. With the first decision of June 2017 now publicly available, this does not appear to have happened – the ball is now with the General Court.

Market delineation in platforms is still evolving. Neither in the EU nor in the US can we discern a full-fletched, coherent analytical competition law approach to multi-sided platform markets. The three stages discerned above show that antitrust authorities and courts are still tackling market delineation in multi-sided markets at different levels of refinement. These are not linear or chronological stages, but are present in different courts or competition authorities throughout these jurisdictions at the same point in time. This indicates that the incorporation of multi-sided markets into antitrust market definition is an ongoing process in both the EU and the US. In particular, the US Supreme Court's *American Express* opinion and dissenting opinion have shown that antitrust is still in search of a well-founded market definition as it applies to platform markets, and that this must be furnished with a well-elaborated theory of harm as it applies to platform markets. Arguably, a court's or an authority's willingness to engage in this discussion will prove more influential on whether or not multi-sided market theory is taken into consideration than the jurisdictional setting in which a particular court or authority finds itself.

Under both EU and US competition law, it should be possible to factor multi-sidedness into the market definition by relying on patterns of substitutability. Analysing the varying stages at which competition law actors find themselves in, the question arises why so many EU and US courts and authorities do not go beyond merely acknowledging the two-sided nature of an antitrust market. Given the relatively recent focus on innovative platform markets in scholarship, which went hand in hand with fast-paced technological developments since the 1990s, this tentative incorporation of two-sided market theory into the antitrust law framework is not surprising. As will be seen below, scholarship has suggested several ways in which platform markets could be delineated for antitrust purposes, and courts and authorities need to rely on trial and error in order to discover which of these methods best meets the system requirements and goals of their respective antitrust laws. The following analysis suggests ways of moving beyond the current state of affairs, providing examples of how antitrust market definition could more efficiently incorporate two-sided market theory.

ii. Doctrinal Views on Multi-sidedness

In antitrust scholarship, the view prevails that antitrust market definition should take a market's multi-sidedness into account – if not always, then in most cases.[139] Ignoring a market's multi-sidedness can lead to a number of wrong assumptions.[140] However, there is no agreement on how to achieve this incorporation of two-sided market theory into competition law or regarding which cases would not require such a change in the analytical framework. It is also an open question as to whether multi-sidedness is not a question of degree.[141] In the following, we trace some major arguments and approaches in current antitrust scholarship, adding additional analysis.

There is a 'conceptual proliferation' in industrial organisation concerning the definition of what constitutes a multi-sided market.[142] What antitrust law requires is a sensible and well-working definition. An overly inclusive definition will capture many instances that do not require the lens of two-sided market theory from an antitrust perspective, while cases of multi-sided platforms may escape an overly restrictive definition. Over- or under-enforcement of competition law may be the consequence.[143]

Markets with the following characteristics are here understood to constitute multi-sided markets: (i) two or more customer groups are present who (ii) need

[139] It has also been suggested that both the delineation of one single market as well as the delineation of separate markets for each of the market sides can be accommodated under the substitutability concept; see S Wismer and A Rasek, 'Market Definition in Multi-Sided Markets' in OECD (ed), *Rethinking Antitrust Tools for Multi-Sided Platforms* (2018) 56 f, www.oecd.org/daf/competition/Rethinking-antitrust-tools-for-multi-sided-platforms-2018.pdf.

[140] See, eg, Wright (n 3) 47–51.

[141] S Holzweber, 'Market Definition for Multi-Sided Platforms: A Legal Reappraisal' (2017) 40 *World Competition* 563, 567.

[142] Auer and Petit (n 5) 432.

[143] On error costs in relation to market definition and innovation, see ch 3, section III.C.

each other in some way, although this does not need to be reciprocal in a strict sense, and who (iii) cannot capture the indirect network effects by themselves and therefore (iv) rely on an intermediary (platform) to capture these indirect network effects.[144] Of these criteria, the second and third are particularly noteworthy, as they require the presence of indirect network effects as well as the customer groups' inability to capture these without the platform acting as an intermediary. The definition encompasses indirect network effects that are one-directional, ie, cases in which one market side needs the other more – or for different purposes – than vice versa. It thus captures a multitude of digital platforms in which the platform provider captures indirect network effects by providing an entirely different service to one user group (eg, online search, social network) compared to the other user group (eg, aggregated user data, targeted online advertising).

Multi-sided markets might appear similar to more traditional vertical relationships in many ways. However, they differ in a decisive respect: vertical agreements are commercial relationships between two or more parties that act on different levels of the distribution chain,[145] thereby fulfilling characteristic (i) of the above definition. They also need each other in some way, as required by characteristic (ii). However, in traditional vertical relationships, the parties are able to enter into a contractual relationship amongst themselves, and there is no intermediary that captures indirect network effects between the parties for them. In some cases, the lines may be blurry. For instance, several authors have concluded that supermarkets may constitute multi-sided markets with consumers on the one market side and suppliers on the other side.[146] In many innovative markets, especially in digital market environments, the question of whether or not they constitute multi-sided markets can be answered in a rather straightforward manner.

Once a definition of what constitutes a multi-sided market is agreed on, the question is how to incorporate this knowledge into the antitrust legal framework – for the present purposes, market definition. A multitude of approaches exist in this respect, several of which have influenced the case law. In as early as 2004, Julian Wright argued that both market sides of a two-sided market must be taken into account in the antitrust analysis, as well as the interdependence that exists between the various market sides.[147] In 2013, Florence Thépot suggested proceeding in two steps when delineating two-sided markets: first, the different market sides are defined; and, second, the platform's two-sidedness is considered and the platform is assessed as its own antitrust market. This allows the analysis to grasp the interconnectedness between the market sides.[148] A similar approach was advocated by

[144] Based on Evans and Schmalensee, 'Antitrust Analysis' (n 20) 409.

[145] Commission Regulation (EU) 330/2010 on the application of Article 101(3) of the Treaty on the Functioning of the European Union to categories of vertical agreements and concerted practices [2010] OJ L102/1 (VABER), art 1(1)(a).

[146] See Auer and Petit (n 5) 439 and, in particular, 460.

[147] Wright (n 3) 61.

[148] Thépot (n 10) 205 f. This has been called 'intermarket analysis'; see Holzweber (n 141) 577 ff.

Joyce Verhaert, who argued that the substitutability on both market sides as well as the platform itself that connects the market sides need to be considered.[149]

David Evans and Michael Noel argue that whether or not a market is multi-sided may have an influence on the ensuing antitrust analysis, but this is not always necessarily the case.[150] The interdependency of demand on the various market sides will affect market definition,[151] and pricing in multi-sided platforms is complex.[152] Market definition in multi-sided platforms should be seen as less of a black-and-white affair and more of a grey area in which strict boundaries can rarely be found.[153] From a legal standpoint, the question is how to deal with this flexible approach – which the dissenting opinion highlighted in *American Express* – for antitrust market definition.

In a paper from 2014, Lapo Filistrucchi and others propose an approach to the delineation of multi-sided markets that differentiates between transaction and non-transaction multi-sided markets. As discussed above (see section II.B), in the case of non-transaction markets – which often provide a service for free to the consumer side – they argue that separate antitrust markets should be defined, even if they are clearly interrelated.[154] This interconnection should then be reflected in the competition law analysis that follows.[155] Transaction markets, on the other hand, which are characterised by a transaction occurring between the two market sides, should be delineated as one comprehensive antitrust market.[156] This approach to transaction multi-sided markets was relied upon by the majority in the US Supreme Court's *American Express* case (see section III.C) and is also supported by other authors.[157] Such a transaction market does not rely on the substitutability criterion that has traditionally been the centre of market delineation.[158] Where there is no direct transaction between the market sides, a separate relevant market should be delineated for each market side, while taking interdependencies into

[149] J Verhaert, 'The Challenges Involved with the Application of Article 102 TFEU to the New Economy: A Case Study of Google' (2014) 35 *European Competition Law Review* 265, 269 f.

[150] DS Evans and M Noel, 'Defining Antitrust Markets When Firms Operate Two-Sided Platforms' [2005] *Columbia Business Law Review* 667, 681, 696.

[151] Evans and Schmalensee, 'Two-Sided Platforms' (n 69) 689.

[152] Evans, 'Antitrust Economics' (n 8) 339 ff.

[153] Evans and Noel (n 150) 697.

[154] Filistrucchi et al (n 39) 298, 300–02. See also L Filistrucchi, 'Market Definition in Multi-Sided Markets' in OECD, *Rethinking Antitrust Tools* (n 139) 42.

[155] Filistrucchi et al (n 39) 322.

[156] ibid 298, 300–02.

[157] S Wismer, C Bongard and A Rasek, 'Multi-Sided Market Economics in Competition Law Enforcement' (2017) 8 *Journal of European Competition Law & Practice* 257, 260.

[158] See the discussion of this 'interrelated markets' test in GA Manne, 'In Defence of the Supreme Court's "Single Market" Definition in *Ohio v American Express*' (2019) 7 *Journal of Antitrust Enforcement* 104, 107, 109 f. In Europe, by contrast, the Court of Justice has held that the credit card market consisted of separate yet interrelated markets; see Case C-67/13 P *Groupement des cartes bancaires v Commission* EU:C:2014:2204, para 79.

account at a later stage.[159] However, it might not be as straightforward as is generally thought to distinguish transaction from non-transaction markets.[160]

Another suggestion in the literature has been to focus on demand substitutability on each market side, an enterprise which would necessarily lead to separate relevant markets for each market side.[161] This would again need to be followed by an analysis of the interconnection between these relevant markets.

Instead of applying two-sided market theory, Giacomo Luchetta suggests framing Google's search engine as an intermediary of personal information that operates in the market for personal user information, with the search engine side as the input market and the advertisers buying this information in order to better target their audience.[162] Google – as the buyer of personal information on the input market – would very likely not have a dominant position on this market.[163] This approach only takes one-directional network effects into account, disregarding any value search engine users might derive from high-quality ads. Another issue that will arise is the question which other market players (eg, social networks that also generate lots of personal user information that is monetised) can be regarded as direct competitors of the search engine based on a similar business model. If the answer is that social networks have different input markets, but substitutable output markets, then this is but a rephrasing of the two-sided market theory: one side of a platform competes with one side of another platform. But the data gathering ('input market') – through whichever channel – may become more shielded from antitrust intervention. While Luchetta's analytical framework seems to solve many lingering market definition questions at first sight, it also raises questions as regards the criterion of demand and supply interchangeability – one of the very premises on which antitrust market definition is based. If one models a two-sided market as a vertical relationship with an upstream and a downstream market, then it is precisely the indirect network effects that multi-sided market theory wants to capture that move out of reach. In addition, the input market will often face the difficulty that users provide data and attention – later used to lure in advertisers – and receive a free service in exchange. Outside the framework of a multi-sided market, it may be more difficult to grasp these non-monetary exchanges as a commercial activity.

Another approach that strongly focuses on data was put forward by Inge Graef. She underlines the high importance of data in innovative platforms and describes how it is used as the essential input in many digital platform markets.[164] She acknowledges that in cases in which data is not used as a commodity, but

[159] Wismer, Bongard and Rasek (n 157) 259.

[160] J-U Franck and M Peitz, 'Market Definition and Market Power in the Platform Economy' (CERRE Report, May 2019) 24.

[161] P Solano Díaz, 'EU Competition Law Needs to Install a Plug-in' (2017) 40 *World Competition* 393, 397 (with reference to the case law).

[162] Luchetta (n 6) 197, 201.

[163] ibid 201 f.

[164] Graef (n 16) 475 ff.

stays within a company for its own business purposes, it cannot be defined as its own relevant product market under current antitrust laws as it lacks a commercial element.[165] However, data represents a valuable specialised asset within digital platforms, and might as such be taken into account as potential competition or as a hypothetical future market.[166] Nicolo Zingales also highlights the importance of data in multi-sided digital platforms, not necessarily as its own product market, but as a key barrier to entry and as an opportunity for important market players to capture neighbouring markets based on their lead in data.[167] Consequently, he suggests regarding user data as a constraining factor on competitive restraints.[168] A 'data market' might also represent a possible solution for a forward-looking market definition framework which incorporates potential competition, and a useful tool within the realm of Article 102 TFEU or section 2 of the Sherman Act.[169]

Whether or not the business model of multi-sided platforms is taken into account at the stage of market delineation may not have a bearing on the end result of a case where the substantive analysis appropriately reflects the multi-sidedness. However, we may lose sight of the economic realities of the market if we ignore a market's multi-sidedness when delineating it for competition law purposes.[170]

iii. Platform Interchangeability: Conceptual Issues

The combination of innovative market environments with platform markets renders antitrust market definition in this area a very complex undertaking, as both of these givens need to be factored in. In the interests of analytical clarity and legal certainty, the simplest possible approach should be taken to two-sided markets. While some authors and courts have suggested that transaction and non-transaction multi-sided markets should be defined differently,[171] what is proposed here is to make this distinction at a later stage, when analysing the antitrust issues at stake.

As Dirk Auer and Nicolas Petit have found, in the recent past 'authorities seemed to intuitively grasp the key findings of the theory of two-sided markets … even though their decisions were not anchored in formal two-sided market reasoning'.[172] While this insight does not replace the need to formally incorporate two-sided market theory into the antitrust law framework, it does show that

[165] ibid 490, 492.
[166] ibid 492, 494 f.
[167] N Zingales, 'Product Market Definition in Online Search and Advertising' (2013) 9 *Competition Law Review* 29, 38 f.
[168] ibid 42.
[169] P Jones Harbour and TI Koslov, 'Section 2 in a Web 2.0 World: An Expanded Vision of Relevant Product Markets' (2010) 76 *Antitrust Law Journal* 769, 773, 785 f.
[170] G Gürkaynak et al, 'Multisided Markets and the Challenge of Incorporating Multisided Considerations into Competition Law Analysis' (2017) 5 *Journal of Antitrust Enforcement* 100, 109.
[171] See already above, sections II.B and III.C.
[172] Auer and Petit (n 5) 457.

authorities are already latently aware of two-sided markets, but perhaps not of all their consequences, as some of the cases analysed above have shown.

Market definition cannot always provide the clear market boundaries that antitrust sometimes requires, and especially not in two-sided markets. However, it can be very insightful to carry out a proper characterisation of the market at issue, including its multi-sidedness.[173] Where reliable quantitative econometric data is not readily available, one has no other option but to resort to qualitatively assessing the relevant market in multi-sided platforms.[174] This is precisely what the European Commission has been doing for a long time, even if its analysis was then not particularly informed by two-sided market theory. The multi-sidedness of a market will also help us to understand how the market works and how market sides can constrain each other's market power. Despite all the uncertainties connected to market definition in multi-sided markets, it remains an important step in the legal analysis.[175]

In some digital platform cases, one may need to focus on the good or service that is sold via those platforms rather than on the digital platform itself.[176] This will then lead to an analysis in which the digital platform constitutes a mere marketplace rather than a market participant. However, if the focus is on the narrower product market – eg, high-quality crystal wine glasses sold on eBay and Amazon Marketplace – we might lose sight of the very relevant fact that trading platforms may be exerting market power over their user sides that goes far beyond a single product. Similarly, focusing on the advertising sold through social media platforms will lose sight of the other user side. Indeed, one might question whether the FTC's non-challenge of Facebook's acquisition of Instagram (2012)[177] and the European Commission's non-challenge of Facebook's acquisition of WhatsApp (2014)[178] might have been such instances in which new kinds of anti-competitive effects were overlooked because they were seen as occurring outside of the relevant market.

When addressing antitrust issues that arise in online platforms, one should question whether we can continue our exclusive reliance on product substitutability for defining antitrust markets on the product level or whether we need to consider platform interchangeability as well. Can one distinguish the antitrust market for a product from the antitrust market for the distribution of the very

[173] A Gebicka and A Heinemann, 'Social Media & Competition Law' (2014) 37 *World Competition* 149, 154; Robertson, 'Delineating Digital Markets' (n 1) 143.

[174] GA Manne and JD Wright, 'Google and the Limits of Antitrust: The Case against the Case against Google' (2011) 34 *Harvard Journal of Law & Public Policy* 171, 196. Qualitative assessment is usually the starting point for any market definition; see PJ Davis and E Garcés, *Quantitative Techniques for Competition and Antitrust Analysis* (Princeton, Princeton University Press, 2009) 166.

[175] See R Podszun, 'The Arbitrariness of Market Definition and an Evolutionary Concept of Markets' (2016) 61 *Antitrust Bulletin* 121, 128 f.

[176] Monopolkommission (n 49) paras 367, 370.

[177] Federal Trade Commission, 'Proposed Acquisition of Instagram by Facebook, Case 121-0121' (22 August 2012).

[178] *Facebook/WhatsApp* (Case COMP/M 7217) Commission Decision of 3 October 2014 [2014] OJ C417/4.

same product – ie, the platform? This certainly departs from traditional antitrust law analysis as discussed in Chapter 2. However, in the light of digital developments over these past two decades, such an enhanced, two-level approach might become worth considering.[179] There are two main options for conceptualising such an approach. The first option assumes that products and platforms need to be assessed separately, thus delineating separate relevant markets for these. The competitive assessment will need to take into account the relationship between the platform and the products. The second option models this interconnectedness of platform and product markets on traditional aftermarket scenarios (on these, see Chapter 7).[180] Such an approach could grasp the interoperability issues related to platforms as gatekeepers. Market definition would then conceptualise platforms in a similar vein as we analyse aftermarkets: as primary and secondary markets that, depending on their relationship, need to be delineated either as a system market or as multiple markets. For instance, a mobile operating system would constitute the primary market that can act as a gatekeeper to a range of secondary markets. The same applies to sharing economy platforms such as ride sharing platforms as primary markets, with the taxi ride constituting the service as the secondary market. Competition issues arising in the secondary markets (eg, regarding app developers, online retailers or Uber drivers) would only be considered problematic if competition in the primary market (eg, mobile operating systems, digital assistants or the Uber platform) does not lead to a competitive environment.[181] This conceptualisation is useful because, like regular primary markets, platforms can have an influence on interoperability or accessibility to the secondary market. Only if there is sufficient interchangeability amongst platforms can we speak of a broader market that encompasses several such platforms, which would then broaden the market definition – just as in 'ordinary' primary and secondary markets. These options are further developed and discussed below (see sections II.A, II.B and II.C.vii in Chapter 10).

IV. Conclusion

Market definition in multi-sided platforms has attracted widespread antitrust attention because of the success of a small number of digital platforms.

[179] See also Mandrescu (n 122) 461 ('platform substitution'); M Bourreau and A de Streel, 'Digital Conglomerates and EU Competition Policy' (CERRE Report, March 2019) 26; J Crémer, Y-A de Montjoye and H Schweitzer, 'Competition Policy for the Digital Era' (Report for DG Competition, 3 April 2019) 47 f ('market for ecosystems' at 48).

[180] Holzweber (n 141) 576; Bourreau and de Streel (n 179) 26.

[181] Here, the fact that Apple does not license its iOS might become relevant – just as with data or trade secrets, something that is kept in-house does not generally constitute a relevant product market; it may constitute a potential market. One might also liken this situation to a proprietary aftermarket as discussed in ch 7, section III.A.

Interestingly, however, the case law on both sides of the Atlantic has primarily dealt with multi-sided platforms in the context of credit cards. While that case law is often applied by analogy, this may not always be feasible.

In multi-sided markets, a remarkable clash between traditional and more dynamic approaches to market definition can be observed. Several stages in the case law were highlighted that represent a growing awareness of current economics literature in the courts. Free services need to be taken into account just like 'regular' market sides in these platform markets, as the business model of multi-sided platforms cannot be understood without them. In a range of cases both from the EU and the US, it could be seen that some courts and authorities – until rather recently – simply ignored the multi-sidedness of a relevant product market. This is not necessarily a negative finding, as for some cases, whether a market is one- or multi-sided will not have any consequences. In other cases, a platform market's multi-sidedness was acknowledged without actually incorporating this finding into the analysis. It is only very recently that some courts and authorities have come to realise how far-reaching the impact of a market's multi-sidedness might be. In *American Express*, the finding of a comprehensive transaction platform market was the key element that decided the case in favour of the defendant.

From a conceptual perspective, inspiration for the delineation of multi-sided markets can come from the delineation of aftermarkets, thus leading to a platform market ('system market') or multiple markets (platform and single market sides).[182] If aftermarkets are relied upon as an inspiration for delineating multi-sided markets, it will be important to recognise the role of the platform itself as an intermediary or a provider of the ecosystem.[183] Therefore, a two-step market analysis may be pertinent in order to properly characterise the relevant market(s) both at the level of the platform and on all market sides. The question of multi-homing on one or all market sides will constitute an important consideration in order to understand competition amongst platforms or ecosystems, as will the impact of big data and big analytics that may provide particular leverage to platform operators that are active in a number of (digital) markets.

[182] See ch 10, sections II.B and II.C.vii.

[183] See above, n 179.

9

Further Issues Concerning Innovation and Market Delineation

I. Standard Economic Tests and Innovation

Antitrust market definition relies to a large degree on industrial (neoclassical) economics. At the time when the market definition tests were formulated, neoclassical economics regarded price as the most important competitive parameter, homogeneous products as the archetype and innovation as a negligible factor.[1] Today, this perception of markets has considerably changed – yet the legal test of how to define a relevant antitrust market has not. Many of the static parameters for market definition were incorporated into the antitrust laws, such as legal presumptions relying on market share thresholds as a measure of market power or concentration ratios in the realm of mergers. Static analysis is still the rule rather than the exception in competition cases.

When writing about industrial economics, a competition lawyer enters dangerous territory. Some may even say that a competition lawyer not trained in economics may wish to abstain from writing about standard economic tests. However, the practical significance of these tests and the fact that they have been relied upon in case law, soft law and legal provisions make it impossible to disregard this issue. Nevertheless, the present chapter should be understood as a lawyer's perspective on antitrust economics.

As David Teece has cautioned, when accounting for innovation, one must be acutely aware of the trap that many economists fall into – preach innovation only to apply static analytical tools.[2] Three economic tools and their applicability in innovative market environments will be considered in the following: the hypothetical monopolist or SSNIP test, market shares and concentration levels.[3] The aim

[1] M Glader, *Innovation Markets and Competition Analysis: EU Competition Law and US Antitrust Law* (Cheltenham, Edward Elgar, 2006) 7.

[2] DJ Teece, 'Favoring Dynamic over Static Competition: Implications for Antitrust Analysis and Policy' in GA Manne and JA Wright (eds), *Competition Policy and Patent Law under Uncertainty: Regulating Innovation* (Cambridge, Cambridge University Press, 2011) 208 f.

[3] For an overview of quantitative economic tools for market definition, see ABA Section of Antitrust Law, *Econometrics: Legal, Practical, and Technical Issues*, 2nd edn (Chicago, ABA Publishing, 2014) 216; PJ Davis and E Garcés, *Quantitative Techniques for Competition and Antitrust Analysis* (Princeton, Princeton University Press, 2009) 169 ff.

is not to prescribe which methods experts in antitrust economics should rely on in providing an antitrust court with evidence, but rather to allow antitrust courts and practitioners to critically reflect on: (1) the limitations of this evidence in the context of dynamic markets; and (2) the problem of basing legal presumptions on these seemingly straightforward measures.

A number of other econometric tools used by competition economists, which are not discussed further here, also rely on prices or markets, such as upward pricing pressure,[4] critical loss analysis[5] or the Lerner index.[6] Concerns relating to these instruments' adaptability to innovative market environments have equally been voiced.[7]

A. The Hypothetical Monopolist Test

Antitrust authorities like to rely on the concept of the hypothetical monopolist or SSNIP test when delineating the relevant market, using economic expertise that is available within the authorities (see sections I.F.ii and II.F.ii in Chapter 2). The test operationalises the product substitutability criterion that courts have relied upon on both sides of the Atlantic,[8] but depends on sometimes misleading assumptions. Courts generally make less use of this formalised framework, adopting a more legalistic or qualitative approach to the issue of market definition instead. However, particularly in the US, one does find plenty of examples in which courts relied on the hypothetical monopolist test. In addition, economic experts will often present the court with opinions relying on this test.

i. From Policy to Case Law

In the US, the hypothetical monopolist test was first embraced in the DoJ's 1982 Merger Guidelines and is still regularly carried out in the context of mergers.[9] The aim of the test is 'to identify a set of products that are reasonably interchangeable

[4] See J Farrell and C Shapiro, 'Antitrust Evaluation of Horizontal Mergers: An Economic Alternative to Market Definition' (2010) 10 *BE Journal of Theoretical Economics* 1, 2.

[5] BC Harris and JJ Simons, 'Focusing Market Definition: How Much Substitution is Necessary?' (1989) 12 *Research in Law and Economics* 207; GJ Werden, 'Demand Elasticities in Antitrust Analysis' (1998) 66 *Antitrust Law Journal* 363.

[6] AP Lerner, 'The Concept of Monopoly and the Measurement of Monopoly Power' (1934) 1 *Review of Economic Studies* 157; JA Ordover, AO Sykes and RD Willig, 'Herfindahl Concentration, Rivalry, and Mergers' (1982) 95 *Harvard Law Review* 1857; *FTC v Swedish Match*, 131 FSupp2d 151, 161 (DDC 2000).

[7] See, eg, RJ van den Bergh and A Giannaccari, 'L'approcio più economico nel diritto comunitario della concorrenza: Il più è troppo o non (ancora) abbastanza?' (2014) XVI *Mercato concorrenza regole* 393, 430.

[8] Similarly for the US, see DF Broder, *US Antitrust Law and Enforcement: A Practice Introduction*, 3rd edn (Oxford, Oxford University Press, 2016) para 5.19.

[9] See section I.F.i in ch 2.

with a product sold by one of the merging firms'.[10] The test analyses whether a hypothetical monopolist would increase the price of a certain product in a small but significant and non-transitory fashion so as to maximise its profit.[11] It looks at the product and the geographical dimension of the relevant market, recognising that a number of factors – eg, transportation costs, language and service availability – can limit the geographical space within which products compete with each other.[12] The result – based on the smallest market principle – is a relevant antitrust market with a product and a geographical dimension.[13]

Through the antitrust agencies, the hypothetical monopolist test has also entered US case law. In *FTC v Staples* (1997), the District Court for the District of Columbia emphasised that it would often refer to the five per cent benchmark in its analysis because this was the one relied upon in the Horizontal Merger Guidelines.[14] In *FTC v Tenet Healthcare* (1998), the District Court for the Eastern District of Missouri discussed evidence relating to market definition within the framework of the hypothetical monopolist test of the Horizontal Merger Guidelines,[15] and on appeal so did the US Court of Appeals for the Eighth Circuit.[16] In another case, *US v SunGard Data Systems* (2001), the District Court for the District of Columbia emphasised that the product substitutability test stemming from *Brown Shoe* and *Cellophane* could be regarded as the same approach followed by the agencies in the Horizontal Merger Guidelines.[17] And in a 2003 Pre-Trial Brief in the *First Data* case, the plaintiff argued that '[t]he D.C. Circuit has long applied the hypothetical monopolist test in defining antitrust markets'.[18] As a case in point, the plaintiff referred to *US v Microsoft* (2001), where it was held that 'plaintiffs must as a threshold matter show that the browser market can be monopolized, *i.e.*, that a hypothetical monopolist in that market could enjoy market power'.[19] In another case, *Heinz* (2001), both the SSNIP test and the Horizontal Merger Guidelines were referred to.[20]

Based on this evidence, one may find that the SSNIP test can rely on a broad consensus in the US,[21] and a similar observation is probably also true for the EU. However, EU competition law continues to regard the SSNIP test as only one

[10] US Department of Justice and Federal Trade Commission, Horizontal Merger Guidelines (19 August 2010) (US Horizontal Merger Guidelines 2010) § 4.1.1.

[11] ibid.

[12] ibid §§ 4.1, 4.2.

[13] A ten Kate and G Niels, 'The Relevant Market: A Concept Still in Search of a Definition' (2009) 5 *Journal of Competition Law & Economics* 297, 304, 307.

[14] *FTC v Staples*, 970 FSupp 1066, 1076 fn 8 (DDC 1997).

[15] *FTC v Tenet Healthcare*, 17 FSupp2d 937, 945 (ED Mo 1998).

[16] *FTC v Tenet Healthcare*, 186 F3d 1045, 1053 fn 11 (8th Cir 1999).

[17] *United States v SunGard Data Systems*, 172 FSupp2d 172, 182 (DDC 2001).

[18] *United States v First Data Corporation*, Case 1:03CV02169 (Plaintiff's Pretrial Brief, DDC 10 December 2003) § V.A.

[19] *United States v Microsoft*, 253 F3d 34, 81 (DC Cir 2001) (emphasis in original).

[20] *FTC v Heinz*, 246 F3d 708, 718 (DC Cir 2001).

[21] MB Coate and JJ Simons, 'In Defense of Market Definition' (2012) 57 *Antitrust Bulletin* 667, 679.

possible method for market delineation. The Commission's 1997 Market Definition Notice relies on a mix of market definition methods including, most importantly, those developed in the case law. The EU Courts have only rarely addressed the applicability of the SSNIP test under EU competition law. When they have, they did so in response to a Commission decision applying (or failing to apply) the SSNIP, and invariably gave the Commission a considerable margin of discretion in the application of this method. In *Swiss Watchmakers* (2010), the General Court noted that judicial review is limited 'in so far as the definition of the relevant market involves complex economic assessments on the part of the Commission'.[22] It went on to cite the Commission's Market Definition Notice at length, and in particular the SSNIP test,[23] thus implicitly accepting this as the Commission's trusted method of market delineation.

In *Telefónica España* (2014), the applicant argued that the General Court had wrongly endorsed the Commission's account of the temporal dimension of the SSNIP – a plea that was subsequently found to be inadmissible before the Court of Justice as it related to a factual question.[24] However, the idea of holding the Commission accountable for its use of the SSNIP test was not directly ruled out by the Court of Justice. In the *Deutsche Börse* merger case (2015), the applicant submitted that the Commission had wrongly conducted the SSNIP test in its analysis by basing it on the wrong price. The General Court engaged with this argument, but found that the Commission had based the SSNIP on the correct price and that '[t]he alleged infringement of the 1997 notice must therefore be rejected'.[25]

In the case of *Topps* (2017), which amongst other things concerned Panini football collectibles, the Commission had insisted in its 2014 decision that it was under no obligation to carry out a SSNIP, while Topps argued that the Commission should have carried out a SSNIP based on the information it had provided the authority with. The General Court found that while the SSNIP was 'a recognised method for defining the market at issue',[26] it was neither compulsory nor the only method for delineating a relevant market. Referring to *Swiss Watchmakers*, it again emphasised that the Commission had 'a certain discretion concerning the definition of the relevant market, in so far as that definition involves complex economic assessments'.[27] It is interesting to note how the General Court de-emphasised the Commission's interpretative freedom through its careful choice

[22] Case T-427/08 *CEAHR v Commission (Swiss Watchmakers)* EU:T:2010:517, para 66.

[23] ibid para 69.

[24] Case C-295/12 P *Telefónica España v Commission* EU:C:2014:2062, paras 87, 89; Case T-336/07 *Telefónica España v Commission* EU:T:2012:172, para 123. This is also the reason why references to the SSNIP are usually found in General Court judgments.

[25] Case T-175/12 *Deutsche Börse v Commission* EU:T:2015:148, para 84.

[26] Case T-699/14 *Topps Europe v Commission* EU:T:2017:2, paras 78, 79, 82 (direct quote). It also held that in that case 'it is not even necessary to adjudicate on the possibility of applying the SSNIP test to children' (para 91).

[27] ibid para 80.

of words in this recent judgment, downgrading it slightly by referring to 'a certain discretion' and thereby highlighting that it was not absolute.

Some economists have argued that although the SSNIP is not always empirically applicable, conceptually it is the only 'appropriate analytical framework for defining relevant market[s] in *all* cases'.[28] This might be true from a seasoned economist's standpoint, but it is questionable whether competition *law* leads to the same conclusion. Bearing in mind the functions that market definition fulfils in competition law, one might doubt whether from a purely legal point of view the SSNIP test is the only viable conceptualisation of market delineation. Some of its weaknesses – both of a general nature and specifically related to innovative markets – are highlighted below. These might require a significant departure from the hypothetical monopolist test as we know it. Market definition needs to evolve with the challenges that new markets pose.

ii. *The Arbitrariness of the SSNIP*

Both the price upon which the SSNIP is based and the price increase it assumes allow for arbitrariness in this test. The price that is used as a benchmark is generally the current price of the final product, but in some circumstances the price increase might also only relate to one component of a final product. In that case, it might be more accurate to model the price increase on the value that a particular company adds to the final product. This possibility is explicitly foreseen in the US Horizontal Merger Guidelines,[29] but not in the EU's Market Definition Notice. However, in some recent cases, the European Commission has equally relied on such an approach.[30]

The price increase that lies at the heart of the SSNIP test – be it a five or 10 per cent increase – is a fabricated number. The choice of the percentage benchmark has been referred to as arbitrary by numerous commentators.[31] Results may vary greatly depending on whether a 10 per cent or only a five per cent price increase is applied. If one could agree on a standard percentage increase to be applied (for instance, five per cent) and applied it uniformly, then this would allow for more precise and consistent market definitions.[32] However, a five per cent price

[28] S Bishop and M Walker, *The Economics of EC Competition Law: Concepts, Application and Measurement*, 3rd edn (London, Sweet & Maxwell, 2010) paras 4-003 (direct quote, emphasis in original), 4-016, 4-024.

[29] US Horizontal Merger Guidelines 2010 (n 10) § 4.1.2 (last sentence before example 8, and example 8).

[30] *Inco/Falconbridge* (Case COMP/M.4000) Commission Decision 2007/163/EC [2007] OJ L72/18, para 379; *Glencore/Xstrata* (Case COMP/M.6541) Commission Decision of 22 November 2012 [2014] OJ C109/1, paras 138, 140.

[31] See, eg, PD Camesasca and RJ van den Bergh, 'Achilles Uncovered: Revisiting the European Commission's 1997 Market Definition Notice' (2002) 47 *Antitrust Bulletin* 143, 150; JB Baker, 'Market Definition: An Analytical Overview' (2007) 74 *Antitrust Law Journal* 129, 146; AP Vassallo, 'Can One (Ever) Accurately Define Markets?' (2017) 13 *Journal of Competition Law & Economics* 261.

[32] See also MA Glick, DJ Cameron and DG Mangum, 'Importing the Merger Guidelines Market Test in Section 2 Cases: Potential Benefits and Limitations' (1997) 42 *Antitrust Bulletin* 121, 136.

increase might have greatly varying effects in different industries, again leading to some arbitrariness.[33] Where different price increases lead to different results and there is no pre-determined level at which to test a SSNIP, it is questionable to what extent the SSNIP should be used at all. The SSNIP therefore has an inherent arbitrariness that cannot easily be overcome.[34] Ultimately, it may be a normative decision which number to apply.[35]

iii. The SSNIP's Heightened Problematic in Innovative Markets

The SSNIP test that EU and US antitrust authorities like to apply – the precise formulation of which slightly differs amongst the two jurisdictions under scrutiny[36] – has had limited success in dynamic market contexts, both because innovative products may often be differentiated and because price may not play the same predominant role as in more static environments.[37] In fact, many industries that heavily rely on IP and/or in which brand loyalty is an important factor will find that even substantial price increases will not push a significant number of customers away.[38] In addition, the SSNIP test does not appear to be very suitable for analysing the degree of competition *for* a market.[39]

The SSNIP test measures interchangeability by focusing on a price increase and is therefore biased towards price competition.[40] In market environments in which other parameters of competition matter, the SSNIP test will often not provide a useful basis for capturing interchangeability patterns.[41] In innovation-intense industries, the SSNIP test could be modified: instead of focusing on price

[33] TA Baker, 'The 1984 Justice Department Guidelines' (1984) 53 *Antitrust Law Journal* 327, 329.

[34] G Niels, H Jenkins and J Kavanagh, *Economics for Competition Lawyers*, 2nd edn (Oxford, Oxford University Press, 2016) § 2.31.

[35] U Schwalbe and D Zimmer, *Kartellrecht und Ökonomie: Moderne ökonomische Ansätze in der europäischen und deutschen Zusammenschlusskontrolle*, 2nd edn (Frankfurt am Main, Verlag Recht und Wirtschaft, 2011) 79.

[36] See D Zimmer, 'The Emancipation of Antitrust from Market-Share-Based Approaches' (2016) 61 *Antitrust Bulletin* 133, 148.

[37] P Crocioni, 'The Hypothetical Monopolist Test: What it Can and Cannot Tell You' (2002) 23 *European Competition Law Review* 354, 362; ML Katz and HA Shelanski, '"Schumpeterian" Competition and Antitrust Policy in High-Tech Markets' (2005) 14 *Competition* 47; Monopolkommission, 'Wettbewerbspolitik: Herausforderung digitale Märkte' (Sondergutachten 68, 2015) para 59.

[38] MA Lemley and MP McKenna, 'Is Pepsi Really a Substitute for Coke? Market Definition in Antitrust and IP' (2012) 100 *Georgetown Law Journal* 2055, 2057. 'Significant' stands for the number of customers that would make a price rise unprofitable under the SSNIP test. This was also the argument of Topps relating to Panini collectibles; see Case T-699/14 *Topps Europe v Commission* EU:T:2017:2.

[39] PA Geroski, 'Competition in Markets and Competition for Markets' (2003) 3 *Journal of Industry, Competition and Trade* 151, 160. Geroski adds that 'competitors in a market impose a different type of discipline on incumbents than competitors for a market' (ibid 162).

[40] R Podszun, 'The Arbitrariness of Market Definition and an Evolutionary Concept of Markets' (2016) 61 *Antitrust Bulletin* 121, 125.

[41] See also R Hartman et al, 'Assessing Market Power in Regimes of Rapid Technological Change' (1993) 2 *Industrial and Corporate Change* 317, 322; A Gebicka and A Heinemann, 'Social Media & Competition Law' (2014) 37 *World Competition* 149, 157.

increases, it could analyse how changes in performance (design, innovation, quality) affect customer behaviour. The decisive question is then 'whether a change in the performance attributes of one commodity would induce substitution to or from another'.[42] If such substitution were to occur, the substitutable technologies or products belong to the same relevant market. This is because 'performance competition is usually the central focus of competitive efforts in these [dynamic] industries',[43] while price competition is often deemed to be secondary. For this reason, it might prove fruitful to base the SSNIP test on a non-price variable such as quality or performance.[44] Conceptually, the SSNIP test can therefore be applied to non-price parameters, even though this makes its practical application much more complex.[45]

The SSNIP assumes a non-transitory price increase and is not overly troubled by temporary price increases, as the latter tend to return to more competitive levels within a short timeframe.[46] If one applies this thinking to innovative and fast-moving markets, the question becomes how long the duration of a price increase needs to be in order to be seen as non-transitory or transitory.[47]

What is particularly relevant for market definition is the extent to which customers would substitute the product for which the price has increased with another product.[48] If only a small fraction were to switch, then the price increase would be profitable. Both quantitative and qualitative evidence can be used in order to estimate demand substitutability. However, in innovative markets that are characterised by considerable differentiation, product substitution is faced with an additional challenge, namely value disparities as regards different possible substitutes.[49] It is not possible to accurately depict these through a SSNIP.

One approach that could increase the SSNIP's applicability in innovative markets is the possibility to include producers into the analysis that could easily switch to producing the product under scrutiny with its individualised features within a short timespan, given the necessary incentive. This would provide a more comprehensive picture of the market landscape, but would significantly modify the SSNIP test by firmly incorporating supply-side substitutability. Currently, the US Horizontal Merger Guidelines only take supply-side substitutability into

[42] Hartman et al (n 41) 334.

[43] C Pleatsikas and DJ Teece, 'The Analysis of Market Definition and Market Power in the Context of Rapid Innovation' (2001) 19 *International Journal of Industrial Organization* 665, 671 f.

[44] DJ Teece and M Coleman, 'The Meaning of Monopoly: Antitrust Analysis in High-Technology Industries' (1998) 43 *Antitrust Bulletin* 801, 853–57; ML Katz and HA Shelanski, 'Mergers and Innovation' (2007) 74 *Antitrust Law Journal* 1, 36.

[45] See Niels, Jenkins and Kavanagh (n 34) § 2.168.

[46] ibid § 2.47.

[47] For conventional markets, competition authorities usually apply a reference period of one to two years; ibid § 2.48.

[48] JB Baker and TF Bresnahan, 'Empirical Methods of Identifying and Measuring Market Power' (1992) 61 *Antitrust Law Journal* 3, 8. This was also emphasised by the General Court in Case T-699/14 *Topps Europe v Commission* EU:T:2017:2, para 91.

[49] Teece, 'Favoring Dynamic over Static Competition' (n 2) 221.

account when determining which suppliers to include in the relevant market,[50] rather than at the stage of market definition. The SSNIP test under the European Commission's Market Definition Notice also focuses on demand-side substitutability, but appears more open to supply-side considerations.[51]

While the cellophane fallacy has taught us that the SSNIP must be performed from the basis of the competitive price, it is far from clear how one should know in advance whether a given price is indeed competitive.[52] Particularly in innovation-driven industries, the additional question is what a competitive price signifies in this market environment of short product cycles, potentially large investment requirements in the beginning but considerably lower running costs, and performance rather than price as an important competitive parameter.

An additional problem is encountered in markets in which companies rely on price discrimination, as this makes the SSNIP inoperable.[53] As digital markets become increasingly versed in dynamic pricing and price discrimination techniques,[54] this complicates market definition based on standard economic tests.

Together, these shortcomings of the SSNIP as regards innovative markets will necessarily have an influence on the antitrust assessment and might unnecessarily hamper innovation. Nevertheless, as has been pointed out by economists, economics has not yet proposed a workable alternative.[55] Competition lawyers should therefore be particularly attentive to the SSNIP test's deficiencies as regards innovative markets when interpreting the results.

iv. Multi-sided Markets

Many standard economic tests that competition law employs cannot simply be transferred to platform markets,[56] amongst them the SSNIP test. While in the US, it is sometimes claimed that the SSNIP test can be applied to define them in a straightforward way,[57] this is a controversial conclusion.

In both the EU and the US, the case law on platform markets predominantly developed in the course of credit card payment networks. In one of these early cases, the SSNIP was applied to multi-sided markets. In *First Data* (2003), the

[50] US Horizontal Merger Guidelines 2010 (n 10) § 5.1.

[51] European Commission, Notice on the definition of relevant market for the purposes of Community competition law [1997] OJ C372/5 (EU Market Definition Notice 1997) paras 20 ff.

[52] LJ White, 'Market Definition and Market Power in Payment Card Networks: Some Comments and Considerations' (2006) 5 *Review of Network Economics* 61, 64; Zimmer (n 36) 150.

[53] ABA Section of Antitrust Law, *Econometrics* (n 3) 12.

[54] See A Ezrachi and ME Stucke, *Virtual Competition: The Promise and Perils of the Algorithm-Driven Economy* (Cambridge, MA, Harvard University Press, 2016) 89 ff.

[55] Bishop and Walker (n 28) § 4-016.

[56] L Filistrucchi, D Geradin and E van Damme, 'Identifying Two-Sided Markets' (2013) 36 *World Competition* 33, 57.

[57] United States in OECD (ed), *Policy Roundtable: Market Definition* (2012) DAF/COMP(2012)19, 329 f; *United States v First Data Corporation* Case 1:03CV02169 (Plaintiff's Pretrial brief, DDC 10 December 2003), § V.C; E Emch and TS Thompson, 'Market Definition and Market Power in Payment Card Networks' (2006) 5 *Review of Network Economics* 45, 47.

DoJ applied the SSNIP test in the already-mentioned case on PIN debit network services, but only to the merchant side of the market.[58] It held that the SSNIP test could be used in two-sided markets and that, when a SSNIP was applied to the merchant side, the test was shown to be useful even when the bank side of the market was also taken into consideration.[59] However, this did not resolve the question of how one market side's response to a SSNIP should be incorporated into a SSNIP test on the other side (if at all).[60]

The European Commission, in its *MasterCard* decision (2007), was asked by MasterCard to view cardholders and merchants as a single demand side and to apply the SSNIP test to this pooled demand.[61] This would have combined the two market sides, which had different interests and certainly also different responses to a price increase or quality decrease, into one demand side. The Commission rejected MasterCard's approach to market definition in two-sided markets, explicitly referring to multi-sided market theory in its analysis.[62]

In the US *Realcomp* case (2009), the administrative law judge applied the SSNIP test to online multi-listing services by realtors. In that case, it was distinguished between the input market (multiple listing services) and the output market (real estate brokerage services).[63] The SSNIP test was applied to those sides of the market separately, although network effects amongst these markets were discussed. Especially in cases in which services can be accessed for free by one group of users (eg, search engine users), while they need to be remunerated by users on the other side of the market (eg, companies wanting to place an ad amongst the search results), one will need to resort to different methods in order to properly delineate the relevant market(s).

In a merger investigation of 2011, the DoJ held that the SSNIP could be expressed through 'a small but significant degradation in the quality of comparative flight search services or increase in price to consumers of these services'[64] in that case, thus showing it was open to adapt the SSNIP to the circumstances of the case – particularly the free nature of the search services.

In *American Express* (2015), the US District Court held that the SSNIP test could be applied to the two-sided market at issue, as was done by the defendant's economic expert. The latter carried out two SSNIP tests, one applying the price increase only to the merchant's switch fee and the other one to the whole fee to be paid by the merchant.[65] Upon appeal, the US Court of Appeals for the

[58] *United States v First Data Corporation* Case 1:03CV02169 (Plaintiff's Pretrial brief, DDC 10 December 2003), § V.B.

[59] ibid § V.C.

[60] RB Hesse and JH Soven, 'Defining Relevant Product Markets in Electronic Payment Network Antitrust Cases' (2006) 73 *Antitrust Law Journal* 709, 716.

[61] *MasterCard* (Cases COMP/34.579, COMP/36.518, COMP/38.580) Commission Decision of 19 December 2007 [2009] OJ C264/8, para 252.

[62] ibid para 257.

[63] *Realcomp II Ltd v FTC*, Case 9320 (ALJ) (10 December 2009) paras 283 ff, 315.

[64] *United States v Google/ITA Software*, Case 1:11-cv-00688 (Complaint, DDC 8 April 2011) para 23.

[65] *United States v American Express*, Case 10-CV-4496 (NGG) (RER) (EDNY 19 February 2015) § III.A.2.a.

Second Circuit emphasised that the District Court should have taken into account the feedback effect on cardholder demand in the case of a price increase to merchants.[66] Upon further appeal, the US Supreme Court held that both market sides needed to be taken into account when defining a market: 'Price increases on one side of the platform likewise do not suggest anticompetitive effects without some evidence that they have increased the overall cost of the platform's services.'[67]

In the case law, one therefore finds examples of each market side being subjected to the SSNIP or both market sides being the subject of a SSNIP. There certainly has been more experience in this regard in the US, which might reflect on that jurisdiction's greater reliance on economics more generally.[68] In the literature, one finds mixed approaches. One group of economists has suggested that the SSNIP test can be applied in two-sided markets, with the main difficulty being that the products on the two market sides are complements rather than substitutes.[69] But it has also been remarked that merely focusing on one market side for a SSNIP necessarily leads to a 'logical conundrum', as this would imply that the overall product offered by the platform is found to be at the same time in as well as outside one and the same product market.[70] Multi-sided markets cannot be artificially split up in this way. Also, when modelling a price increase on one market side, one cannot ignore resulting feedback effects from the other market side.[71] It would be misguided 'as a matter of economics to ignore significant demand interdependencies among the multiple platform sides'.[72] Competitive constraints upon a platform trying to profitably raise prices may come from various sides: single-sided firms on either platform side, as well as other platforms that target the same or either platform side.[73]

The SSNIP test relies on substitutability patterns in the event of a price increase in a one-sided market. Within the realm of multi-sided platforms, two factors complicate this analysis. First of all, many innovative platform markets know one market side which ostensibly receives services for free (or against 'payment' of a different kind, for instance, the divulging of data). In that case, the applicability of the SSNIP has to be questioned, for one cannot readily apply the test to non-monetary market sides. A percentage price increase on the free user side will have no effect – zero remains zero. Indeed, it has been found that any price-based assessment will encounter great difficulties where multi-sided markets are

[66] *United States v American Express*, 838 F3d 179, 200 (2d Cir 2016).

[67] *Ohio v American Express*, 585 US ___ (2018), 138 SCt 2274, 2286 (2018).

[68] For a short discussion of economics and US antitrust law, see ch 2, section IV.B.i.

[69] Niels, Jenkins and Kavanagh (n 34) § 2.148.

[70] Emch and Thompson (n 57) 54.

[71] DS Evans and R Schmalensee, 'The Antitrust Analysis of Multisided Platform Businesses' in RD Blair and DD Sokol (eds), *The Oxford Handbook of International Antitrust Economics*, vol 1 (Oxford, Oxford University Press, 2015) 421.

[72] ibid 424.

[73] ibid.

concerned,[74] even if neither side receives services for free. It is also questionable whether decreases in quality will herald the same result as price increases, even if this alternative measurement is sometimes suggested.[75] Second, the demand on all market sides is intricately intertwined, so an effect on one market side will regularly also have an effect on the other one.[76] Indeed, price increases on one market side may have several linked effects on either or both sides.[77] This makes it impossible to carry out the SSNIP on each market side without considering the platform market as a whole. However, if a platform market is conceptualised as one single relevant antitrust market, rather than defining each market side as its own relevant market, then one may decide to use the overall price charged to both sides in order to capture the platform's price increase.[78] The question then arises as to how to model the price increase of five to 10 per cent, ie, to which market side this will be allocated. In the business model underlying multi-sided markets, it is upon the platform to decide which price to allocate to which side. When modelling a price increase and each market side's reactions to that price increase, one will nonetheless need to decide on which side the price increases by which percentage. This influences the entire analysis, including the outcome. There could also be a danger of unnecessarily finding microscopic markets when trying to apply the SSNIP test in multi-sided market environments; for instance, in the area of online search, one might end up with narrow markets for search terms and/or organic search results.[79]

B. Market Shares and Alternatives in Innovative Settings

Within competition law, market shares and market share thresholds are frequently used as a screen to separate plausibly unproblematic cases from probably

[74] F Thépot, 'Market Power in Online Search and Social Networking: A Matter of Two-Sided Markets' (2013) 36 *World Competition* 195, 216; R Bitetti, 'Google, Competition Policy and the Owl of Minerva' in M Kovač and A-S Vandenberghe (eds), *Economic Evidence in EU Competition Law* (Cambridge, Intersentia, 2016) 309.

[75] See OECD (ed), *Policy Roundtable: The Role and Measurement of Quality in Competition Analysis* (2013) DAF/COMP(2013)17, 8.

[76] On the SSNIP test within multi-sided markets, see also DS Evans and R Schmalensee, 'Markets with Two-Sided Platforms' in ABA Section of Antitrust Law (ed), *Issues in Competition Law and Policy* (Chicago, ABA Publishing, 2008) 689; L Filistrucchi et al, 'Market Definition in Two-Sided Markets: Theory and Practice' (2014) 10 *Journal of Competition Law & Economics* 293, 331–33; T Hoppner, 'Defining Markets for Multi-Sided Platforms: The Case of Search Engines' (2015) 38 *World Competition* 349, 350.

[77] For an overview of these possible price effects, see M Blaschczok, *Kartellrecht in zweiseitigen Wirtschaftszweigen: Eine Untersuchung vor dem Hintergrund der ökonomischen Forschung zu 'two-sided markets'* (Baden-Baden, Nomos, 2015) 72.

[78] This approach is suggested by Emch and Thompson (n 57) 54. For a critical appraisal thereof, see Blaschczok (n 77) 82–85.

[79] See GA Manne and JD Wright, 'Google and the Limits of Antitrust: The Case against the Case against Google' (2011) 34 *Harvard Journal of Law & Public Policy* 171, 201.

problematic cases.[80] They form part of a structural market analysis and are relied upon as indicators of a company's market power, or to provide safe harbours based on an assumption of the absence of such market power. Due to the characteristics of innovative market environments (see section II in Chapter 3), market shares may not be a good indicator of a company's market position in such markets. In addition, market shares require a reliable delineation of the relevant market.[81]

i. Market Share Thresholds in the Case Law as Remnants of a Static Analysis

EU competition law has a long tradition of relying on market shares in order to inform antitrust analysis. In *AKZO* (1991), the Court of Justice established a positive presumption of market dominance for the purposes of Article 102 TFEU starting from market shares of 50 per cent.[82] In that same year, the Commission underlined in the merger of *Tetra Pak/Alfa-Laval* that while high market shares strongly point to a position of market dominance – particularly in that case, in which Tetra Pak had a market share exceeding 90 per cent on the relevant market – 'in certain rare circumstances even such a high market share may not necessarily result in dominance'.[83] This was an early instance in which the Commission hinted at the possibility that market shares may not be conclusive when technology-intense markets are at issue. Despite this, the *AKZO* precedent continues to establish a strong, albeit rebuttable presumption of market dominance.

Market shares as an indication of monopoly power have also been extensively discussed under section 2 of the Sherman Act. In *Alcoa* (1945), Judge Hand stated with regard to the relationship between market power and a company's market share of over 90 per cent that such a high market share certainly 'is enough to constitute a monopoly; it is doubtful whether sixty or sixty-four percent would be enough; and certainly thirty-three per cent is not'.[84] This relates well to Abba Lerner's view that high market shares – and thus market concentration – may not always indicate power over a market.[85]

Although market shares have been used as a shorthand for market power assessments in the US,[86] the US judicature has retreated from strong structural

[80] A Coscelli and G Edwards, 'Dominance and Market Power in EU Competition Law Enforcement' in I Lianos and D Geradin (eds), *Handbook on European Competition Law* (Cheltenham, Edward Elgar, 2013) 357.

[81] P Areeda, 'Justice's Mergers Guidelines: The General Theory' (1983) 71 *California Law Review* 303, 307.

[82] Case C-62/86 *AKZO Chemie v Commission* EU:C:1991:286, para 60.

[83] *Tetra Pak/Alfa-Laval* (Case IV/M.068) Commission Decision 91/535/EEC [1991] OJ L290/35, § 3.3.

[84] *United States v Aluminum Co of America (Alcoa)*, 148 F2d 416, 424 (2d Cir 1945). *Alcoa* was later specifically endorsed by the US Supreme Court; see *American Tobacco v United States*, 328 US 781, 811 ff (1946).

[85] Lerner (n 6) 165 f.

[86] TE Kauper, 'The Problem of Market Definition under EC Competition Law' (1997) 20 *Fordham International Law Journal* 1682, 1685.

presumptions.[87] In 1974, the US Supreme Court acknowledged in *General Dynamics* that past market shares do not always give an accurate picture of how a market is expected to develop.[88] In addition, what matters is not so much a company's market share, but its ability to maintain that market share over time.[89] This is even more the case in innovative market environments. Overall, market share presumptions under US antitrust law are today 'only a shadow of [their] prior self'.[90]

Two cases involving Microsoft – one from the US and one from the EU – will be relied upon to demonstrate how market shares may be interpreted in innovative market environments. In *US v Microsoft* (2001), the US Court of Appeals for the District of Columbia Circuit was confronted with Microsoft's argument that it was operating in a Schumpeterian market, implying that an innovation-sensitive way of analysing the case was called for. However, the Court found that even in markets that may be radically affected by innovations overthrowing current markets, it was still in a position to assess the suspected anti-competitive behaviour.[91] It noted that while in such highly dynamic industries, innovation may have 'already rendered the anticompetitive conduct obsolete (although by no means harmless)',[92] antitrust enforcement should nevertheless take place. The case then moved on to the issue of market shares, and it was found that Microsoft had an impressively high market share on the relevant product market. Microsoft urged that 'even a predominant market share does not by itself indicate monopoly power' in such a dynamic market.[93] As structural entry barriers protected Microsoft's future market position, the Court concluded that the high market share was indicative of market power in this case.[94]

A decade later, the *Microsoft/Skype* merger showed that the EU's General Court was willing to take a market's innovation dimension into account when interpreting the meaning of market shares for market power. Microsoft wanted to acquire Skype on the video communications market. In its merger assessment, the Commission found that this was 'a nascent and dynamic sector and market shares can change quickly within a short period of time',[95] leading it to conclude that in this particular environment, 'market shares are not the best proxy to evaluate the market power … and they only give a preliminary indication of the competitive situation in these dynamic markets'.[96] In late 2013, the General Court accepted this finding based on the fact that the market at issue was 'characterised by short innovation cycles in which large market shares may turn out to be ephemeral. In such

[87] Coate and Simons (n 21) 673 fn 10.
[88] *United States v General Dynamics Corp*, 415 US 486, 498 ff (1974).
[89] *United States v Syufy Enterprises*, 903 F2d 659, 665 f (9th Cir 1990).
[90] Coate and Simons (n 21) 691.
[91] *United States v Microsoft*, 253 F3d 34, 49 f (DC Cir 2001).
[92] ibid 49.
[93] ibid 54.
[94] ibid 54–56.
[95] *Microsoft/Skype* (Case COMP/M.6281) Commission Decision of 7 October 2011, para 78.
[96] ibid para 99.

a dynamic context, high market shares are not necessarily indicative of market power and, therefore, of lasting damage to competition.'[97] As a consequence, the acquisition was not opposed.

It is interesting to note how differently the authorities and courts dealt with the issue of market shares and their relationship with market power in these innovative markets. While one might suggest that perhaps time has taught the antitrust enforcers a lesson, giving them a more realistic view of the limited market power insights that can be gained from high market shares in dynamic markets, the discrepancy more likely stems from the fact that market shares need to be seen and interpreted in their particular market context. In *US v Microsoft*, the structural entry barriers allowed the conclusion that the high market share was stable and could thus be maintained over time, while in *Microsoft/Skype*, the market was seen as highly dynamic and therefore the high market share did not provide the kind of stability that an inference of market power would require. Rather than reflecting different analyses based on different moments in time or approaches differing according to the jurisdiction under scrutiny, it is the particular market context that appears to have been decisive in these cases.

ii. Market Share Thresholds in Policy Documents – and Alternatives

Market share thresholds as policy benchmarks are relied upon in both the EU and the US. The European Commission has been particularly active in establishing market share thresholds. Its legally binding Block Exemption Regulations, which exempt agreements from scrutiny under Article 101(3) TFEU, foresee many market share thresholds. In order to come under the safe harbour of block exemption, the Vertical Agreements Block Exemption Regulation only applies to agreements where the supplier and the buyer do not exceed a market share of 30 per cent on any affected relevant market.[98] The Technology Transfer Block Exemption Regulation (TTBER) applies to licensing agreements between competitors up to a combined market share of 20 per cent and between non-competitors up to an individual market share of 30 per cent.[99] The Specialisation Block Exemption Regulation foresees an aggregate market share threshold of 20 per cent for all parties involved,[100] and the R&D Block Exemption Regulation imposes no

[97] Case T-79/12 *Cisco Systems & Messagenet v Commission* EU:T:2013:635, para 69. Recently, see *Apple/Shazam* (Case M.8788) Commission Decision of 6 September 2018 [2018] OJ C417/4, para 162.

[98] Commission Regulation (EU) 330/2010 on the application of Article 101(3) of the Treaty on the Functioning of the European Union to categories of vertical agreements and concerted practices [2010] OJ L102/1 (VABER), art 3(1).

[99] Commission Regulation (EU) 316/2014 on the application of Article 101(3) of the Treaty on the Functioning of the European Union to categories of technology transfer agreements [2014] OJ L93/17 (TTBER), art 3.

[100] Commission Regulation (EU) 1218/2010 on the application of Article 101(3) of the Treaty on the Functioning of the European Union to certain categories of specialisation agreements [2010] OJ L335/43 (Specialisation BER), art 3.

market share threshold on R&D agreements between non-competitors, but a 25 per cent market share threshold if the R&D agreement is concluded between competitors.[101] These market share thresholds are combined with black lists that determine which types of clauses are never allowed to be included in agreements wanting to benefit from the safe harbour of block exemption, and grey lists which may be included in an agreement, but must be individually assessed, outside of block exemption.[102]

The Commission has also established market share thresholds in its non-binding (soft law) competition guidance, for instance, an aggregate market share threshold of five per cent, combined with a turnover threshold of €40 million, in its Notice on Effect on Trade,[103] and an aggregate market share threshold of 10 per cent for competitors, but an individual market share threshold of 15 per cent for non-competitors in its *De Minimis* Notice.[104] These Commission documents are based on the assumption that companies with low market shares do not enjoy substantial market power, and their anti-competitive behaviour would thus not greatly impact competition on the relevant market. However, the soundness of this assumption may be questioned, especially in the context of innovative markets in which market shares are less and less indicative.

The volatility of market shares in product markets that rely heavily on IPRs can constitute an issue under the TTBER, making market shares in the context of licensing agreements fairly unworkable in practice.[105] The Commission solved this by foreseeing that licensing agreements with market shares that are below the levels foreseen by the TTBER at the time the licensing agreement is entered into, but that in due course go beyond the market share thresholds foreseen in Article 3 TTBER, will remain subject to the exemptions for two consecutive calendar years.[106] Similar solutions have been adopted under other BERs.[107]

The number of times the Commission sets out market share thresholds in its guidance indicates that it believes in the added value of calculating market shares based on the finding of a relevant market. Parties also hold the Commission

[101] Commission Regulation (EU) 1217/2010 on the application of Article 101(3) of the Treaty on the Functioning of the European Union to certain categories of research and development agreements [2010] OJ L335/36 (R&D BER), art 4(1), 4(2)(a) and (b).

[102] VABER (n 98) art 4 (black list), art 5 (grey list); R&D BER (n 101) art 5 (black list), art 6 (grey list); Specialisation BER (n 100) art 4 (black list); TTBER (n 99) art 4 (black list), art 5 (grey list).

[103] European Commission, Guidelines on the effect on trade concept contained in Articles 81 and 82 of the Treaty [2004] OJ C101/81, para 52(a).

[104] European Commission, Notice on agreements of minor importance which do not appreciably restrict competition under Article 101(1) of the Treaty on the Functioning of the European Union [2014] OJ C291/1 (*De Minimis* Notice) para 8.

[105] SD Anderman, 'The New EC Competition Law Framework for Technology Transfer and IP Licensing' in J Drexl (ed), *Research Handbook on Intellectual Property and Competition Law* (Cheltenham, Edward Elgar, 2008) 113, 117 f.

[106] TTBER (n 99) art 8(e).

[107] VABER (n 98) art 7(d) and (e); R&D BER (n 101) art 7(d) and (e); Specialisation BER (n 100) art 5(d) and (e).

accountable in cases in which it does not adhere to its own system of market share thresholds.[108] As far as can be seen, the EU Courts have never challenged the Commission's frequent reliance on market shares in its policy documents.

In its Horizontal Merger Guidelines, the Commission assures that it will take market conditions into account when interpreting market shares for merger analysis, particularly in market environments which are innovation-driven.[109] In its Non-horizontal Merger Guidelines, the Commission states that it will normally not view mergers that do not lead to significant concentration as problematic, unless it is known that one of the companies involved will expand shortly – for instance, thanks to a recent innovation.[110]

On the US side, starting with the 1982 Horizontal Merger Guidelines, the antitrust agencies have relied on safe harbours based on market shares.[111] The IP Guidelines contain three rules on safety zones, ie, safe harbours within which companies can generally rest assured that their licensing agreements will not be challenged based on antitrust rules. The safety zone rules differentiate based on the type of market concerned: if the relevant market is a product market, then the agencies will not challenge licensing restraints if (a) the restraint is not per se anti-competitive, and (b) the combined market share of the licensor and its licensees does not exceed 20 per cent on any affected relevant product market.[112] They therefore combine a market share threshold with a black list, just like the block exemption regulations under EU competition law. The safety zone rules for technology and innovation markets differ considerably from the general rule for traditional product markets. Both of these safety zone rules foresee that a restraint may not be per se anti-competitive, but they do not contain any market share threshold. For technology markets, the agencies will not challenge licensing restraints if at least four independently controlled technologies are present on the relevant market on top of the technologies controlled by the parties to the licensing agreement, and those technologies are considered viable substitutes for the licensed technology. For R&D markets, the agencies will not challenge licensing restraints if at least four independently controlled entities are present on the relevant market on top of the parties to the licensing agreement, and those entities 'possess the required specialized assets or characteristics and the incentive to engage in research and development

[108] See, for instance, Case T-199/08 *Ziegler v Commission* EU:T:2011:285; Case C-439/11 P *Ziegler v Commission* EU:C:2013:513. On the *Ziegler* case, see ch 2, section II.B.

[109] European Commission, Guidelines on the assessment of horizontal mergers under the Council Regulation on the control of concentrations between undertakings [2004] OJ C31/5, para 15.

[110] European Commission, Guidelines on the assessment of non-horizontal mergers under the Council Regulation on the control of concentrations between undertakings [2008] OJ C265/6, paras 25, 26(a).

[111] Coate and Simons (n 21) 674. Arguing that these structural presumptions are generally working well, see J Kwoka, 'The Structural Presumption and the Safe Harbor in Merger Review: False Positives or Unwarranted Concerns?' (2018) 81 *Antitrust Law Journal* 837.

[112] US Department of Justice and Federal Trade Commission, Antitrust Guidelines for the Licensing of Intellectual Property (14 January 2017) (US IP Guidelines 2017) § 4.3.

that is a close substitute of the research and development activities of the parties to the licensing agreement'.[113] In areas in which the agencies expect that market shares do not provide any valuable insight into sustainable market power, they have therefore replaced market share thresholds with another technique, namely that of counting competing technologies or companies. This alternative method equally requires market definition in order to be applied, for it is not possible to otherwise determine what is within and what is outside the relevant markets, ie, what technology or which competitor might constitute a competitive constraint. However, the reason for delineating the market under these circumstances clearly lies in market characterisation. Nevertheless, it avoids any inference of market power from market shares.

The US Competitor Collaboration Guidelines foresee two safety zones within which companies do not normally need to expect antitrust enforcement.[114] The first one is a general safety zone that applies to collaborations in which all participants have an aggregate market share of up to 20 per cent. Agreements that are analysed as a per se offence or which the agencies would challenge without engaging in a detailed market analysis cannot benefit from this safety zone.[115] This general safety zone applies to any product or technology market; contrary to the IP Guidelines, there are no specific rules for technology markets. For R&D agreements, the Guidelines foresee a special safety zone that only applies to innovation markets. The agencies will not challenge competitor collaborations in this area if at least three independently controlled entities are present on the relevant market on top of the participants to the collaboration, and those entities 'possess the required specialized assets or characteristics and the incentive to engage in R&D that is a close substitute for the R&D activity of the collaboration'.[116] The collaboration may also not be illegal per se or be an agreement that the agencies would challenge without engaging in a detailed market analysis. When establishing substitutability between R&D efforts, the agencies will take the following factors into account:

> [T]he nature, scope, and magnitude of the R&D efforts; their access to financial support; their access to intellectual property, skilled personnel, or other specialized assets; their timing; and their ability, either acting alone or through others, to successfully commercialize innovations.[117]

It has been held that the R&D-specific safety zone could be invoked by R&D collaborations with a market share of roughly 25 per cent and thus provides a

[113] ibid.

[114] Federal Trade Commission and US Department of Justice, Antitrust Guidelines for Collaborations among Competitors (April 2000) (US CC Guidelines) § 4.1.

[115] ibid § 4.2.

[116] ibid § 4.3.

[117] ibid. An almost identically worded paragraph can be found in the US IP Guidelines 2017 (n 112) § 4.3.

more generous safe harbour than the general safety zone.[118] However, in fact, market shares are irrelevant for the special safety zone; the number of alternative R&D efforts alone is decisive, regardless of their respective market shares. One R&D effort may be close to marketability, while another may only be nascent. This focus on alternative R&D activities is a useful point of reference because the primary concern in innovative markets is the number of innovators that are conducting R&D, not their current market shares. Indeed, as long as a sufficient number of innovators are present in a (future/potential) market and innovation remains an important competitive parameter, competition concerns should be limited in the absence of plainly anti-competitive agreements. Such an approach reflects the limited value of determining (unstable) market shares in markets that are prone to paradigm shifts.

iii. The Use of Market Shares in Innovative Markets

Market share thresholds may be simplistic analytical tools, as market definition with its in/out mentality can never accurately depict the continuum that distinguishes real markets.[119] For this reason, economics cannot tell us at what point this continuum should be broken up into a relevant antitrust market; that is the task of the legal framework for antitrust market definition.[120] Against this background, steadfast presumptions of market power based on market shares may be problematic.[121] Safe harbour rules based on market shares may be less so, as they do not lead to antitrust liability, but merely decide which cases need to be assessed more fully.[122]

Louis Kaplow has argued that calculating market shares 'is conceptually flawed' because, depending on the specific market environment, different market shares may indicate greatly varying degrees of market power.[123] Kaplow also insists that in order to make any assertions as to the meaning of market shares, one must necessarily have in mind some sort of standard reference market which allows us to make assumptions about the significance of a certain market share.[124] However, as was shown in the two *Microsoft* cases contrasted above (see section III.B in Chapter 6),

[118] S DeSanti and W Cohen, 'Competition to Innovate: Strategies for Proper Antitrust Assessments' in R Cooper Dreyfuss, D Leenheer Zimmerman and H First (eds), *Expanding the Boundaries of Intellectual Property: Innovation Policy for the Knowledge Society* (Oxford, Oxford University Press, 2001) 337.

[119] Zimmer (n 36) 150.

[120] DF Turner, 'The Role of the "Market Concept" in Antitrust Law' (1980) 49 *Antitrust Law Journal* 1145, 1147 f.

[121] See D Cameron and MA Glick, 'Market Share and Market Power in Merger and Monopolization Cases' (1996) 17 *Managerial and Decision Economics* 193.

[122] Kauper (n 86) 1691.

[123] L Kaplow, 'Market Definition' in Blair and Sokol (n 71) 361.

[124] L Kaplow, 'Why (Ever) Define Markets?' (2010) 124 *Harvard Law Review* 438, 459 ff, in particular 462.

courts take this dependence on the specific market environment into account and are capable of interpreting market shares in their specific context. If we successfully take into account the specific market context when inferring market power (if any) from certain market shares, then this defies the purpose of any standard reference market.

If the beautiful simplicity of market shares has come under attack as a general matter, then this critique is increased in dynamic markets. Market shares in innovative environments may be subject to sudden and important change and innovative environments call for special care when interpreting market shares. This also means that previously calculated market shares may not provide a good basis for antitrust analysis in these markets,[125] while market shares in future markets will often be impossible to calculate or highly speculative. At the same time, it can be difficult to calculate market shares in cases in which a newcomer places a new and superior product on the market.[126]

In platform markets, the application of market share thresholds is particularly difficult. To begin with, it is unclear whether each market side needs to be considered separately for this purpose or whether the platform market as such needs to be considered. However, at a more fundamental level, one will need to ask to what extent market shares can be applied to multi-sided markets at all.[127] They may require an altogether alternative benchmark.

The heavy reliance on market shares that is found in EU competition law, accompanied by the relatively low threshold for intervention, is often traced back to its ordoliberal roots.[128] By contrast, one encounters limited reliance on market shares in US antitrust law. Both the IP Guidelines and the Competitor Collaboration Guidelines are more sensitive to the limited value of market shares in innovative markets, and focus on the number of innovators present rather than on their market shares. The IP Guidelines' safety zone also takes into account the number of available technology substitutes. This, it would seem, is a viable alternative because it shows possible sources of competition – and therefore competitive restraints – rather than focusing on current market share levels that might be subject to significant change in the near future. In addition, this approach focuses on credible innovation efforts rather than forecasting their ultimate success. In the light of innovation's uncertainty, this appears to be a commendable approach.

In the context of dynamic markets, antitrust enforcers should be wary of drawing quick conclusions from market shares, instead taking into account the market's particular characteristics when interpreting them. Competition policy based on market shares may want to consider using alternative benchmarks in innovative markets. Where, as under EU competition law, presumptions of market power rely

[125] GA Hay, 'Innovations in Antitrust Enforcement' (1995) 64 *Antitrust Law Journal* 7, 12 f.
[126] U Schwalbe and F Maier-Rigaud, 'Background Note' in OECD, *Market Definition* (n 57) 58.
[127] Monopolkommission (n 37) paras 56 f.
[128] C Ahlborn and C Grave, 'Walter Eucken and Ordoliberalism: An Introduction from a Consumer Welfare Perspective' (2006) 2 *Competition Policy International* 197, 207.

on market share thresholds, this should be rethought in the context of innovation. At the very least, the fact that a market environment is highly dynamic should constitute a good reason for rebutting such a presumption in order to make way for a more detailed analysis.

C. Concentration Levels

Antitrust authorities rely on the Herfindahl-Hirschman Index (HHI) to assess market concentration, particularly within the context of horizontal mergers.[129] By adding the squares of companies' market shares, the HHI gives relatively more importance to big players, while smaller companies are given less weight. The delta, ie, the change in market concentration that occurs through the merger,[130] is used as a first indication of how the merger might structurally affect competition. The importance of mavericks cannot be depicted in HHI levels, even if they are likely to disrupt the market with their innovations in the future.

Under the 1982 US Merger Guidelines, the HHI was first used as 'a structural proxy for competition'.[131] Today, the HHI still forms part of the Horizontal Merger Guidelines 2010. However, the agencies note that concentration levels will be given more importance in markets in which market shares have remained stable over time, while concentration levels will be less significant where market shares have changed significantly.[132] This is in line with an innovation-sensitive approach.

In *General Dynamics* (1974), the Supreme Court cautioned that neither market shares nor concentration levels were definitive indicators that a merger was to be regarded as anti-competitive.[133] In a more recent case before the US Court of Appeals for the Fifth Circuit, *Chicago Bridge & Iron Company* (2008), the merging parties argued that the FTC should have calculated the HHI on an annual basis rather than over a multiple-year range. The Court held that concentration levels constituted one amongst a number of indicators demonstrating market concentration and needed to be assessed in their particular market context.[134] It also remarked that the merging parties too heavily relied on HHIs, especially as they were not a mandatory analytical instrument.[135]

In the early European case of *France v Commission* (also known as *Kali & Salz*, 1997), Advocate General Tesauro remarked that contrary to other antitrust

[129] JB Baker, 'Market Concentration in the Antitrust Analysis of Horizontal Mergers' in KN Hylton (ed), *Antitrust Law and Economics* (Cheltenham, Edward Elgar, 2010).

[130] This is calculated by subtracting the pre-merger HHI from the post-merger HHI.

[131] Coate and Simons (n 21) 676. For the 1982 Guidelines, see US Department of Justice, 1982 Merger Guidelines (1982) § III.A.

[132] US Horizontal Merger Guidelines 2010 (n 10) § 5.3.

[133] *United States v General Dynamics Corp*, 415 US 486, 498 (1974).

[134] *Chicago Bridge & Iron Company v FTC*, 515 F3d 447, 466 (5th Cir 2008).

[135] ibid 467.

jurisdictions, the EU did not at that time rely on presumptions linking certain market shares to unlawfulness under the merger laws.[136] The EU's Horizontal Merger Guidelines of 2004 subsequently introduced concentration levels to EU merger analysis, outlining that concentration levels such as the HHI would be relied upon as 'useful first indications of the market structure and of the competitive importance of both the merging parties and their competitors'.[137] However, HHI levels and the delta do not constitute presumptions.[138] The EU Courts, including Advocates General, have only rarely alluded to the HHI and generally only when prompted to do so by the parties.[139] In one instance, *Sun Chemical Group* (2007), the General Court noted that as the Commission had not calculated the HHI, it had to be established whether it had thereby disregarded its own Horizontal Merger Guidelines.[140] In another case, the General Court cautioned that 'market shares may only be used as indicia of competition concerns to the extent that the market to which those shares relate has been defined beforehand. The same is true of the HHI to which the applicants also refer'.[141]

Already in the late 1980s, the HHI was criticised as 'a new set of crude, unqualified rules for determining the legality of horizontal mergers'.[142] As an economist asserted at that time, while the HHI measures seem intuitively correct, no theory or reliable econometric evidence shows that the HHI is useful for predicting whether increased concentration will ultimately lead to anti-competitive effects.[143] More worryingly, economists have cautioned that relying on concentration measures in order to ascertain whether a merger would have anti-competitive effects may be particularly erroneous in the presence of differentiated products and possible unilateral – as opposed to coordinated – effects.[144] One cannot assume that the same HHI thresholds should equally apply to all markets either.[145]

The HHI, which we have seen both EU and US antitrust authorities calculate in order to measure changes in concentration that a horizontal merger would

[136] Joined Cases C-68/94 and C-30/95 *France v Commission (Kali & Salz)* EU:C:1997:54, Opinion of AG Tesauro, para 7 (referring to the HHI and the 1984 US Horizontal Merger Guidelines in fn 10).

[137] Guidelines on the assessment of horizontal mergers under the Council Regulation on the control of concentrations between undertakings (n 109) para 14.

[138] ibid para 21.

[139] A search on curia.eu (search terms: HHI or Herfindahl) only prompted seven results (last searched 10 May 2019), and in most of these the Court did not engage with the HHI argument.

[140] Case T-282/06 *Sun Chemical Group v Commission* EU:T:2007:203, paras 133 f. The Court subsequently carried out the HHI calculations, assessed them against the Horizontal Merger Guidelines and found that the Commission had not disregarded its own Guidelines (ibid paras 137–41).

[141] Case T-79/12 *Cisco Systems & Messagenet v Commission* EU:T:2013:635, para 65.

[142] RS Markovits, 'International Competition, Market Definition, and the Appropriate Way to Analyze the Legality of Horizontal Mergers under the Clayton Act: A Positive Analysis and Critique of Both the Traditional Market-Oriented Approach and the Justice Department's Horizontal Merger Guidelines' (1988) 64 *Chicago-Kent Law Review* 745, 806.

[143] FM Fisher, 'Horizontal Mergers: Triage and Treatment' (1987) 1 *Journal of Economic Perspectives* 23, 31.

[144] Farrell and Shapiro (n 4) 1.

[145] MA Glick and D Campbell, 'Market Definition and Concentration: One Size Does Not Fit All' (2007) 52 *Antitrust Bulletin* 229, 234.

bring about, is only meaningful if the relevant market has been correctly defined and all important competitors on this market are identified.[146] However, it 'can deceive analysts with its false air of precision',[147] particularly in innovative environments. If innovative markets are defined too narrowly, then the HHI levels will exaggerate the concentration – and thus the antitrust concern – on the market.[148] However, even where all these conditions are fulfilled, it remains questionable why certain levels of concentration are deemed a threat for competition rather than others and where the cut-off point should be.[149] Against the background of the problematic nature of market shares in innovative market environments (see section I.B above), it is obvious that summing squared market shares in order to obtain concentration levels will not alleviate these concerns. This calls the significance of HHI levels into question, particularly in dynamic markets. It is therefore a positive turn that courts have been cautious when assessing concentration levels in horizontal merger cases.

II. Innovation and Geographical Market Definition

While the present study focuses on product market definition, a few observations shall be made on the relevant geographical market in innovative environments. These merely intend to provide some food for thought and perhaps an incentive for further research.

The relevant geographical market constitutes the second dimension of market definition. Whereas the product market situates alleged anti-competitive conduct within the economic sphere, the geographical market situates it in space. Lerner suggested that 'objects having the same physical characteristics are not the same goods if they are at different places'.[150] An interesting aspect of market definition in innovative environments is the fact that geography is often a rather accommodating feature in these markets. While findings from spatial economics can lead to very fragmented geographical markets in certain goods (eg, atomised geographical markets for local bakeries) or services (eg, small geographical markets for cinemas in a given town),[151] the opposite is the case for many technology markets. As know-how, knowledge and innovations incorporated into IPRs can easily

[146] Similarly, see S Calkins, 'The New Merger Guidelines and the Herfindahl-Hirschman Index' (1983) 71 *California Law Review* 402, 404 fn 14.

[147] AC Hruska, 'A Broad Market Approach to Antitrust Product Market Definition in Innovative Industries' (1992) 102 *Yale Law Journal* 305, 313.

[148] Pleatsikas and Teece (n 43) 672.

[149] See L Kaplow, 'Market Definition Alchemy' (2012) 57 *Antitrust Bulletin* 915, 932.

[150] Lerner (n 6) 166.

[151] J-F Thisse, 'Does Space Matter for Economic Theory? From von Thünen to Krugman through Launhardt and Hotelling: How to Model Competition across Space' (Graz Schumpeter Lectures, Graz, 10 December 2014).

spread thanks to them being non-physical, they can effortlessly cross borders and permeate new geographical spaces. As a consequence, geographical markets tend to be quite wide or even global for technology.[152] There is an increasing number of companies that are 'born global' in the sector of information and communications technology.[153] Transport costs that could constrain a customer's choice[154] are often irrelevant where innovation is concerned – at least before the innovation is incorporated into physical goods.

The internet represents a special case in terms of geographical market definition, as it is not in itself tangible.[155] Some have suggested that the internet may constitute a geographical market in its own right.[156] For network industries such as telecommunications or air transport, geographical market delineation may differ significantly from a more traditional market definition approach, as it will often be directly linked to the product in question – for instance, a flight between city A and city B.[157]

The US Supreme Court advocates a pragmatic and fact-based approach to defining the relevant geographical market, holding that the geographical market should conform to commercial realities.[158] In the US Horizontal Merger Guidelines, the agencies describe the geographical market as the area of competition affected by the merger,[159] while the Supreme Court has referred to it as the 'area of effective competition'.[160] Particularly in technology-intensive industries, the geographical market may be global.[161] For instance, when Western Digital acquired Hitachi in 2012, a merger that concerned hard disk drives, the FTC found the relevant geographical market to be worldwide.[162] The same had been concluded by the European Commission.[163]

[152] Similarly for innovation markets, see RJ Gilbert and SC Sunshine, 'The Use of Innovation Markets: A Reply to Hay, Rapp, and Hoerner' (1995) 63 *Antitrust Law Journal* 75, 81.

[153] L Sleuwaegen, I de Voldere and E Pennings, 'The Implications of Globalization for the Definition of the Relevant Geographic Market in Competition and Competitiveness Analysis' (Ref Ares(2014)77066, January 2001), 7.

[154] Schwalbe and Maier-Rigaud (n 126) 31.

[155] J Kagan, 'Bricks, Mortar, and Google: Defining the Relevant Antitrust Market for Internet-Based Companies' (2011) 55 *New York Law School Law Review* 271, 282.

[156] CC Eblen, 'Defining the Geographic Market in Modern Commerce: The Effect of Globalization and E-Commerce on *Tampa Electric* and Its Progeny' (2004) 56 *Baylor Law Review* 49, 79.

[157] P Larouche, 'Relevant Market Definition in Network Industries: Air Transport and Telecommunications' (2000) 1 *Journal of Network Industries* 407, 442–43.

[158] *Brown Shoe v United States*, 370 US 294, 336 (1962) (with references to cases and legislative materials).

[159] US Horizontal Merger Guidelines 2010 (n 10) § 4.2.

[160] *Standard Oil v United States*, 337 US 293, 299 fn 5 (1949).

[161] United States, 'Geographic Market Definition' in OECD (ed), *Working Party No 3 on Co-operation and Enforcement* (8 November 2016) DAF/COMP/WP3/WD(2016)49, para 25.

[162] *Western Digital/Hitachi*, 153 FTC ___ (5 March 2012) § I.CC.

[163] *Western Digital Ireland/Hitachi* (Case COMP/M.6203) Commission Decision of 23 November 2011 [2013] OJ C241/11, para 389 (for hard disk drives; see paras 390 ff for a different conclusion on external hard disk drives, namely that this market was only EEA-wide in terms of geographical scope).

In a recent case before the US District Court for the Northern District of California, *US v Bazaarvoice* (2014), the parties had argued that the relevant geographical market for the product in question – reviewing and rating (R&R) software platforms for online sellers – should be defined as worldwide because 'technology knows no borders'.[164] The Court did not agree, holding instead that the US was to be regarded as the relevant geographical market. It based this conclusion on several factors, amongst other things that the licences required for R&R platforms were often geographically limited, that customers frequently used different R&R platform providers in different territories, that US customer reviews were not interchangeable with reviews from other territories, for instance because the products sold were different, and that language and cultural issues added to this territorial barrier between the US and other countries.[165] The Court also pondered the application of antitrust to highly dynamic markets more generally, but ultimately held that it was not the 'Court's role to weigh in on this debate'.[166] Instead, its duty was to 'assess the alleged antitrust violations presented, irrespective of the dynamism of the market at issue'.[167]

In the EU, the geographical market is understood as an area in which conditions of competition are sufficiently homogeneous.[168] With the progressing completion of the single market, geographical markets that used to be considered national are becoming broader EU-wide markets.[169] The manufacturing of active pharmaceutical ingredients has even been considered to be a worldwide market.[170] This illustrates how different levels of the same production chain may constitute geographical markets that are not congruent. The Commission has also emphasised that international trade flows only lead to the finding of a broad geographical market if those flows are actually capable of restraining market power.[171] Overall, the Commission has found that 'technological innovation drives the move towards a wider geographical market'.[172] This is evidenced by geographical markets which, in 2012/13, were found to be EEA-wide or wider in 61 per cent of the European Commission's merger cases.[173]

The European Commission in its Technology Transfer Guidelines has set out some principles for geographical market delineation. Under the TTBER, it

[164] *United States v Bazaarvoice*, Case 13-cv-00133-WHO (ND Cal 8 January 2014) paras 93, 120.

[165] ibid paras 120–25. This analysis was also supplemented by the findings of the government's economic expert, Carl Shapiro; ibid paras 126–47.

[166] ibid § III.

[167] ibid (citation omitted).

[168] See on this Case 27/76 *United Brands v Commission* EU:C:1978:22, para 53; EU Market Definition Notice 1997 (n 51) para 8.

[169] *Ciba-Geigy/Sandoz* (Case IV/M.737) Commission Decision 97/469/EC [1997] OJ L201/1, paras 47 f.

[170] ibid para 50.

[171] *Arsenal/DSP* (Case COMP/M.5153) Commission Decision of 9 January 2009 [2009] OJ C227/24, para 49.

[172] European Commission, 'Market Definition in a Globalised World' [March 2015] *Competition Policy Brief* 1, 3.

[173] ibid 2.

will assume that the technology market's geographical market is identical with the geographical market of the relevant product market. However, outside of the TTBER, the Commission might consider the technology market's geographical scope separately from that of the product market.[174] Indeed, the incorporeal nature of know-how might lead to a much wider geographical market.

An issue in the EU that could favour narrower geographical markets also in innovation-intense markets is the fact that IP protection is largely still granted on a national basis and may therefore not be the same in all Member States. This allows for varying degrees of protection that go hand in hand with heterogeneous competitive circumstances.[175] For instance, interesting issues have arisen in the area of the exhaustion of copyright or trademark rights and how this relates to competition law, more specifically pricing abuses and the relevant geographical market. As IPRs are territorial in nature, they allow an IPR owner to price its products differently depending on the market environment and to exclude competition from parallel imports.[176] The single market imperative, which governs EU commercial law at large, has had an important influence on this particular area of the IP/antitrust intersection.[177] In the EU, IP owners can no longer separate national markets based on their IP rights, as the principle of EU-wide exhaustion means that IP-bearing products marketed in the EU can freely circulate within the whole EU.[178] However, price differentiation is still possible with relation to IP-bearing products placed on the market outside of the EU. This has led to the question of whether prices that are perceived as excessive in view of the prices charged outside of the EU can be attacked under EU competition law.[179] In *Micro Leader* (1999), different prices were charged for the French-language versions of Microsoft's copyrighted software in Canada and in France. Micro Leader relied on the market of software in its complaint, with market sectors including word processing, spreadsheets and operating systems.[180] The Court annulled the Commission decision rejecting Micro Leader's complaint, finding that under exceptional circumstances, a claim of excessive pricing might have merit.[181] This possibility means that in cases in which IP owners excessively engage in this type

[174] European Commission, Guidelines on the application of Article 101 of the Treaty on the Functioning of the European Union to technology transfer agreements [2014] OJ C89/3, para 89.

[175] See also Glader, *Innovation Markets* (n 1) 114.

[176] Nothing else was done in *Consten and Grundig*, where the GINT trademark was assigned in order to prevent parallel imports; see Joined Cases 56 and 58/64 *Consten and Grundig v Commission* EU:C:1966:41.

[177] See H Ullrich, 'IP-Antitrust in Context: Approaches to International Rules on Restrictive Uses of Intellectual Property Rights' (2003) 48 *Antitrust Bulletin* 837, 842 f.

[178] On exhaustion of IPRs, see Case C-355/96 *Silhouette v Hartlauer* EU:C:1998:374; Case T-198/98 *Micro Leader v Commission* EU:T:1999:341; Joined Cases C-414 to C-416/99 *Zino Davidoff and Levi Strauss* EU:C:2001:617.

[179] T Heide, 'Trade Marks and Competition Law after Davidoff' (2003) 25 *European Intellectual Property Review* 163, 165.

[180] Case T-198/98 *Micro Leader v Commission* EU:T:1999:341, para 43.

[181] ibid para 56; Heide (n 179) 165 (calling this theoretical possibility problematic).

of price differentiation, the relevant geographical market can go beyond the EU and thus include countries that are not subject to the EU-wide exhaustion rule.

III. Conclusion

As economic tests for antitrust market delineation were developed for static market environments with price as their central competitive parameter, it appears that the traditional tools may be ill-suited to delineate dynamic markets. Nevertheless, the conceptual underpinnings of the hypothetical monopolist test can continue to provide valuable insights for a more qualitative market analysis that is geared towards market characterisation. Obviously, this limits the possibility of relying on strict market share thresholds, market power presumptions based on market shares and concentration ratios. For competition law, it is important to recognise the limits of economic analysis when it comes to dynamic markets.

In terms of the geographical market definition, cases in innovative markets may often rely on global geographical markets. However, there are also important factors that may limit an innovative market's geographical scope, such as observed licensing and business practices, language and cultural issues, and possibly also limited IP exhaustion that enables the re-erection of borders even where the market characteristics as such would lend themselves to a global market.

Conclusions on Part II: Accounting for Innovation When Delineating Markets

Part II uncovered the many ways in which antitrust authorities, courts and legal scholarship – in the EU and the US – have tried to meet the challenges of dynamic markets when delineating antitrust markets. As innovative markets are so closely connected, European and US antitrust authorities and courts often deal with nearly identical cases involving the same parties. This also leads to increased communication among the competition authorities, which can be a source of legal convergence. At the same time, the procedural differences between the two jurisdictions that continue to exist give the authorities and courts different roles to play. While the European Commission directly adopts infringement decisions that can be reviewed by the EU Courts in the case of an appeal, the US agencies need to resort to the courts in order to get their cases through. This, it appears, has also impacted upon these authorities' self-perception. The European Commission is both policy- and decision-maker, a dual role that can sometimes conflict.

In some instances, eg, proprietary aftermarkets or the use of the hypothetical monopolist test in innovative markets, EU competition law and US antitrust law are still worlds apart, and any tentative sign of convergence (such as the *Kodak* case) has not prevailed in practice. In multi-sided markets, the recent cases in the EU and the US are not entirely in sync in developing a new methodology for delineating these platform markets. In *American Express*, the US Supreme Court anchored its relevant market in the transactional nature of the market at issue, while the European Commission mostly steered clear of multi-sided market insights in recent cases such as *Google Shopping*.

Building on the comparative insights on market delineation in dynamic market contexts, Part III will reconceptualise market delineation so as to provide guidelines for gauging substitutability in these non-traditional markets.

PART III

Reconceptualising the Legal Framework for Delineating Antitrust Markets in Dynamic Contexts

10

An Antitrust Framework for Delineating Dynamic Markets

The present chapter develops two options for model guidance based on the legal understanding of market definition, as discussed in section II.A in Chapter 1. In many jurisdictions, this understanding may require a redefinition of the role of economic experts in competition law cases: their role should be understood as that of assisting in the fact-finding, not in the application of the law to these facts. Such a legal approach may not necessarily have great repercussions on day-to-day market definition in practice because '[t]he majority of market definitions are not based on complex economic data, but are instead built on facts, opinions, and logic'.[1]

In innovative market environments, the relevant market will often have to be delineated repeatedly, reflecting changing circumstances. In addition, one cannot regularly rely on previous market definitions because of the dynamics of these markets.[2] Against the background of the findings made throughout this study, this conclusion can only be underlined. The guidance options here developed cannot provide a remedy to this – it is inherent in the sometimes fast-moving, often uncertain nature of innovation. Instead, it is vital to raise antitrust enforcers' awareness in this respect. A careful market characterisation can contribute to a better understanding of a dynamic market's evolution.

After setting out the typology that the market delineation framework relies upon (section I), the two guidance options are stated in a comprehensive fashion (sections II.A and II.B) and subsequently discussed (section II.C). Thereafter, the question is addressed which approach a legislator or an enforcement agency may wish to adopt (section III).

[1] M Sousa Ferro, 'Judicial Review: Do European Courts Care about Market Definition?' (2015) 6 *Journal of European Competition Law & Practice* 400, 410.

[2] Monopolkommission, 'Wettbewerbspolitik: Herausforderung digitale Märkte' (Sondergutachten 68, 2015) para 28.

I. A Typology for Reconceptualising the Market Definition Framework

A three-pronged typology guides the following development of options for market definition guidance in innovative market environments. As market definition fulfils several intertwined functions in the various areas of competition law (see Chapter 2), one can easily envision many different policy approaches to a reconceptualisation of market definition in innovation, each with its own – sometimes implicit, sometimes explicit – agenda. The typology relied on here somewhat resembles a stereotype and quite strongly states the policies pursued by each policy type. This serves to highlight our understanding of competition policy. In order to reframe antitrust market definition, we need to understand the policy approaches that implicitly underlie such a reconceptualisation in order to assess whether implementation of these new approaches is worthwhile or whether it is in line with a jurisdiction's policy goals. The typology provides this policy background information, even if it only represents three positions on an entire spectrum of possible competition policy approaches.

Type A views market *power* as market definition's primary function, and this is reflected in its approach to market definition in a traditional sense. It subscribes to a static policy approach that one could perceive as more conventional, with the continued application of longstanding antitrust tools as one of its underlying assumptions. While being aware of the possibility of over-enforcement, it is more concerned about under-enforcement of the antitrust laws. It wants to maintain the option of investigating cases, possibly at the expense of deterring innovation (which ultimately results in consumer harm, although this certainly is not type A's intention). Type A acknowledges the benefits that innovation can have for consumer welfare, but due to its static focus, it is more intrigued by the short-term gains of lower prices for consumers. This fits well with its overall more static perspective. Type A also treasures legal certainty that companies subjected to the antitrust laws can rely on and assumes that this can be achieved with its static approach.

Going ahead, it is conceivable that traditionalists of the type A approach will maintain the upper hand. Frequently, they can rely on settled case law and will not have to question well-accepted doctrines in order to mould their approach in a more innovation-sensitive way. At the same time, this means that their ability to reshape market definition is limited, as they need to rely on static tools that provide price-based market delineations, allow for the computation of seemingly clear market shares and the application of clear-cut market share thresholds. Despite these limitations, there exists some wiggle room within which a type A approach allows us to take innovation-based considerations into account. While type A therefore represents a more static approach, it in no way negates the importance of innovation, but rather wants to foster it within an analytical framework to which competition law has grown well accustomed.

Type B is focused on the market *characterisation* function of market definition, which acts as a basis for informing the antitrust theory of harm, but can also inform market power assessments in a less traditional sense. It follows a more dynamic policy approach,[3] with a very obvious innovation agenda. As such, it treasures innovation in the long term more than low prices or improved quality in the short term. This fits well within its overall more dynamic perspective. It is very aware of the fast-moving nature of innovative markets. On the other hand, it also realises the importance of network effects and consumer lock-in for the further development of innovative markets. It is not opposed to departing from more traditional antitrust analytical frameworks. On the contrary, it is keen on adapting the antitrust law framework to new market settings and believes in the power of economics to provide us with insights into the workings of competition and innovation. Type B guidance is based on the view that market delineation in innovative environments needs to cater to contextuality and improve our understanding of the competitive process.[4] This again highlights its dynamic nature.

What if type B were to gradually achieve acceptance in the competition community? This would certainly entail a less straightforward role for market definition in the grand scheme of antitrust. However, it may also re-invigorate market definition's standing and be more realistic, assigning a role to market delineation that it can fulfil rather than placing hopes and aspirations for clear-cut answers in an analytical tool that the latter simply cannot deliver. After decades of entrusting market definition with such a role, it will be difficult for competition lawyers – both academics and practitioners – to come to terms with this new role for market definition. Eventually, however, this allows us to look at competition cases with renewed vigour, framing antitrust theories that are anchored in market realities rather than market fantasies. The model guidance that is suggested as type B may help competition law in reconceptualising market definition in this way: as a legal concept with a more realistically defined task.

Some will interject that there is no valid legal basis for this limited role of market definition as proposed under type B, that the legal provisions – as was seen in sections I and II in Chapter 2 – require the reliance on market definition. This may be true. But at the same time, it has always been the courts which filled this requirement for market definition with meaning. While this might require some judicial effort, the courts may in the future also assign such a more realistic role to market definition, particularly in dynamic market contexts. While in the past it was argued that most cases turned on market definition, in the future it might then

[3] Arguing for a focus on market characterisation within dynamic markets, see A Capobianco and A Nyeso, 'Challenges for Competition Law Enforcement and Policy in the Digital Economy' (2018) 9 *Journal of European Competition Law & Practice* 19, 24; G Colangelo and M Maggiolino, 'Applying Two-Sided Markets Theory: The *MasterCard* and *American Express* Decisions' (2018) 14 *Journal of Competition Law & Economics* 115, 115, 124; J Furman et al, 'Unlocking Digital Competition' (Report of the Digital Competition Expert Panel, March 2019) para 1.43.

[4] See already MB Coate and JJ Simons, 'In Defense of Market Definition' (2012) 57 *Antitrust Bulletin* 667, 708; and ch 4, section I.D above.

be that most cases will turn on the antitrust theory of harm – which can only be properly understood in the light of a diligent market characterisation.

A third approach shall briefly be mentioned: type C is not so much a policy approach to market definition as a general negation of the need for or utility of market definition for competition law analysis, particularly in innovative markets. This coincides with the approaches of Louis Kaplow, Richard Markovits and others discussed in section IV.A in Chapter 2. For type C, no set of market definition guidelines will be developed in the following. Bearing in mind the many challenges and pitfalls of delineating innovative markets for antitrust law purposes, the sheer simplicity of type C can seem appealing. However, this is something of a dead-end road. The current legal frameworks for antitrust law in both the US and the EU rely on market definition to fulfil many functions (section III in Chapter 2). A policy approach that takes itself out of the realm of the current legal framework may not be very viable, at least for the foreseeable future.

These three stylised policy types allow for many different positions on the spectrum of antitrust policies towards market delineation. It would be a vast overgeneralisation to state that most competition lawyers will identify with a type A approach, most competition economists with a type B approach and most competition law and economics scholars with a type C approach. Many practitioners and scholars will see themselves on some sort of middle ground between the extremes, and in many cases this might be the way forward. The idea behind these types is also not to advocate the one while belittling the other – instead, the pros and cons of both types A and B shall be discussed. Only type C is excluded from the analysis, as this study does not subscribe to its tenets. By lifting the policy assumptions underlying each type out into the open, we can start to appreciate that every tweak in our antitrust market definition guidance also reflects a certain policy stance, serves a certain goal and is not as neutral as we sometimes like to believe.

II. Options for Market Definition Guidance in the Presence of Innovation

In order to delineate dynamic markets with as much legal certainty as possible, market delineation must build upon 'a coherent underlying framework'.[5] In the following, two such frameworks are developed based on the type A and type B policy approaches outlined above. For each of the aspects of market definition in innovative environments discussed in Part II, two different ways of reconceptualising market definition in this setting will thus be developed. In doing so, the rich comparative insights on market definition in the EU and the US uncovered in Part II can be built upon.

[5] FW McElroy, 'Alternatives to the US Antitrust Agency Approach to Market Definition' (1996) 11 *Review of Industrial Organization* 511, 511.

The guidance intends to operationalise the legal parameters of market definition in innovative market contexts. These legal parameters inform the attorney/judge/competition practitioner when delineating a relevant product market. As such, the guidance does not prescribe in any detail what a competition economist may wish to rely on when presenting an expert opinion on one of these parameters; it does not suggest which evidence may be used before a court or competition authority.[6] It is for the economic expert to implement this guidance. But his or her choice of modelling may then be scrutinised by the judge or competition authority against the benchmarks of the guidance.

A. Type A Guidance: Market Delineation with a Focus on Market Power

§ 1. Antitrust Product Markets in Dynamic Contexts

a. An antitrust product[7] market represents that area of competition within which company behaviour is assessed under the competition laws. It allows for a subsequent assessment of market power within that market.
b. A dynamic market environment is one which exhibits several of the following characteristics: short product cycles, fast-moving market environment, high importance of R&D and intellectual property rights (IPRs) (including know-how), prominence of direct and/or indirect network effects, performance as an important parameter of competition and the use of big data analytics in the design and/or provision of the product.

§ 2. Innovative Products

a. When delineating a relevant product market in innovative market environments, one must focus on the principal product under investigation (focal point). Due attention needs to be paid to the possibilities of substitution that innovation might have brought about, and to asymmetric substitutability.
b. In innovative markets, actual consumer behaviour is an important factor when delineating the relevant antitrust product market from a demand perspective. Supply substitutability is carefully assessed if its effects are as immediate as those of demand substitutability.

[6] Such evidence may consist of previous substitution patterns, quantitative tests, customers' and competitors' views, consumer preferences, entry barriers and switching costs, different customer groups and price discrimination; see D Hildebrand, *The Role of Economic Analysis in EU Competition Law: The European School*, 4th edn (Alphen aan den Rijn, Kluwer Law International, 2016) 212.

[7] The nature of a product is to be understood broadly, ie, as constituting a good, a service or a technology.

§ 3. *Product Differentiation and Innovation*

a. Substitutability amongst online and offline products is analysed against the background of actual substitutability, with the latter being determinative for a finding of separate or converged market spheres.
b. As long as a product's functionalities are evolving within the market environment, they are not determinative of the relevant product market.
c. Whether products distributed through to-pay and freemium business models form part of the same relevant product market depends on actual consumer behaviour.

§ 4. *Time Horizon*

a. The time horizon applied for analysing a relevant product market must reflect market realities (average product and innovation cycles, customer response etc), without taking into account developments at an indeterminate time in the future.

§ 5. *Future Markets*

a. Future markets are relevant product markets the emergence of which can be assumed based on nearly completed R&D and/or pending regulatory approval.

§ 6. *Potential Competition*

a. Potential competition in innovative markets is taken into account after the stage of market definition, reflecting real and credible possibilities of market entry that can exert competitive pressure on both current and future product markets.

§ 7. *Innovative Product Markets and Innovation Competition*

a. Innovation competition is 'competition among firms seeking to develop the same new product or process'.[8] Unless innovation competition can be captured through future markets, it is assessed after the stage of market definition.
b. Where R&D is far advanced, regulatory market approval imminent or an innovation has in any other way nearly reached the stage of market entry in an observable way, innovation competition can be incorporated into market definition as any other product/technology (future market).

[8] JB Baker, 'Beyond Schumpeter vs Arrow: How Antitrust Fosters Innovation' (2007) 74 *Antitrust Law Journal* 575, 579.

c. Factors to consider when assessing whether two research endeavours are in a competitive relationship with each other include the nature, scope and magnitude of R&D efforts, each company's access to specialised assets such as financial and human resources, data or know-how/intellectual property rights, the timing of R&D efforts and each company's ability to successfully commercialise its innovations. These factors may be useful in order to delineate markets in innovative environments.

§ 8. *Intellectual Property Rights*

a. IPRs may constitute their own relevant antitrust market. However, ownership of an IPR does not automatically equate market power.
b. An IPR delimits its own antitrust market where there are no viable substitutes for the IPR from the customers' point of view. Supply-side substitutability is taken into account where IP protection and customer behaviour do not exclude successful substitution from the supply side.
c. Where an IP-protected technology is licensed rather than merely embodied in a product, the resulting relevant market consists of the IPR in question and its available substitutes (technology market).
d. In technology markets and other markets delineated by an IPR, market share thresholds can generally be applied.
e. In standard-essential patents (SEPs), a separate technology market consisting of the SEP in question may be found.
f. Where an IPR does not constitute its own relevant antitrust market (non-IP market), the IPR may be relied upon as an important factor in delineating a relevant antitrust market, eg, as a barrier to entry and expansion.

§ 9. *Aftermarkets*

a. An aftermarket (secondary market) is a market the demand for which derives from a primary product.
b. The primary and secondary market together constitute a relevant antitrust market (system market) where competition on the primary market constrains market power on the secondary market. This is particularly the case where customers have the ability to engage in lifecycle costing before acquiring a primary product.
c. Where lifecycle costing is not possible or where a policy change affects already-existing customers which are thereby locked in, separate markets for the primary products on the one side and each secondary product related to a primary product on the other side need to be delineated (multiple markets).
d. For market delineation, it is irrelevant whether aftermarkets are foreclosed by relying on IPRs, contractual or technical tying.
e. Aftermarkets derived from franchise agreements can be delineated in line with the above.

§ 10. Multi-sided Markets

a. Multi-sided markets are platforms with two or more interconnected market sides. They are characterised by the presence of network effects.
b. Where a multi-sided platform has the primary aim of enabling direct transactions amongst the market sides, the relevant antitrust product market is the platform market, defined as a combination of the different market sides and the platform. As a precondition, (i) competitive conditions need to resemble each other on all market sides, and (ii) the platform's activity by definition encompasses all market sides.[9] Cross-platform externalities (network effects) are taken into consideration in this definition.
c. Where the conditions mentioned in point b. are not fulfilled, each market side constitutes its own relevant antitrust product market. Subsequently, the network effects at play are carefully analysed in the substantive antitrust analysis.
d. Where one market side generates revenue, while another does not (the 'free' market side), it shall be the first that is decisive for market definition. A free market side on its own does not constitute a relevant antitrust market, but may influence a connected relevant market.
e. The SSNIP test can be applied to each market side, thus determining which is its relevant product market. The interdependence amongst market sides should be reflected in this analysis.
f. Market share thresholds are applied to each relevant market, as defined above. Legal instruments incorporating market share thresholds may lower these thresholds where a multi-sided platform is much more significant than its competitors.

§ 11. Economic Tests as Legal Benchmarks

a. When carrying out a hypothetical monopolist test in innovative markets (the SSNIP test), special attention should be paid to the importance of factors other than price.
b. Where market shares are calculated in innovative markets, special attention should be paid to that market's particular characteristics when calculating and interpreting these market shares or applying market share thresholds.
c. Special attention should be paid to an innovative market's particular characteristics when calculating and interpreting concentration levels.

[9] Based on S Wismer, C Bongard and A Rasek, 'Multi-Sided Market Economics in Competition Law Enforcement' (2017) 8 *Journal of European Competition Law & Practice* 257, 260.

B. Type B Guidance: Market Delineation with a Focus on Market Characterisation

§ 1. Antitrust Product Markets in Dynamic Contexts

a. An antitrust product[10] market represents that area of competition within which company behaviour is assessed under the competition laws. A comprehensive characterisation of a dynamic market is essential for the ensuing antitrust analysis and, in particular, for the antitrust theory of harm and a successful set of remedies.

b. A dynamic market environment is one which exhibits several of the following characteristics: short product cycles, fast-moving market environment, high importance of R&D and IPRs (including know-how), prominence of direct and/or indirect network effects, performance as an important parameter of competition and the use of big data analytics in the design and/or provision of the product.

c. Antitrust markets in innovative environments need to be grounded in market realities as they present themselves, not necessarily as they are presented by parties under investigation or their competitors. Any evidence produced by (third) parties and their economic or legal experts shall be assessed against the background of possible intellectual capture. At the stage of market delineation, policy objectives (eg, promotion of innovation, ease of enforcement) do not enter the analysis.

§ 2. Innovative Products

a. When delineating a relevant product market in dynamic market contexts, one must focus on the principal product under investigation (focal point). Due attention needs to be paid to the possibilities of substitution that innovation might have brought about and to asymmetric substitutability.

b. In innovative markets, actual consumer behaviour is the decisive factor when delineating the relevant antitrust product market from a demand perspective. At the same time, supply substitutability needs to be carefully assessed. For both demand and supply substitutability, the effect that algorithms have on substitutability is taken into account.[11]

[10] The nature of a product is to be understood broadly, ie, as constituting a good, a service or a technology.

[11] Some algorithm-based applications may restrict the available substitutes, while others might enlarge them.

c. In order to discern the importance of network effects for the delineation of an innovative relevant market, actual customer behaviour is assessed. No stringent rational behaviour is assumed for either customers or suppliers.
d. Whether a platform of any kind acts as a gatekeeper for interoperable goods or services from a supplier's or a customer's perspective, possibly leading to a narrowly defined market, is analysed based on the framework for aftermarkets (see § 9). From the demand perspective, issues of interoperability are analysed based on actual customer behaviour and possible capture.
e. In innovative market environments, contextuality is particularly relevant to the delineation of the relevant antitrust market. A full understanding of the competitive relationship amongst innovators is necessary in order to fully characterise the relevant product market.

§ 3. *Product Differentiation and Innovation*

a. Substitutability amongst online and offline products is analysed against the background of actual substitutability, with the latter being determinative for a finding of separate or converged market spheres.
b. As long as a product's functionalities are evolving within the market environment, they are not determinative of the relevant product market. Innovative potential is considered in assessing functionalities.
c. Whether products distributed through to-pay and freemium business models form part of the same relevant product market depends on actual consumer behaviour.

§ 4. *Time Horizon*

a. The time horizon applied for analysing a relevant product market must reflect market realities (average product and innovation cycles, customer response etc).

§ 5. *Future Markets*

a. Future markets are relevant product markets the emergence of which can be assumed, particularly based on nearly completed R&D and/or pending regulatory approval. Based on a comprehensive market characterisation, the emergence of future markets must go beyond mere plausibility. The time horizon within which the emergence of a future market is expected depends on the respective market environment.

§ 6. *Potential Competition*

a. Potential competition in innovative markets is taken into account after the stage of market definition, reflecting real and credible possibilities of market entry that can exert competitive pressure both on current and future product markets. Competitive pressure by potential competitors can be factored into the analysis by regarding the latter as (future) market participants.

§ 7. *Innovation Competition*

a. Innovation competition is 'competition among firms seeking to develop the same new product or process'.[12] The concept relates to the competitive pressure that is exerted in innovative industries to win the competitive race for a new market. It implies a certain independence from current markets and competition for rather than in a market.

b. As a parameter of competition, innovation competition generally needs to be assessed outside of clearly defined antitrust markets. For instance, it can be usefully addressed when characterising a particular antitrust market or it may relate to a specific theory of harm. Only if the resulting innovations are considered imminent can they be assessed as future markets or potential competition. In such cases, the inherent uncertainty in innovative markets needs to be factored into the market definition.

c. Factors to consider when assessing whether two research endeavours are in a competitive relationship with each other include the nature, scope and magnitude of R&D efforts, each company's access to specialised assets such as financial and human resources, data or know-how/IPRs, the timing of R&D efforts and each company's ability to successfully commercialise its innovations. These factors may be useful benchmarks in the assessment of future markets, of potential competition and of innovation competition.

§ 8. *Intellectual Property Rights*

a. IPRs may constitute an antitrust product market subject to the substitutability rules set out below. Otherwise, they may constitute a factor in establishing and characterising the relevant product market. There is no presumption that IPRs constitute their own antitrust market.

b. Irrespective of the type of IPR at issue, a market delineated by the IPR in question is established where there are no conceivable substitutes for the IPR or the IP protected product from the customers' point of view or from the supply side. Substitutability goes beyond IP-protected subject matter. The degree to which differentiated products are substitutable with the IPR in question is assessed. In addition, it is assessed whether IP markets are subject to considerable competitive pressure from potential competition and/or the

[12] Baker, 'Beyond Schumpeter vs Arrow' (n 8) 579.

constant threat of innovation. Where these considerations are not such as to alter the market delineation, they can nevertheless be taken into account in the substantive analysis.

c. Where an IPR is or conceivably can be licensed, the relevant upstream product market (the technology market or licensing market) consists of the IPR in question and its available substitutes. Substitutability takes the subject matter of the IPR as a starting point, ie, a particular copyrighted work, a patented invention, a protected design or a trademarked sign.[13] However, eventually, it may go beyond it. Substitutability is first assessed from the point of view of the licensee, ie, the customer. In this respect, a SSNIP test can be employed. The availability of substitutes also needs to take imminent market entry into account, for instance, where an IPR has already been applied for but not yet registered or where a generic pharmaceutical manufacturer is preparing its market entry because of the imminent expiry of a patent. Licensing markets may be found alongside product markets incorporating the IP-protected technology or work.

d. The fast-moving nature of many licensing markets and IP-delineated markets means that market share thresholds cannot be applied in a straightforward manner. Instead, where a legal instrument does not explicitly foresee alternatives, the presence of a total of three viable alternatives[14] independently owned by competitors will be seen as equivalent to not surpassing a 30 per cent market share threshold, the presence of a total of five viable alternatives owned by competitors will be seen as equivalent to not surpassing a 20 per cent market share threshold, and the presence of a total of eight viable alternatives owned by competitors will be seen as equivalent to not surpassing a 10 per cent market share threshold.

e. Markets consisting of SEPs are delineated in the same way as conventional licensing markets. Factors to consider when assessing substitutability are the success of the standard that the SEP relates to and possible alternative technologies of any kind.

f. Where an IPR does not constitute its own relevant antitrust market (non-IP market), it may be relied upon as an important factor in characterising a relevant antitrust market, eg, as a barrier to entry and expansion.

§ 9. *Aftermarkets*

a. An aftermarket (secondary market) is a market the demand for which derives from a primary product.

[13] It is acknowledged that while technology markets typically relate to patents, the licensing of copyright and other IPRs raises very similar issues.

[14] Alternatives are understood to be either alternative technologies, works or inventions; these need not be IP-protected.

b. The primary products on the one hand and each secondary product relating to a primary product on the other hand each constitute their own relevant antitrust market (multiple markets) where competition on the primary market does not constrain market power on the secondary market. Factors to take into account in this assessment are customers' information costs, switching costs, their reliance on lifecycle pricing at the time of buying the primary product and the timeliness of customer response in the case of price rises in the secondary market.[15]
c. Primary and secondary markets together may be delineated as a system market where the primary market's competitiveness constrains market power in the secondary market, based on the factors set out in point b.
d. For market definition, it is irrelevant whether aftermarkets are foreclosed by relying on IPRs, contractual or technical tying.
e. Aftermarkets derived from franchise agreements can be delineated in line with the above.

§ 10. Multi-sided Markets

a. Multi-sided markets are markets in which (i) two or more customer groups are present who (ii) need each other in some way and (iii) rely on an intermediary (platform) to capture indirect network effects amongst each other that they cannot capture by themselves.[16]
b. In order to characterise the relevant antitrust market in multi-sided market environments, one distinguishes between the platform (primary product) and the market sides (secondary products).
c. The platform as the primary market and its market sides together constitute a relevant antitrust market (system or platform market) where workable competition at the platform level or in one of the market sides significantly constrains market power on the primary market and all secondary markets. This is particularly the case where customers can and do engage in lifecycle costing specific to platforms (as outlined in point d. below) before engaging with a primary product. Cross-platform externalities (network effects) are taken into consideration in this definition.
d. Factors to take into account as lifecycle costing in multi-sided markets include price, customers' information costs, switching costs, reliance on parameters such as quality, interoperability, data protection, data portability etc at the time of acquiring/using the primary product, and the timeliness of customer

[15] This is a condensed and amalgamated version of the *Kodak* factors and the *EFIM* criteria. On these cases, see ch 7.

[16] This definition is based on DS Evans and R Schmalensee, 'The Antitrust Analysis of Multisided Platform Businesses' in RD Blair and DD Sokol (eds), *The Oxford Handbook of International Antitrust Economics*, vol 1 (Oxford, Oxford University Press, 2015) 409.

response in the case of price rises, quality decreases etc in the platform's provision of services on one market side.[17] Multi-homing of each market side can also prove relevant.

e. Where lifecycle costing relating to a platform, based on the factors in point d. above, is either not possible or not taken into account by customers, or where a policy change affects already-existing customers which are thereby locked in, separate markets for the primary platform on the one hand and each market side related to a platform on the other hand need to be delineated (multiple markets). Where multiple markets are delineated, cross-platform externalities (network effects) need to be taken into consideration in the subsequent antitrust analysis.

f. The fact that one of the market sides involved does not directly generate any revenue ('free' market side) is not relevant for market delineation. Such a non-price market side is incorporated into the market definition like all other market sides. However, the exercise of market characterisation should begin on that market side which generates the revenue and possibly subsidises the 'free' market side.

g. The SSNIP test in its traditional form is not as such applied to multi-sided markets. Instead, substitutability is assessed for each market side and then combined to an overall understanding of the multi-sided market at issue.

h. Market share thresholds are not applied to multi-sided markets; instead, instruments containing market share thresholds should foresee alternatives.[18] In the event that this is not the case, for platform markets the presence of a total of three viable competing platforms will be seen as equivalent to not surpassing a 30 per cent market share threshold, the presence of a total of five viable platforms will be seen as equivalent to not surpassing a 20 per cent market share threshold, and the presence of a total of eight viable platforms will be seen as equivalent to not surpassing a 10 per cent market share threshold. For multiple markets, the presence of a total of three viable competitors for the primary and each of the secondary product sides of which the platform market is comprised will be seen as equivalent to not surpassing a 30 per cent market share threshold, the presence of a total of five viable competitors each will be seen as equivalent to not surpassing a 20 per cent market share threshold, and the presence of a total of eight viable competitors will be seen as equivalent to not surpassing a 10 per cent market share threshold.

[17] Essentially, this is a condensed and amalgamated version of the *Kodak* factors and the *EFIM* criteria. On these cases, see ch 7. These factors constitute a moveable system that allows for an overall appreciation of whether or not customers engage in lifecycle costing specific to platforms.

[18] Such alternatives can, for instance, relate to the number of competitors on a specific market or market side.

§ 11. Economic Tests as Legal Benchmarks

a. The SSNIP test is only conceptually applied in innovative markets. Substitutability between products and suppliers needs to be assessed based on (limited) interchangeability with other products, against the background of an insight into the differentiated nature of the products at issue.
b. In markets characterised by high innovation, market shares are not relied upon as a proxy for market power. Market share thresholds foreseen in policy instruments or legal provisions should provide for alternative tests. In the event that this is not the case, the presence of a total of three viable competitors on the relevant market will be seen as equivalent to not surpassing a 30 per cent market share threshold, the presence of a total of five viable competitors will be seen as equivalent to not surpassing a 20 per cent market share threshold, and the presence of a total of eight viable competitors will be seen as equivalent to not surpassing a 10 per cent market share threshold.
c. Concentration levels cannot be applied as such to horizontal mergers in highly innovative markets. Instead, a qualitative assessment of the merger's influence on the post-merger competitive environment needs to be carried out, with special regard to the innovation dimension.

C. Discussing the Guidance Options: Depicting Innovation in Antitrust Markets

Innovative companies are faced with possible antitrust intervention in the EU, the US and beyond. Bearing in mind the efficiencies – including dynamic efficiencies – that antitrust convergence related to innovation-intense industries could adduce, it is tempting to propose 'one-size-fits-all' model guidelines for delineating antitrust relevant markets in dynamic market contexts. However, jurisdictions embrace different policy approaches. The dividing line is not necessarily the Atlantic, as courts, competition authorities and legislators may all have different policy objectives (or *no* policy objectives, but simply the application of the law) in mind. For this reason, the present study relies on a dual approach to showcase options that are available when rethinking antitrust market definition in the light of innovative market environments. The proposals provide two analytical frameworks for legally determining the relevant antitrust product market in the presence of innovation, each with a slightly different nuance based on the underlying policy premises. While both frameworks necessarily contain some economics, they are primarily based on an analytical framing of substitutability, as this is the legal benchmark for antitrust market definition that has emerged in the case law. These options are not as such ready to be incorporated into a national framework, for adjustments specific to each jurisdiction will need to be made in light of the legal requirements, case law and policy objectives pursued. The options presented are not restricted to one area of antitrust – as the US Horizontal Merger Guidelines are – but equally apply to all areas covered by competition law: anti-competitive agreements, unilateral anti-competitive behaviour and mergers. Although market delineation will differ amongst those three areas of competition law due to their particular outlook on markets, it should nevertheless rely on the same methodology. The style that was followed in the guidance more closely resembles the European Commission's 1997 Notice, as this allows for a more rule-like application; however, this may just as well be adapted.

We may not need new antitrust rules in order to do justice to innovative market environments,[19] but we do need to take the particular characteristics of innovative markets into account when delineating and assessing these markets. The guidelines developed above draw on the rich experience with delineating dynamic markets that has developed in the US and in the EU (see Part II), without necessarily following approaches found in the one rather than in the other jurisdiction. The typology upon which the two sets of guidelines are based flows from the two main functions that are ascribed to market definition as an analytical tool in antitrust: the market power function ('type A') and the market characterisation function ('type B'). In the following, the two guidance options are discussed, especially as concerns the question how they relate to each other.

i. General Aspects

A first, general section of the model guidance addresses the question what a relevant antitrust market represents (§ 1). It contains an inclusive definition as it understands the relevant market as the 'area of competition within which company behaviour is assessed under the competition laws'. A footnote clarifies that 'product' is understood as a comprehensive term encompassing goods, services and also technology. For type A, the main purpose of delineating the market is to inform subsequent market power assessments. It thus requires clearly identifiable market boundaries. By stating this approach to market definition right at the outset, a type A approach naturally makes itself susceptible to the line of criticism which has focused on market definition's relationship with market power.[20] Type B, on the other hand, emphasises the importance of market characterisation and cites this as its main purpose. It can more easily accommodate the ever-changing nature of dynamic markets and, in particular, the growing integration of markets based on digitisation and innovation. However, by de-emphasising the relevant market's relationship with market power assessments, type B guidance opens up questions as to how market share thresholds and other market share-based tests should be carried out under this guidance, a topic which is covered by § 11 of the guidance.

[19] M Jaeger, 'Perspective of the Judiciary' (12th GCLC Annual Conference, Brussels, 27 January 2017).

[20] See ch 2, section IV.A.

Both guidelines then establish in identical terms what they understand by an innovative market environment, linking back to the characteristics of innovative markets discussed in section II in Chapter 3. These characteristics include short product cycles, a fast-moving market environment, the high importance of R&D and intellectual property rights (including know-how), the prominence of direct and/or indirect network effects, performance as an important parameter of competition, and the use of big data analytics in the design and/or provision of the product under scrutiny. Neither guidance requires the presence of a certain number of these characteristics, and there is no ranking amongst them. Instead, these characteristics represent a moveable system that, overall, indicates whether a market is more innovative/dynamic or more static. Having a moveable system rather than a clear set of criteria induces some flexibility which is necessary in order to accommodate a wide range of innovative industries that may exhibit these characteristics with different intensities, as can be gathered from the discussion of different kinds of innovative industries in Chapter 4. However, such a moveable system also leads to a certain degree of uncertainty when deciding whether a market environment is sufficiently dynamic to require the application of the innovation-specific market definition rules – and subsequently perhaps also innovation-specific substantive rules.

Type B guidance adds a third paragraph on policy to § 1, emphasising how market definition needs to be informed by actual market realities rather than representing 'constructed market realities' as sometimes presented by parties, their (economic or legal) experts or their lobbyists. This is a very sensitive issue and one of high importance as it relates to lobbying efforts, intellectual capture and, ultimately, to the question of whether companies with the necessary financial means can make antitrust law believe in their constructed market realities. As shown in section I.A in Chapter 4, it is of high practical relevance and can, particularly where the substantive analysis heavily rests on market definition, be decisive for a case. Furthermore, type B guidance urges competition authorities to be objective in their market delineation; at the stage of market definition, policy objectives – such as the promotion of innovation or the ease of enforcement – should not play a role. This links back to the understanding underlying this study that market delineation is not a suitable instrument in order to promote antitrust's policy goals.

ii. Innovative Products and Product Differentiation

§ 2 of the guidance has innovative products at its heart, thus attempting to capture the diverse nature of products in innovative environments that were outlined in Chapter 4. Both types of guidance emphasise the importance of relying on a focal point in delineating a relevant market. This highlights that markets may need to be seen differently depending on the question at hand, for instance, as regards one-way substitutability that is specifically mentioned.[21] However, this does not

[21] See already ch 4, section I.D.

justify different approaches to market delineation depending on the area of antitrust. Both type A and type B underline the importance of relying on actual rather than on hypothetical or merely anecdotal consumer behaviour when delineating the relevant market, an insight that is well accepted but not always followed. Particularly in the area of brand loyalty, following this guidance could lead to the finding of narrow relevant markets based on irrational consumer behaviour. The guidance mentions the possibility that innovation might have given rise to further possibilities of substitution, and such facts should be taken into account. Type A guidance goes on to state that both demand and supply substitutability are to be relied upon, although the importance of the latter is limited to circumstances where its effects are as immediate as that of demand substitutability. This is in line with the focus on demand substitutability that is palpable in the case law and in the policy documents from the EU and the US.[22] Type B proposes looking at both demand and supply substitutability in equal measure, thus particularly acknowledging the fact that it is usually the supplier who introduces product or process innovations. Type B also suggests that any effects the use of algorithms may have on substitutability should be closely scrutinised, thus paving the way for an approach that is more sensitive to data-driven algorithms possibly influencing consumer choice. As (self-learning) algorithms are incorporated into an increasing number of devices and applications, the effects of such an approach might be significant. As a footnote in the guidance explains, algorithm-based applications may either restrict the availability of substitutes or may increase their number. Particular examples, of course, are the digital assistants and the IoT applications discussed in section I.B.iii in Chapter 4.

Type B guidance adds three paragraphs to the section on innovative product markets, which relate to three important dimensions for understanding the complexity of many innovative markets: network effects, interoperability and contextuality as a broader notion. This is in line with the close market characterisation that type B wants to achieve. As regards network effects, the guidance relies on insights from behavioural economics by not assuming rational consumer behaviour. On interoperability and digital platforms, the guidance relies on the analytical framework that is developed for aftermarkets in § 9 of the type B guidance. This means that digital platforms and their aftermarkets – eg, operating software and its application store distributing third-party applications – may constitute multiple markets where competition on the platform market does not constrain market power on the market for application stores; otherwise they may be seen as a system/platform market. This can lead to the delineation of narrow markets and can solve interoperability issues in those jurisdictions in which companies with strong market positions have a special responsibility under the antitrust laws. However, the insight that high market shares in innovative industries are not very significant may contradict such findings. The section also highlights

[22] See ch 2, sections I.A.i and I.F.i for the US; and ch 2, sections II.A.i and II.F.ii for the EU.

consumer capture that might limit demand substitutability, thus further narrowing possibly broader relevant markets. Finally, type B guidance draws attention to the importance of market realities and the legal and economic context of a certain relevant product market. Particularly where innovative industries are concerned, a full understanding of the current competitive relationships, ongoing R&D endeavours and possible futuristic scenarios grounded in present markets is important to characterise the relevant market for competition law purposes. Bearing in mind the technology-driven integration of markets, the possible limitation of choices through algorithms and the market environments sketched in section I.B.iii in Chapter 4, it becomes clear that this is a complex assessment.

§ 3 turns to ways of delineating innovative, differentiated markets for antitrust purposes. Both type A and type B guidance highlight three central aspects related to innovative product differentiation. The first is the online/offline paradigm, for which both types of guidance propose relying on actual substitutability in order to assess whether the online and offline markets compete with each other. Here, market realities may shift over time as online and offline markets become more integrated, as was shown in section II.B in Chapter 4. Challenging questions as to geographical market delineation may also arise in this context, but are not focused on here. Second, reference is made to product functionalities and their importance for market delineation. Here, both sets of guidelines acknowledge that a product's functionalities do not determine the relevant product market as long as they are still evolving, thus referring to one of the innate characteristics of innovative markets. Type B adds that innovative potential ought to be taken into account when assessing functionalities, thus asking for a characterisation of the market under scrutiny. However, an analysis of innovative potential is often more easily carried out with hindsight than prospectively. Overall, functionalities may help in a closer characterisation of the market. Finally, freemium models in which one encounters a slimmed-down free version of a product and a to-pay-for enhanced version (with more functionalities, without advertisements etc) need to be assessed based on consumer behaviour, not simply with regard to the business model. As a point of departure, the analysis must start from the focal product being investigated (as suggested under § 2 of the guidance); in the case of platform markets, this may yield different results depending on which market side is being investigated, particularly under a type A approach.

iii. The Time Horizon and Future Markets

§ 4 establishes that the time horizon for analysing a relevant product market must reflect market realities, in particular average product and innovation cycles, customer response and so on. The guidance needs to remain rather vague in this respect due to the different market realities that may be affected by this (see section III in Chapter 4). Type A cautions that future developments should only be taken into account to a limited extent, thus focusing more on currently observable or foreseeable developments. This is due to the fact that type A guidance is focused on market power assessments, which are very much anchored in present markets rather than hinging on uncertain future market developments.

However, for a market characterisation under type B, possible future market developments may be of more relevance, for they might indicate that some uncertainty as to innovation is present in the market – thus perhaps destabilising any cartel or other anti-competitive behaviour in the making.

In dynamic market contexts, an argument can be made for longer time horizons, thus preferring long-term dynamic efficiency gains over short-term allocative efficiency gains. Ultimately, it will be important for competition authorities and courts to develop a coherent approach in this respect, particularly as no abstract guidelines can be given in this regard.

§ 5 of the guidance relates to future markets. Both type A and type B guidance understand future markets to be relevant product markets which will emerge with high probability, based on R&D endeavours that have nearly reached completion where regulatory approval is pending. Future markets can be based on current products or on still-emerging products.[23] While the guidance does not mention pharmaceuticals by name, the reference to regulatory approval is specifically tailored to pharmaceuticals and other products for which such market approval is required. To this, type B guidance adds that a detailed market characterisation can improve our estimate of whether a future market is indeed likely to emerge and that this must go beyond mere plausibility, ie, some form of proof is required to substantiate such an expectation. These future markets can then be relied upon in the same way as current product markets. They are reminiscent of the US innovation market approach, but are more confined and thus perhaps more realistic in their scope of application. Type B guidance also includes a reference to the time horizon that should be applied to future markets, noting that this is case-dependent and should be judged with reference to the specific market environment. As type B guidance includes a comprehensive market characterisation, it should also enable competition law to assess what the appropriate time horizon is in a particular market setting.

iv. Potential Competition and Innovation Competition

Moving beyond future markets, § 6 of the guidance concerns potential competition. This is a dimension of competition that needs to be applied after the relevant antitrust market has been delineated. Potential competition is understood to consist of credible possibilities of market entry, thus anchoring potential competition in market realities. Potential competition is different from future markets in that it relates to credible entry in current or future markets rather than the emergence of such a market. However, there is often a floating transition from one

[23] See ch 4, section IV.

to the other. Type B guidance adds that potential competitors may be included in the analysis as market participants, as is indeed the case under the US Horizontal Merger Guidelines. This, however, does not change the delineation of the relevant product market as such. As type B guidance does not rely on market shares (see § 11 of that guidance), this inclusion will not shift market shares and thereby directly influence the analysis of market power. However, it may allow competition law to regard potential competitors as close to the current/future market under investigation, and one may consider including them as viable competitors under § 10h or § 11b of the type B guidance.

Innovation competition is then considered in § 7 of the guidance. Based on Baker, it is defined as 'competition among firms seeking to develop the same new product or process'.[24] Type A guidance states right away that innovation competition may need to be considered after a relevant market has been delineated, unless it can be captured as a future market or, perhaps, as potential competition relating to a current market or a future market (§ 7a). Type A guidance also highlights how innovation efforts that are already very far advanced can be grasped as future product or technology markets, rather than resorting to the notion of innovation competition (§ 7b). Innovation efforts are far advanced where regulatory market approval is imminent or where the innovation has in any other way nearly reached the stage of market entry. Type A guidance thus tries to capture innovation competition through palpable markets rather than resorting to the transient concept of innovation competition. This can be understood against the background of its focus on market power.

Type B guidance rests on the same initial definition of innovation competition (§ 7a), but then further characterises this concept as relating to competition for the market rather than competition in the market. It also alludes to the competitive pressure that is exerted in competition for the market. Further, type B guidance in § 7b explicitly notes that innovation competition takes place outside of clearly defined antitrust markets, that innovation is inherently uncertain and that innovation competition is more helpful as a tool to characterise a market rather than delineate it. For this very reason, innovation competition can greatly contribute to a better understanding of the relevant market under type B guidance. In order to delineate future markets or even ascertain potential competition related to current or future markets, type B guidance requires a temporal proximity.

Both sets of guidance then list the same factors that help in an assessment of the proximity of two innovation efforts (§ 7c). These are the nature, scope and magnitude of R&D efforts, each company's access to specialised assets such as financial and human resources, data or know-how/IPRs, the timing of R&D efforts and each company's ability to successfully commercialise its innovations. All these factors – except access to data – can be found in the current EU and US guidance.

[24] Baker, 'Beyond Schumpeter vs Arrow' (n 8) 579.

While the type B approach lists these factors with a view to closely characterising a market, the type A approach does so with the intention of basing a market delineation on them.[25] The reference to data reflects the importance of data in digital markets and necessarily includes the ability to analyse big data.

v. Intellectual Property Rights

§ 8 of the guidance turns to IPRs and their role in market delineation. Type A acknowledges that IPRs may delineate a relevant market, but that ownership of an IPR does not automatically indicate market power. This phrasing captures the current stance in many jurisdictions that theoretically acknowledges that IPRs do not in themselves delineate antitrust markets, while at the same time often following the intuition that in practice this is nevertheless the case (see section II in Chapter 6). Type B guidance is more cautious in its approach, referring to the necessity to rely on the substitutability rules it sets out in order to decide whether an IPR constitutes its own antitrust product market. Type B guidance also suggests that IPRs can instead be relied upon as a factor to establish and characterise a relevant product market, thus highlighting the significance of market characterisation. Importantly, there is no presumption that IPRs constitute their own antitrust product market.

Type A guidance considers substitutability for an IPR from a customer's perspective to be determinative for the question as to whether an IPR delimits its own antitrust market, and points to supply-side substitutability as a further source of competition where IP protection does not exclude successful substitution from the supply side. Type B guidance also emphasises the importance of the customer's perspective for conceivable substitutes, but in equal measure considers supply-side substitution for the IPR or the IP-protected product. It highlights that substitutability must be understood as going beyond the IP-protected subject matter. An IPR should therefore not be used as the basis of the relevant market definition, for there may be substitutes available that do not benefit from IP protection. Type B points to differentiation induced by the IPR and to possible competitive pressure from potential competition or the constant threat of innovation. This paragraph of type B guidance is phrased in a Schumpeterian way and focuses on market characterisation rather than on a clear demarcation of the relevant market, allowing for a depiction of the various competitive forces acting on markets in which IPRs play a significant role.

The licensing of technology is then defined as a technology market by the type A approach (§ 8c), with all available substitutes of the IPR forming part of such a market. It is also explained that this technology market is distinct from product markets that simply incorporate IP-protected technology. Type B guidance contains a somewhat closer characterisation of technology markets and

[25] In this sense, one could say that the type A approach has not yet quite overcome innovation markets.

suggests that these could also be referred to as licensing markets understood as upstream markets. The guidance includes IPRs that are currently licensed as well as such that could conceivably be licensed, thus allowing market definitions to be centred on IPRs that are currently only used in-house, but could potentially be viably commercialised. It considers that substitutability for licensing markets should be looked at from a perspective that is broader than the subject matter of the IPR; substitutes need not be IP-protected. A SSNIP test can be relied upon in order to discern a licensee's perception of substitutability. Imminent market entry is a factor to be taken into account, particularly where an IPR has been applied for but has not yet been entered into the register. This is a future market approach applied to technology markets. It also points to the possibility that a licensing market exists alongside a product market in which the IPR at issue is incorporated, thus allowing for the parallel finding of downstream product markets and upstream licensing markets.

Type A guidance insists that market share thresholds can be applied to technology markets as well as to other markets delineated by an IPR (§ 8d), thus holding on to the well-accepted market share thresholds relied on in many jurisdictions. Type B guidance, on the other hand, does not use market share thresholds for licensing and other IP-delineated markets. Instead, type B guidance proposes relying on the number of viable alternatives that are present on the market, modelled on several US policy instruments. Where a legal instrument does not explicitly foresee alternatives for market share thresholds, type B guidance proposes that the presence of a certain number of viable alternatives – be they works, inventions or technologies of any kind – owned by competitors should be seen as equivalent to not surpassing a certain market share threshold. Here, of course, the exact number of viable alternatives and their translation into market share thresholds in a traditional sense are up for discussion.

For SEPs, type A guidance articulates a cautious positive presumption of a separate technology market (§ 8e). This may facilitate the finding of market power. Type B guidance insists on treating SEPs as any other licensing market, pointing to the fact that the SEP's success will often determine its substitutability from the customer perspective. Finally, type A guidance considers that where an IPR does not constitute or delineate its own relevant market, it may nevertheless constitute an important factor in the market delineation exercise, for instance, as a barrier to entry or expansion (§ 8f). Similarly, type B guidance underlines IPRs' importance for closely characterising a market.

vi. Aftermarkets

Aftermarkets or secondary markets are considered in § 9 of the guidance options. These are markets in which demand is derived from a primary product. The type A approach, with its emphasis on traditional tools for market delineation, focuses on the option of lifecycle costing and, where this option is available, delineates a system market which consists of the combined primary and secondary markets.

Only where such lifecycle costing is not possible, or where there is a policy change on the company side that affects existing customers, does type A consider delineating multiple markets, ie, separate markets for the primary product and each secondary product related to the latter. This is more in line with what US Courts of Appeals have adjudicated, even after *Kodak*.[26] The type B approach, on the other hand, implements insights from behavioural economics: it only delineates a comprehensive system market where customers actually engage in lifecycle costing rather than letting the possibility to do so suffice. Otherwise, type B relies on multiple markets in which market power may be more easily ascertained. The latter approach is more in line with European case law (*EFIM*) and some US case law (*Kodak* and *Virtual*).[27]

There certainly is a policy implication at stake when choosing one approach over the other, relating to the question whether lucrative aftermarkets should be reserved to the innovator that conquered the primary market. Might such an extension encourage further innovation in the long run or should third parties be allowed easy access to these aftermarkets for the immediate benefit of consumers?

Interestingly, the system market approach in aftermarkets strays away from that basic conceptualisation of market delineation: substitutability. It groups together two aspects of a market – a primary and a secondary product – which are not interchangeable with each other, but which are complements instead. This is due to the particular nature of competition in aftermarkets.

vii. Multi-sided Platforms

§ 10 sets out market definition options for multi-sided platforms. Both the type A and the type B approach begin by setting out what they understand by a multi-sided market. While type A defines multi-sided markets as platforms with two or more interconnected market sides on which network effects are present, type B guidance relies on David Evans and Richard Schmalensee in defining platforms, highlighting the presence of two or more customer groups that need each other in some way and that rely on a platform to capture indirect network effects amongst each other.

Concerning the method for delineating the relevant antitrust market under each approach, type A guidance again opts for a more traditional route that is also geared towards an assessment of market power rather than towards the characterisation of the market. For transactional platforms, ie, platforms that have the primary aim of enabling direct transactions amongst the market sides involved, type A guidance will delineate a comprehensive platform market consisting of the platform and its different market sides (§ 10b). However, this is only possible where

[26] On this case, see ch 7.
[27] On these cases, see ch 7.

conditions of competition are very much alike for all market sides. In particular, the guidance mentions two conditions previously developed in the literature, namely that: (i) competitive conditions resemble each other on all market sides; and (ii) the platform's activity by definition encompasses all market sides.[28] Therefore, this market definition approach may rarely be put into practice.[29] In all other cases (§ 10c), type A guidance assumes that each market side constitutes a separate relevant antitrust market. The network effects then need to be taken into account at the stage of the substantive analysis. The platform's role under this approach is left open.

Type A guidance does not regard free market sides on their own as antitrust markets (§ 10d), although they may be incorporated into a broader market definition or play a role in the substantive analysis. In many jurisdictions, the stance regarding 'free' market sides is now changing; some have even introduced specific legislation in that regard.[30] Therefore, this part of the guidance may be gradually adapted to depict an evolving type A approach. As it tries to stick with the traditional market definition tools that it has gotten accustomed to, type A wants to keep relying on the SSNIP test when delineating single market sides (§ 10e), as well as relying on market share thresholds (§ 10f). In the end, whether a platform market or single markets are delineated may not make a big difference to the outcome of an antitrust analysis, as network effects are merely taken into account at different stages.

One of the recommendations of the German Monopoly Commission when assessing digital markets was to further develop the legal framework in order to better analyse the special characteristics of multi-sided platforms.[31] The type B approach to defining multi-sided platforms for antitrust purposes does just that. It is more nuanced than the type A guidance and relies on an approach similar to that deployed in aftermarkets.[32] However, it must be recognised that in aftermarkets, demand is derived from a primary product, while in multi-sided markets, this relationship is not as straightforward. Instead, in multi-sided markets, one market side's demand is often derived from another market side, but can only be satisfied via the primary product (ie, the platform). Nevertheless, in order to keep with a vocabulary that competition lawyers have become accustomed to, type B guidance

[28] Wismer, Bongard and Rasek (n 9) 260.

[29] Similarly, see J-U Franck and M Peitz, 'Market Definition and Market Power in the Platform Economy' (CERRE Report, May 2019) 38.

[30] See the German example mentioned in ch 8, section II.A.

[31] Monopolkommission (n 2) para 477.

[32] Generally, see M Bourreau and A de Streel, 'Digital Conglomerates and EU Competition Policy' (CERRE Report, March 2019) 26. For a similar conceptualisation based on aftermarket insights, see S Holzweber, 'Market Definition for Multi-Sided Platforms: A Legal Reappraisal' (2017) 40 *World Competition* 563, 576. Holzweber's approach does not position the platform itself within the relevant market; instead, it combines the market sides to a single relevant market. As the platform captures the positive externalities among the market sides, the present study includes it in its conceptualisation.

distinguishes between the platform as the primary product and the market sides as the secondary products (§ 10b).

Just as is the case for aftermarkets, type B delineates a combined system market (or platform market) where workable competition at the platform level or in one of the market sides significantly constrains market power on the secondary market(s) (§ 10c). This resembles a moveable system that can become highly complex, thus reflecting the complexity of market realities in these cases. In particular, the platform market approach requires a taking-into-account of lifecycle costing specific to platforms, ie, one needs to assess whether customers can and do engage in lifecycle costing based on the factors set out in § 10c before engaging with a primary product. Cross-platform externalities (network effects) also need to be taken into consideration in this definition. As is the case in aftermarkets, this approach to market characterisation is not based on the criterion of substitutability amongst the market sides.

Type B guidance sets out factors which can help in establishing whether customers on one market side take into account 'lifecycle costing' specific to multi-sided platforms when 'acquiring'[33] the primary product (§ 10d). Such factors go beyond price to include customers' information costs, switching costs, reliance on parameters such as quality, interoperability, data protection and data portability at the time of acquiring or using the primary product, and the timeliness of customer response in the case of price rises, quality decreases etc in the platform's provision of services on one market side. Under type B guidance, multiple markets – ie, a market consisting only of the platform and single markets for each market side – are delineated where no lifecycle costing is relied upon by customers, either because lifecycle costing is not possible or because customers do not make use of it in practice, or because of a policy change that affects existing customers that are thereby locked in (§ 10e). This necessarily leads to narrower antitrust markets, reflecting the need to take competition amongst the platforms (primary products) as such into account when applying the competition laws. § 10d also refers to multi-homing, ie, the practice of customers on either market side to use more than one platform service. This may prove relevant in order to establish substitutability (or the lack thereof).

As is generally acknowledged today (see section II.B in Chapter 8), the fact that one of the market sides is 'free' in that it does not directly generate any revenue is not relevant for market delineation under the type B approach (§ 10f). Free market sides are incorporated into the market definition just like any other market sides. However, it is suggested that market definition begin on that market side which generates the revenue and possibly subsidises the free market side. Thereby, market characterisation can gain an understanding of what might be the driving force in a market.

[33] Acquiring is understood in a broad sense, also encompassing the non-monetary acquisition of a service, for instance, through the use of a free search engine or social networking site in exchange for user data and/or attention.

The type B approach does not regard the SSNIP test in its conventional format as applicable to multi-sided platforms (§ 10g). It proposes assessing substitutability for each market side and then combining this to an overall understanding of the multi-sided market at issue. This will not necessarily lead to a clearly delineated market as is required by clear-cut market share thresholds. For this reason, § 10h of the type B guidance replaces traditional market share thresholds with the number of viable competitors present on a platform market or on multiple markets – unless a legal instrument foresees specific alternatives. What constitutes a viable competitor, and which time horizon should be employed when judging this viability, will undoubtedly be a contentious issue. Nevertheless, this approach – inspired by the US IP Guidelines – can serve as a substitute for market share thresholds, while at the same time leading to informative discussions on the innovations at stake and possible future scenarios. The precise number of viable competitors also remains debatable and insights from competition economics may be useful in setting these numbers.

The market conceptualisation of the type B approach is an attempt at characterising the market and, as such, cannot be the basis for a purely structural application of competition law. In particular, a platform market may only compete with another platform market on one market side.[34]

viii. Economic Tests

§ 11 is very brief, restricting itself to guidance relating to economic tests as far as these might be relied upon in the legal framework. It should be understood as guidance for lawyers, not necessarily for economists and their expert testimony or opinion.

The type A approach tries to salvage the traditional economic tools while at the same time adapting them to innovative market environments and their particular market characteristics. When carrying out the SSNIP test, it urges the paying of attention to factors other than price in innovative markets (§ 11a). It does not give any further guidance on how this could be done, thereby leaving considerable leeway to antitrust authorities in this respect. While type A relies on market shares and HHI concentration levels in a traditional way, it also cautions that the characteristics of innovative markets need to be taken into account when calculating and interpreting market shares or applying market share thresholds (§ 11b) and when calculating and interpreting HHI concentration levels (§ 11c).

This is contrasted by guidance under a type B approach, which only wants to apply the SSNIP test conceptually in innovative markets (§ 11a). In particular, substitutability between products and suppliers needs to be assessed based on (limited) interchangeability with other products and against the background of an

[34] See on this Stigler Committee on Digital Platforms, 'Final Report' (September 2019) 91.

insight into the differentiated nature of the products at issue. Again, this helps to closely characterise the market without drawing any distinct market boundaries. Furthermore, type B guidance does not want to rely on market shares as a proxy for market power when it comes to very dynamic markets. Market share thresholds should be replaced by alternatives (§ 11b). Where soft law or legal provisions do not foresee alternatives to market share thresholds, type B guidance proposes that the presence of a certain number of viable competitors on the relevant market should equate to a certain market share threshold. Type B guidance does not regard HHI concentration levels as applicable to horizontal mergers in highly innovative markets either. Instead, it proposes carrying out a qualitative assessment of the merger's influence on the post-merger competitive environment, with special regard to the innovation dimension (§ 11c). This is in line with its view that market definition primarily serves to closely characterise the market – and the competitive harm – under scrutiny.

III. Choosing a Market Definition Framework

Under a type A approach to market delineation, well-established methods are maintained to the extent possible, while at the same time introducing a more innovation-sensitive approach. This is well in line with an approach that requires clear market boundaries for its structural analysis, particularly in order to assess market power based on market shares, changes in concentration levels and the like. As was previously noted, this policy stance entails that type A's ability to reconceptualise market definition against the background of innovation considerations is limited. That said, one does find some leeway that can bring innovation-based considerations on board. This is particularly noticeable in relation to the online/offline paradigm, the scepticism towards the continued reliance on functionalities in innovative markets, and the consideration of freemium and other business models. However, a certain reticence to modernise the concept of innovation competition is also palpable, based on the far-reaching changes that this would bring to traditional structural market analysis. In 'free' market sides within the framework of multi-sided platforms, it was seen how jurisdictions such as Germany have recently undergone a slight paradigm shift in accepting how these can be considered antitrust markets in their own right. This is not reflected in the type A model guidelines, as indeed many other jurisdictions have not yet found a way to accommodate such insights within a more traditional take on competition law.

Jurisdictions favouring an approach to antitrust market definition along the lines of type A will need to make fewer changes to their analytical tools, and market definition can continue to fulfil its role as a basis for market power assessment, even though market power analysis based on market definition/shares has rightly been called into question both by scholars and by enforcers. Where market definition based on a type A analysis yields too narrow an antitrust market, an innovation-specific substantive analysis can possibly remedy this, but needs to be clearly formulated so as to achieve uniformity and predictability.[35]

The adoption of a type B approach to market definition would bring about palpable change to antitrust legal analysis. In many respects, it is a market characterisation more than a market delineation. The type B approach would not (need to) yield the neatly defined antitrust markets that a more traditional approach requires. Acknowledging the dynamic nature of innovative markets, it allows for a better understanding of the market environments at issue through a close market characterisation, while at the same time understanding that a market definition might consist of a continuum of substitutions within which it is not necessarily helpful to draw artificial market boundaries. This means that a lot of the structural analysis that builds upon clearly delineated antitrust markets will require an overhaul.[36] As daunting as such a prospect may seem, it might serve to re-invigorate market definition's standing and would certainly be more realistic, assigning a role to market delineation that it can fulfil rather than requiring clear-cut answers from an analytical tool which, by design, it cannot deliver. After decades of entrusting market definition with such a major role, it will be difficult for competition lawyers – both academics and practitioners – to come to terms with this limited role of market definition. Eventually, however, this will allow for renewed vigour when assessing competition law cases, framing antitrust theories that are anchored in market realities rather than 'constructed realities' shaped by the parties who might benefit from them. The type B model guidelines may help reconceptualise market definition in this way: as a legal concept with a more realistically defined task. As was seen, a type B approach to market definition repeatedly emphasises the role of market characterisation that falls to market definition, as well as the agnostic role that market definition should play in understanding market realities, be they market dynamics, patterns of competition or innovation cycles. For this very reason, the type B approach relies on a close characterisation of the market based on an understanding of actual demand and supply substitutability, also in the context of algorithms. It explicitly refers to network effects and interoperability issues, which require a good understanding in innovative markets, but also points to contextuality more generally. It links innovation competition to the substantive analysis and highlights the fluent transition between current markets, future markets, potential competition and innovation competition. With regard to IPRs, it regards these as possible demarcation points for relevant markets, but not necessarily so. It particularly does not presume any kind of market power as flowing

[35] On innovation-specific substantive antitrust analysis, see P Ibáñez Colomo, 'Beyond the "More Economics-Based Approach": A Legal Perspective on Article 102 TFEU Case Law' (2016) 53 *CML Rev* 709; B Mäihäniemi, 'The Role of Innovation in the Analysis of Abuse of Dominance in Digital Markets: The Analysis of Chosen Practices of Google Search' (2017) 1 *Market and Competition Law Review* 111; N Petit, 'Significant Impediment to Industry Innovation: A Novel Theory of Harm in EU Merger Control?' (ICLE White Paper 2017-1, 2017).

[36] This overhaul should not occur in the ways foreshadowed by Kaplow, Markovits and others; what is at stake is a legal conceptualisation of antitrust concepts and analysis, not an economics-based one.

from SEPs without further qualification. As regards multi-sided platforms, the type B approach adopts a conceptualisation based on aftermarket analysis, which insists on the importance of looking at platform competition and substitutability in addition to the competition and substitutability that may exist for each individual market side. It also proposes factors that can help understand 'lifecycle costing' as applicable to multi-sided platforms, particularly as these factors often relate to non-price considerations. Throughout type B proposals for reconceptualising market definition, proxies like market shares and market share thresholds are regarded with considerable scepticism. They are replaced with an approach that has been advocated in the US Competitor Collaboration Guidelines, namely the number of viable competitors.

IV. The Guidance Options as a Way Towards Convergence in Market Definition?

Companies that are active in several jurisdictions face different antitrust laws in every single one of them. As important differences remain between EU competition and US antitrust law,[37] this might deter companies from engaging in behaviour that would be seen as pro-competitive and, one might add, innovation-enhancing.[38] Differing analytical bases for competition law are also highly relevant because of the jurisdictional reach of competition law regimes:[39] US antitrust law relies on the effects doctrine, meaning that any commercial behaviour that has effects on US territory will be caught, irrespective of where a company is incorporated.[40] Similarly, EU competition law applies to any anti-competitive behaviour that is implemented within the EU or has qualified effects in its territory.[41] The EU Merger Regulation relies on turnover thresholds that have a European and a global aspect, irrespective of where the merging companies are incorporated.[42] For companies active in innovative industries, which can easily affect both

[37] See, eg, K Czapracka, *Intellectual Property and the Limits of Antitrust: A Comparative Study of US and EU Approaches* (Cheltenham, Edward Elgar, 2009) 3.

[38] WE Kovacic, 'Competition Policy in the European Union and the United States: Convergence or Divergence?' in X Vives (ed), *Competition Policy in the EU: Fifty Years on from the Treaty of Rome* (Oxford, Oxford University Press, 2009) 315.

[39] See also DJ Gerber, 'Global Competition Law Convergence: Potential Roles for Economics' in T Eisenberg and G Ramello (eds), *Comparative Law and Economics* (Cheltenham, Edward Elgar, 2016) 211.

[40] *United States v Aluminum Co of America (Alcoa)*, 148 F2d 416, 443 (2d Cir 1945); *United States v Aluminum Co of America (Alcoa)*, 377 US 271, 273 (1964); *Hartford Fire Insurance Co v California*, 509 US 764, 796 (1993).

[41] Joined Cases 89, 104, 114, 116, 117 and 125 to 129/85 *Ahlström and Others v Commission (Wood Pulp)* EU:C:1993:120, paras 12–18; Case T-102/96 *Gencor v Commission* EU:T:1999:65, paras 78 ff, in particular paras 90 ff; Case T-286/09 *Intel v Commission* EU:T:2014:547, para 236, upheld in Case C-413/14 P *Intel v Commission* EU:C:2017:632, paras 40–47.

[42] Council Regulation (EC) 139/2004 on the control of concentrations between undertakings [2004] OJ L24/1 (EU Merger Regulation), art 1.

EU and US markets (and far beyond), the question of regulatory convergence is therefore one of high relevance, not least as it concerns market definition. Some innovators might even argue that antitrust convergence is more important for their business than the 'right' antitrust solution. By being active in multiple jurisdictions, they find themselves in something of a prisoner's dilemma, caught between different antitrust jurisdictions. Antitrust convergence could be key to solving that dilemma.

Convergence is an important theme in comparative competition law. However, the present study does not directly aim at providing a uniform analytical framework for market delineation; rather, it bases its two proposals for the future development of market definition in dynamic contexts on a comparative legal analysis of these two jurisdictions. The guidelines, then, show possibilities from different policy approaches without proposing a harmonisation of the competition laws. The guidelines may serve as a blueprint upon which competition law and policy can base its discussion of a new understanding of market definition where innovative markets are concerned. This may, ultimately, lead to some harmonisation and thus convergence. Through its longitudinal study, the research also uncovered points of convergence that have already been reached.

Today, convergence is sometimes seen as 'the only viable strategy for developing competition law' due to this area's inherently international dimension.[43] This may be particularly true in the innovation context. Bearing in mind the limits imposed upon legal convergence by legal culture, the guidelines developed in this study are unlikely to lead to comprehensive convergence. However, they may serve as an important point of convergence if the same policy approach is chosen and subsequently applied in the same manner.[44]

One word of caution is in order: the present study is limited to antitrust market definition. Any convergence that is perceived in this context, or any scope for further convergence that is highlighted, is limited to this particular legal concept. Market definition as an analytical tool of antitrust law may not have the same functions in every jurisdiction. This can limit the scope for convergence. A much broader effort would need to be undertaken to achieve an overall harmonisation of the antitrust laws – if indeed one deems it a worthwhile aim.

Another caveat when speaking of the possibility of leading to greater convergence amongst competition laws relates to legal culture.[45] Especially in the

[43] DJ Gerber, 'Comparative Law and Global Regulatory Convergence: The Example of Competition Law' in M Adams and J Bomhoff (eds), *Practice and Theory in Comparative Law* (Cambridge, Cambridge University Press, 2012) 121.

[44] On the need for a point of convergence to anchor processes of convergence, see Gerber, 'Global Competition Law Convergence' (n 39) 208.

[45] On the importance of legal culture, the limitations that comparative law is confronted with in this respect and the consequences that the postmodern school wanted to derive from these insights, see P Legrand, 'The Impossibility of "Legal Transplants"' (1997) 4 *Maastricht Journal of European and Comparative Law* 111, 115 f; P Legrand, 'Negative Comparative Law' (2015) 10 *Journal of Comparative Law* 405; D Nelken, 'Using the Concept of Legal Culture' (2004) 29 *Australian Journal of Legal Philosophy* 1; D Nelken, 'Using Legal Culture: Purposes and Problems' (2010) 5 *Journal of Comparative Law* 1.

well-established competition law regimes under investigation, a new set of guidelines will not start with a clean slate, but will be read and applied against the background of previous case law and experience. It would need to be interpreted in its inherently national (or supranational) context. In new competition law regimes, the prevailing policies and circumstances will equally influence how such a new set of guidelines would be interpreted.[46] Even if the same model guidelines were to be adopted by different jurisdictions, they would be interpreted against the background of the specific competition law culture. Lawyers in different jurisdictions may attribute different (legal) value to such guidance, the (legal and factual) circumstances found in the country have their bearing, and the prevailing understanding of competition law will leave its mark. Where competition authorities embark on constructive dialogue and cooperation, and where competition courts intently engage with judgments from other jurisdictions, a shared understanding may be attainable.

V. Conclusion

The present chapter set out options of model guidance that can further develop our understanding of market delineation in innovative market environments, pointing to areas that need to be considered when delineating innovative markets in practice, and ways how the relevant market can be conceptualised.

The first guidance option ('type A') focuses on the function of market definition that informs market power assessments and is more traditional and static in nature, but values innovation as one of the goals of competition law. Its proposals for model guidance (see section II.A above) reflect this approach, emphasising the delineation of markets based on conventional price-based tools as far as possible and continuing to rely on the SSNIP test, market shares and concentration levels. Innovation considerations are only able to enter this approach to the extent that this does not interfere with its focus on clear market boundaries. Otherwise, innovation is more of an afterthought – and can re-enter the antitrust stage when the substantive analysis is carried out. While such an approach may well reflect the views held by many courts, it is not shared by all. Several scholars and competition authorities have forcefully argued that market power assessments and other structural analyses based on relevant markets with bright lines as market boundaries should be a thing of the past. There is considerable merit to these views and they principally allow for two conclusions: the first is to abandon market definition altogether – or at least to reduce competition law's dependence on it – while the second is to redefine the role of market definition and how relevant markets are relied upon.

For a critique of postmodernism, see BS Markesinis, *Comparative Law in the Courtroom and the Classroom* (Oxford, Hart Publishing, 2003) 51 f.

[46] Generally, see A Ezrachi, 'Sponge' (2017) 5 *Journal of Antitrust Enforcement* 49, 59, 75.

The second proposal ('type B'; see section II.B above) showcases how the relevant market can be delineated by relying on a more innovation-sensitive approach, despite the inherently static conception of a relevant market which needs to be anchored in time. Adopting a type B approach requires a considerable change in the role that is assigned to the relevant market. This is called the characterisation function of market definition, which acts as a basis for informing the antitrust theory of harm and designing successful antitrust remedies. In addition, such a characterisation can inform market power assessments in a less traditional sense. While this is a role that the relevant market already fulfils in many jurisdictions, it will become increasingly important as we move away from structural market power assessments based on market shares and similar considerations, instead embracing an analysis that relies on a thorough understanding of the competitive forces at play. A type B approach would adduce an in-depth understanding of the relevant market that the judge or the competition enforcer can rely on. It does so by providing a framework within which the judge or authority can assess which factors and circumstances need to be taken into account, and how specific market situations can be understood from an antitrust perspective. Ultimately, this is for the judge or the competition authority to decide, with economic experts clarifying questions of fact rather than questions of the law. Market characterisation provides a very detailed analysis of the relevant market. However, what it does not do is to provide the clear demarcation lines that lawyers frequently look for – essentially hoping to encounter a predetermined playing field very much like the referee in tennis. This, it is submitted, is an expectation which market definition simply cannot fulfil, especially not in the innovative market environments that lie at the heart of this study.

Both type A and type B guidelines have the potential to improve the competition laws' ability of fostering rather than thwarting innovation, albeit with quite different starting points. One important caveat to this is that legal culture might not readily allow for the adoption of (parts of) these guidelines in certain jurisdictions, due to prevailing legal interpretations, traditions and preconceptions. And even if the same guidance were to be implemented in the seemingly same fashion in several jurisdictions, this does not guarantee that its application would lead to similar outcomes. The guidelines should therefore not be seen as an instrument for harmonisation, but can perhaps act as a point of convergence and a basis for discussion. In a global market context, one can easily recognise the appeal that a global antitrust culture and the promise of convergence can have for innovators.

11

Reflections

Is Market Definition Too Big to Fail – or is it Failing Innovation?

Market definition would not be so important if it were not so important.[1]

The present study has scrutinised different legal and economic approaches to antitrust market delineation in innovative market environments in the EU and the US, in the law, the case law and in policy instruments. The deconstruction of these two jurisdictions' antitrust market definition frameworks brought to light the many intricacies of each framework. These were viewed against the rich case law in both jurisdictions and the manifold soft law guidance developed by their respective competition authorities. When setting out market definition's functions in a comparative fashion, the point was made that market definition continues to be an indispensable element of the antitrust legal framework in both jurisdictions under scrutiny. Market definition is so deeply entrenched in antitrust analysis that it is, in practical terms, almost impossible to conceive of an antitrust framework without market definition as an integral analytical step. It plays manifold roles – both in terms of market power assessment and as regards market characterisation – that cannot simply be erased or replaced by other tools. However, where market definition clings on to its traditional static conception, with its archetypes of homogeneous products and long product cycles, it will often not do justice to dynamic market realities. This is a dichotomy that competition law is constantly confronted with. While one might be tempted to argue that this issue only affects competition law frameworks that aim at furthering innovation, this would be too narrow a point of view. Indeed, the relevant market should be seen as an agnostic analytical tool without any particular policy objective in mind. Regardless of the specific market environment, it is indispensable to accurately delineate the relevant market, ie, that part of commerce in which companies' behaviour is assessed from a competition law perspective.

The special difficulty in dynamic market environments is that their characteristics make them particularly elusive for market definition, while market

[1] DS Evans, 'Lightening up on Market Definition' in E Elhauge (ed), *Research Handbook on the Economics of Antitrust Law* (Cheltenham, Edward Elgar, 2012) 76.

delineations that were relied upon in the past only give a limited insight into today's market dynamics. The interplay of the different facets of innovative markets that affect antitrust market definition, as analysed in Part II, put market delineation to the ultimate test. In the face of the numerous challenges that dynamic market environments present antitrust market definition with – their fast-moving nature, short product cycles, inherent uncertainty, vast differentiation opportunities, network effects, unpredictable paradigm shifts and strong reliance on competitive parameters other than price – it needs to be asked whether, as a matter of principle, innovative markets which are inherently dynamic can be captured by a static concept such as market definition at all. This dilemma also becomes apparent when considering the fluent transitions between current markets, future markets, potential competition and innovation competition. While the first two of this quartet were considered as forming part of market delineation proper, the last two were considered as dimensions that are outside of market definition, but that can form part of the substantive analysis. However, where disruptive innovation is considered, the importance of these last two will by far outweigh the first two. At the same time, one might argue that legal certainty as such is also a static concept, and to provide the former it is worthwhile to try and capture inherently dynamic markets with an intrinsically static concept. Applying the law requires us to define its sphere of application, and for antitrust law this is done based on a conceptualisation of the relevant market.

While it was argued that market definition should serve as an agnostic analytical tool, not every jurisdiction may wish to see market definition in this light. What if a jurisdiction instead wished to further innovation as a policy goal from the outset? This would not be easily achievable. We have seen that the relationship between innovation and the structure of competition is not yet fully understood; in fact, it probably is the case that this relationship is highly case-dependent, not adhering to steadfast rules. One can therefore not confidently state that a certain antitrust intervention – or non-intervention – will let innovation thrive to the benefit of consumers. In fact, it is frequently far from clear whether antitrust intervention that was intended to increase consumer welfare did not, in the long run, do more harm than good – one merely needs to think of the *Microsoft* or *IBM* cases litigated on both sides of the pond. This question is ultimately for the substantive legal analysis to decide. Market definition is not a good place to position such policy goals, even if it will often be relied upon as a policy tool to capture certain market behaviour. Particularly where novel theories of harm are developed, it is important to realise that they necessarily need to relate to a well-understood and well-characterised relevant market. This is certainly an issue that will play a role before the General Court in the ongoing *Google* cases in Europe. Market definition was also the decisive issue in the US Supreme Court's 2018 landmark case on multi-sided markets, *American Express*.

Naturally, the promotion of innovation may constitute a policy goal that an antitrust jurisdiction may wish to further. If such a policy goal should not – and indeed cannot – form part of an antitrust market definition framework, it is quite reasonable that antitrust enforcers will ask what (if any) positive effect on innovation we can purport to bring about by reconceptualising the antitrust market definition framework in innovative market environments in one of the ways proposed here. There are two interrelated answers to this question. One lies in the fact that an in-depth discussion on market definition in these innovative markets can help us to better conceptualise how these markets can be defined as relevant antitrust markets in as consistent a fashion as possible. This is because by focusing on the various parameters of innovative markets that might play into market definition, one can build a comprehensive picture of how competition works, thus ultimately also leading to an understanding of the relevant market at stake. The second and closely related answer is that a comprehensive assessment of the many dimensions of innovative markets enables insights into the particular complexities of these markets that antitrust law needs to take into account, not only at the stage of market delineation but also in the competitive assessment that builds on the latter. While providing a more solid basis for establishing market boundaries, a circumspect market characterisation will also improve our understanding of the theories of harm connected to these markets. Particularly in new types of markets, it is indispensable to link the theory of harm to the specificities of the relevant market. Overall, therefore, while no direct effects on innovation will be harnessed by carrying out an innovation-conscious market delineation, the analysis will certainly enable competition law enforcers – be they courts, agencies or private actors – to rely on a streamlined framework that can adduce more legal certainty in the inherently uncertain world of innovation. This may in turn encourage innovators in their endeavours.

The two types of policy approaches that were explored in order to reconceptualise the market definition framework in a more innovation-sensitive light represent two positions on an entire spectrum of possibilities. They constitute options and should not be understood as straightforward policy recommendations, in particular against the legal culture caveat voiced in this study. However, going beyond reconceptualisation, it is pertinent to ask what competition law would stand to lose if it were to give up on market definition in innovative market environments altogether, thus adopting the third policy approach which was not further pursued in this study. Without a doubt, it would stand to lose out on what competitive innovative markets have to offer. Although it was seen that economics is not yet decided when it comes to the market structure that is most conducive to innovation, a competition law regime that arbitrarily applies its rules without properly understanding the market environment in which it is operating, without trying to capture the essence of the competitive relationships at play, cannot be beneficial to innovation. Without a predictable framework for market delineation, competition law is not connected to market realities in the way that the competition law provisions of the EU and the US have foreseen. Well-administrable and predictable rules on market delineation can infuse this legal concept with legal certainty that is particularly needed in innovative market contexts, despite the scope for interpretation that always remains open. Of course, here lies the crux of the problem: how can we come up with such rules? The two approaches proposed in this study hope to incite a discussion to this effect, while acknowledging that no easy solutions may be found.

The complexities of reality – in this case, of the actual markets in a business sense – do not allow us to formulate any 'simple bright-line test' that always delivers an accurate relevant market. Antitrust law requires 'an analytical construct … that helps make sense of the dynamic situation, evaluate the various forces at play, and reach an informed and reasoned decision'.[2] For our purposes, this analytical construct is that of antitrust market definition. While it necessarily oversimplifies many of the complex economic realities, it strives to give antitrust law some important background for the legal analysis. As such, the relevant market was understood as a legal concept throughout this study. Although it is a concept that has its roots in economics and is now shared amongst law and economics, it takes on a specific legal conception within the context of antitrust. There was thus only a limited role for economics in the research endeavour here pursued. Just like the line umpires in tennis, the role of economists was seen as that of fact providers – with the referee (in competition law, the judge) having the ultimate authority to decide. In tennis, of course, the court consists of predetermined demarcation lines and tennis players only disagree about whether a ball was in or outside the court. In competition law, the demarcation lines of the playing field need to be negotiated in every case. As an added complexity, the market behaviour in question has frequently already occurred at the point in time when the demarcation lines are being questioned. Companies will naturally try to convince the judge or the competition authority that the market boundary needs to be drawn in a certain way so as to exclude their market behaviour from antitrust scrutiny from the outset (or include their competitor's). This also raises the issue of intellectual capture through which companies with the necessary financial means may try and sway antitrust opinion in their favour. Market delineation, of course, is a particularly rewarding area for such activities, as an antitrust case may stand and fall with the market delineation.

An intriguing question linked to the preceding paragraph, and an issue that was touched upon on a number of occasions throughout this study, is who has the power to delineate antitrust markets? Can the legislator provide sufficiently clear guidance so as to minimise opportunism when it comes to parties providing constructed market realities? Is it courts that are called upon to elucidate sometimes ambiguous legal provisions on market definition, thus acting as the ultimate authority of market definition? Are economists the true actors in market definition issues, providing evidence for the parties or perhaps directly to the court? Can competition authorities provide unbiased guidance despite the policy objectives that they pursue, both on a general level and in individual cases? And to what

[2] D Cameron, MA Glick and D Mangum, 'Good Riddance to Market Definition?' (2012) 57 *Antitrust Bulletin* 719, 722.

extent is there intellectual capture that influences the perception of market realities, as portrayed above? With market definition having far-reaching consequences for the substantive analysis, these are not philosophical questions, but issues that come to bear on the outcome of individual cases, on the competition landscape and, ultimately, on innovation efforts. The main actors for market delineation depend on a number of factors: the jurisdiction in which a case is situated, whether the case is brought through a private claim or through public enforcement, how proactive a court/competition authority is, whether a competition authority's decision requires court approval or whether it can only be challenged in court, the intensity of judicial review, the role of economic experts and their cultural capital etc. These captivating questions must be left for further research. Their answer could add an important layer of understanding to the issue of market definition seen from a comparative perspective.

Another crucial question to be explored from a comparative point of view is whether and how convergence of all or some parts of the market definition framework could increase legal certainty and promote innovation to the benefit of consumers. While providing two possible options for further developing the market definition framework in the presence of innovative markets, this study has cautiously argued that convergence amongst antitrust jurisdictions as concerns market delineation may be beneficial in both respects. Such convergence may benefit innovators, even if it is not the optimal solution towards which antitrust jurisdictions converge. The simple fact of convergence may suffice in order to lower costs for innovators and free up resources to further invest in R&D or disseminate existing innovations. And although there is no clear correlation between R&D expenditure and innovation, one can easily see how regulatory convergence would benefit innovators who simultaneously market their innovations in several jurisdictions. While some commentators are critical of the ability of even the EU and the US to converge in antitrust matters[3] – let alone when more culturally diverse jurisdictions are included in the mix – such an opportunity for innovators, and ultimately for consumer welfare, should not be foregone lightly. In the long run, innovation benefits consumers, even if it sometimes comes at a cost in the short run, such as higher prices. The comparative analysis in this study deconstructed EU and US market definition as it concerns many innovation-related aspects. In doing so, it uncovered areas in which certain aspects of market delineation are already converging. Especially where new market definition issues are addressed, such as in multi-sided platforms, convergence from the outset may still be a viable option and worthy of being considered. Where market definition issues are already quite settled, such as in IPRs or aftermarkets, it may require some regulatory effort to bring about convergence in these areas.

[3] DJ Gifford and RT Kudrle, 'Antitrust Goals, Procedures, and Policies in the US and the EU' (2017) 62 *Antitrust Bulletin* 239.

Apart from the innovation/competition interface which remains topical, the criticism that a number of influential scholars confront antitrust market definition with is particularly forceful in the context of innovative markets because the characteristics of innovative markets increase the uncertainties associated with market definition. One of the implications stemming from the difficulties of reliable market definition in innovation-intensive industries could be that in certain innovative environments, market definition will have to focus much more on its market characterisation function and play a much less prominent role for market power assessments. In the US, a number of scholars – and to a certain extent also the agencies, eg, in the Horizontal Merger Guidelines 2010 – are already putting forth arguments that call for a less prominent role for market definition. For instance, they argue that dynamic markets are frequently in a state of flux, making the analytical tools of market definition and market structure inept in these circumstances.[4] In some respects, the legal framework of US antitrust law already allows market definition to play a less decisive role. There are only few market share thresholds in the US that would necessarily require market definition, and market concentration might have to be cautiously assessed outside of the HHI framework in innovative industries. The US Guidelines on Intellectual Property and on Competitor Collaboration outline possibilities to account for innovation competition and the reduced significance of market shares when defining safe harbours. The US Horizontal Merger Guidelines also express the view that relevant markets do not need to be precisely defined. However, recent case law shows that while the agencies may be ready to (partially) abandon market definition, the US courts are not. Cases continue to be won and lost based on market delineation, or the lack of such a delineation.

In the EU, the whole system of Commission guidance relating to Article 101 TFEU relies heavily on market share thresholds. While these are soft law instruments, they nevertheless inform Commission practice and would therefore need to be substantially revised in order to account for a less pronounced role of market definition. The block exemption regulations with their manifold market share thresholds represent enforceable law and would also need to be revised. The EU Courts have repeatedly underlined the significance of market definition, even determining a positive presumption of market dominance for companies with a market share above 50 per cent. Also, the EU Courts do not generally interfere with the Commission's complex economic assessments. From a practical perspective, market definition remains a necessary analytical step for competition law analysis. As such, it appears that it is a concept that is too big to fail (at least for now).

Based on these insights, it is unlikely that one can simply declare market definition obsolete in dynamic markets – or indeed in any type of market.

[4] See HA Shelanski, 'Information, Innovation, and Competition Policy for the Internet' (2013) 161 *University of Pennsylvania Law Review* 1663, 1670.

Therefore, market definition may need to be reconceptualised along the lines shown in this study in order to make it a more useful tool in innovative markets. Only a minority of innovative settings will be such that one cannot meaningfully characterise the current/future relevant product market and will thus have to rely on other tools for the purposes of antitrust law analysis. One can also note the high number of merger cases in dynamic markets in which the European Commission left the precise market delineation open, being satisfied that no competitive harm would arise even under the narrowest possible market definition.

Reducing the importance of market definition while exclusively focusing on innovation competition would require nothing less than a considerable effort on the part of antitrust courts and authorities.[5] It also remains questionable whether this would represent a desirable goal. Market definition provides a framework within which a company's market behaviour can be assessed. It does not act as a straitjacket, as other factors can come into play during later stages of the antitrust analysis. However, it does provide guiding principles that tell us which of the market realities antitrust law needs to pay particular attention to. It provides the boundaries of the playing field, to use the tennis analogy employed at the very outset of this study. Even if innovative markets make these lines somewhat more blurry, they remain useful. Without the help of market definition, it would seem that antitrust analysis could look anywhere and nowhere, that parties would not have to reach a consensus on which field it is they are playing on – and eventually each pass would get lost in the myriad of possibilities. This might appear to be the ideal solution for companies wishing to dodge antitrust scrutiny. From a consumer welfare perspective, it is the worst possible outcome.

Some reflections on the comparative insights that have come to light over the course of this study are also in order. Comparative competition law most frequently compares EU competition and US antitrust law, and in the context of innovation this is almost a necessity. However, the inclusion of further jurisdictions such as China and Japan – if feasible despite the language barrier – is highly desirable in order to enrich the insights and the cultural breadth that can be attained. There is thus considerable scope for further, more multicultural research in this area. Notwithstanding the mostly monocultural nature of the present study, it has been seen that innovation is not always considered in the same or even a similar way when antitrust markets are delineated. In proprietary aftermarkets, for instance, current legal practice leans towards much narrower relevant markets in the EU based on an understanding that may be inspired by behavioural economics, while the predominant US approach relies on a Chicagoan approach of laissez-faire. In other areas, such as multi-sided markets, new market delineation methodology is simultaneously developing in both jurisdictions (one merely need to think of the

[5] Equally emphasising the role of the EU Courts for assigning a new role to market delineation, see H Schweitzer et al, 'Modernisierung der Missbrauchsaufsicht für marktmächtige Unternehmen: Endbericht' (Report for the Federal Ministry for Economic Affairs and Energy, 29 August 2018) 168.

pending *Google* appeals in the EU and the *American Express* case recently decided by the US Supreme Court), presenting a unique chance to tackle market definition in these innovative markets in a harmonious way. Merger cases that simultaneously come before the EU and US authorities give comparative law an additional opportunity to assess the status quo of convergence or divergence in this field. In the area of innovation, these parallel cases also show how innovators are affected by both antitrust jurisdictions simultaneously, and most likely also by antitrust jurisdictions beyond the two included here.

As markets mature, they typically 'outgrow' the characteristics of innovative markets, in particular the short product cycles and superseding market dynamics.[6] Does this mean we can dispense with attempts to more realistically delineate them at their earlier stage? It is submitted here that this is not the case; quite the contrary. Even if some markets become more traditional – and thus perhaps static – over time and can then be delineated along the more familiar lines of traditional market definition, new markets will continue to emerge to which this does not apply. It would not only be conceptually misguided to apply overcome market definition frameworks to these dynamic cases, but would also mean that we cannot be certain that we properly understand these markets, their competitive dynamics and the possible competitive harm that is occurring in them. In addition, we might be finding ourselves at a time in which the nature of competition is fundamentally changing based on innovations such as self-learning algorithms, digitisation and data harvesting,[7] with the consequence that it is time to think about new ways of conceptualising competition, competition law and what the latter can and should achieve.

If it is accepted that market definition is, at least for now, too big to fail, then efforts must be made to enhance our understanding of innovative market environments and to better depict them through the lens of antitrust market definition – be it based on a more traditional approach (referred to as type A in this study) or a more dynamic approach (referred to as type B in this study). Going forward, the function of market definition within the antitrust laws might need to concentrate on market characterisation. This is a point of view that can be brought into line with many of the judgments that have dealt with market definition – although in some instances, in particular where market share thresholds and concentration ratios are concerned, some creativity is required. Here, it will be for the courts to take the lead. An adapted market definition framework can be a useful tool for a more insightful antitrust analysis of dynamic market contexts. The comparative analysis carried out in the present study hopes to serve as a basis for an enhanced mutual understanding and as a foundation for future ambitions of competition

[6] L Peeperkorn and E Paulis, 'Competition and Innovation: Two Horses Pulling the Same Cart' in P Lugard and L Hancher (eds), *On the Merits: Current Issues in Competition Law and Policy – Liber Amicorum Peter Plompen* (Cambridge, Intersentia, 2005) 24.

[7] A Ezrachi and ME Stucke, *Virtual Competition: The Promise and Perils of the Algorithm-Driven Economy* (Cambridge, MA, Harvard University Press, 2016).

law convergence. This may prove beneficial to innovators and, ultimately, to consumers. Market definition may be too big to fail, but it need not be failing innovation.

By increasing our understanding of dynamic markets, of the antitrust theories of harm that relate to them and of the way in which competition remedies may succeed in curbing anti-competitive outcomes in these markets, the relevant market concept can serve its purpose. The proposals developed in this study may serve as a basis for achieving an antitrust market delineation that does justice to innovative market environments, while at the same time maintaining the integrity of the antitrust legal framework. However, in order to harness its potential in dynamic markets, the focus of market delineation in these environments may need to increasingly shift to its market characterisation function.

BIBLIOGRAPHY

ABA Section of Antitrust Law, *Market Definition in Antitrust: Theory and Case Studies* (Chicago, ABA Publishing, 2012)

——, *Market Power Handbook: Competition Law and Economic Foundations*, 2nd edn (Chicago, ABA Publishing, 2012)

——, *Econometrics: Legal, Practical, and Technical Issues*, 2nd edn (Chicago, ABA Publishing, 2014)

Abbott, AF, 'Intellectual Property Licensing and Antitrust Policy: A Comparative Perspective' (2003) 34 *Law & Policy in International Business* 801

Abbott, AF, Michel, S and Irizarry, A, 'The Right Balance of Competition Policy and Intellectual Property Law: A Federal Trade Commission Perspective' in P Marsden (ed), *Handbook of Research in Trans-Atlantic Antitrust* (Cheltenham, Edward Elgar, 2006) 356–97

Ackermann, T, 'European Competition Law' in K Riesenhuber (ed), *European Legal Method* (Cambridge, Intersentia, 2017) 513–35

Ahlborn, C and Grave, C, 'Walter Eucken and Ordoliberalism: An Introduction from a Consumer Welfare Perspective' (2006) 2 *Competition Policy International* 197

Ahlborn, C, Grave, C, and Padilla, AJ, 'Competition Policy in the New Economy: Is European Competition Law up to the Challenge?' (2001) 22 *European Competition Law Review* 156

Alexiadis, P and Guerrero Pérez, A, 'European Commission Proposes Stricter EU Antitrust Rules on Technology Transfer' (2013) 35 *European Intellectual Property Review* 415

Allen, JA, *Scientific Innovation and Industrial Prosperity* (Amsterdam, Elsevier, 1967)

Anderman, SD, 'Substantial Convergence: The US Influence on the Development of the Regulatory Framework for IP Licensing in the EC' in P Marsden (ed), *Handbook of Research in Transatlantic Antitrust* (Cheltenham, Edward Elgar, 2006) 335–55

——, 'The New EC Competition Law Framework for Technology Transfer and IP Licensing' in J Drexl (ed), *Research Handbook on Intellectual Property and Competition Law* (Cheltenham, Edward Elgar, 2008) 107–38

——, 'The Interface between Intellectual Property Rights and EU Competition Law' in A Ohly (ed), *Common Principles of European Intellectual Property Law* (Tübingen, Mohr Siebeck, 2012) 241–46

——, 'Innovation, IPRs and EU Competition Law' in I Lianos and D Geradin (eds), *Handbook on European Competition Law* (Cheltenham, Edward Elgar, 2013) 561–87

Anderman, SD and Kallaugher, J, *Technology Transfer and the New EU Competition Rules: Intellectual Property Licensing after Modernisation* (Oxford, Oxford University Press, 2006)

Anderman, SD and Schmidt, HSK, *EU Competition Law and Intellectual Property Rights: The Regulation of Innovation*, 2nd edn (Oxford, Oxford University Press, 2011)

Andrews, P, 'Aftermarket Power in the Computer Services Market: The Digital Undertaking' (1998) 19 *European Competition Law Review* 176

Andriychuk, O, 'The Dialectics of Competition Law: Sketching the Ordo-Austrian Approach to Antitrust' (2012) 35 *World Competition* 355

Antitrust Modernization Commission, 'Report and Recommendations' (April 2007)

Areeda, P, 'Justice's Mergers Guidelines: The General Theory' (1983) 71 *California Law Review* 303

——, 'Market Definition and Horizontal Restraints' (1983) 52 *Antitrust Law Journal* 553

Areeda, P and Hovenkamp, HJ, *Antitrust Law: An Analysis of Antitrust Principles and Their Application*, 4th edn (Wolters Kluwer 2017)

Arewa, OB, 'YouTube, UGC, and Digital Music: Competing Business and Cultural Models in the Internet Age' (2010) 104 *Northwestern University Law Review* 431

Armstrong, M, 'Competition in Two-Sided Markets' (2006) 37 *RAND Journal of Economics* 668

Armstrong, M and Huck, S, 'Behavioral Economics and Antitrust' in RD Blair and DD Sokol (eds), *The Oxford Handbook of International Antitrust Economics*, vol 1 (Oxford, Oxford University Press, 2015) 205–28

Arrow, KJ, 'Economic Welfare and the Allocation of Resources to Invention' in National Bureau of Economic Research (ed), *The Rate and Direction of Inventive Activity* (Princeton, Princeton University Press, 1962) 609–25

Audretsch DB, Baumol, WJ and Burke, AE, 'Competition Policy in Dynamic Markets' (2001) 19 *International Journal of Industrial Organization* 613

Auer, D and Petit, N, 'Two-Sided Markets and the Challenge of Turning Economic Theory into Antitrust Policy' (2015) 60 *Antitrust Bulletin* 426

Ayres, I, 'A Private Revolution: Markovits and Markets' (1988) 64 *Chicago-Kent Law Review* 861

Bain, JS, *Price Theory* (New York, Holt, Rinehart & Winston, 1952)

——, *Barriers to New Competition* (Cambridge, MA, Harvard University Press, 1956)

Baker, JB, 'Product Differentiation through Space and Time: Some Antitrust Policy Issues' (1997) 42 *Antitrust Bulletin* 177

——, 'Beyond *Schumpeter vs. Arrow*: How Antitrust Fosters Innovation' (2007) 74 *Antitrust Law Journal* 575

——, 'Market Definition: An Analytical Overview' (2007) 74 *Antitrust Law Journal* 129

——, 'Market Concentration in the Antitrust Analysis of Horizontal Mergers' in KN Hylton (ed), *Antitrust Law and Economics* (Cheltenham, Edward Elgar, 2010) 234–60

——, 'Taking the Error out of "Error Cost" Analysis: What's Wrong with Antitrust's Right' (2015) 80 *Antitrust Law Journal* 1

Baker, JB and Bresnahan, TF, 'Empirical Methods of Identifying and Measuring Market Power' (1992) 61 *Antitrust Law Journal* 3

——, 'Economic Evidence in Antitrust: Defining Markets and Measuring Market Power' in P Buccirossi (ed), *Handbook of Antitrust Economics* (Cambridge, MA, MIT Press, 2008) 1–42

Baker, JB and Shapiro, C, 'Reinvigorating Horizontal Merger Enforcement' in R Pitofsky (ed), *How the Chicago School Overshot the Mark: The Effect of Conservative Economic Analysis on U.S. Antitrust* (Oxford, Oxford University Press, 2008) 235–88

Baker, TA, 'The 1984 Justice Department Guidelines' (1984) 53 *Antitrust Law Journal* 327

Balto, DA and Pitofsky, R, 'Antitrust and High-Tech Industries: The New Challenge' (1998) 43 *Antitrust Bulletin* 583

Barton, JH, 'The Balance between Intellectual Property Rights and Competition: Paradigms in the Information Sector' (1997) 18 *European Competition Law Review* 440

Batchelor, B and Jenkins, T, 'Commission Consults on Revisions to the Competition Rules on Technology Transfer Regime: Proposes Tightening of the Rules' (2013) 34 *European Competition Law Review* 348

Bauer, JP, 'Antitrust Implications of Aftermarkets' (2007) 52 *Antitrust Bulletin* 31

Baxter, WF, 'The Definition and Measurement of Market Power in Industries Characterized by Rapidly Developing and Changing Technologies' (1984) 53 *Antitrust Law Journal* 717

Baye, MR, 'Market Definition and Unilateral Competitive Effects in Online Retail Markets' (2008) 4 *Journal of Competition Law & Economics* 639

Beaton-Wells, C, 'Mergers without Markets? Unilateral Effects Analysis in the United States and Its Prospects in Australia' (2006) 34 *Australian Business Law Review* 186

Beckner, CF III and Salop, SC, 'Decision Theory and Antitrust Rules' (1999) 67 *Antitrust Law Journal* 41

Bergqvist, C, 'Google and the Search for a Theory of Harm' (2018) 39 *European Competition Law Review* 149

Bernard, K, 'Innovation Market Theory and Practice: An Analysis and Proposal for Reform' (2011) 7 *Competition Policy International* 159

Bishop, S and Walker, M, *The Economics of EC Competition Law: Concepts, Application and Measurement*, 3rd edn (London, Sweet & Maxwell, 2010)

Bitetti, R, 'Google, Competition Policy and the Owl of Minerva' in M Kovač and A-S Vandenberghe (eds), *Economic Evidence in EU Competition Law* (Cambridge, Intersentia, 2016) 295–325

Blair, RD and Carruthers, CK, 'The Economics of Monopoly Power in Antitrust' in KN Hylton (ed), *Antitrust Law and Economics* (Cheltenham, Edward Elgar, 2010) 64–81

Blaschczok, M, *Kartellrecht in zweiseitigen Wirtschaftszweigen: Eine Untersuchung vor dem Hintergrund der ökonomischen Forschung zu 'two-sided markets'* (Baden-Baden, Nomos, 2015)

Boettke, PJ, 'Austrian School of Economics' in DR Henderson (ed), *The Concise Encyclopedia of Economics*, 2nd edn (Carmel, Liberty Fund, 2008)

Bohannan, C and Hovenkamp, HJ, *Creation without Restraint: Promoting Liberty and Rivalry in Innovation* (Oxford, Oxford University Press, 2012)

Borenstein, S, MacKie-Mason, JK and Netz, JS, 'Exercising Market Power in Proprietary Aftermarkets' (2000) 9 *Journal of Economics & Management Strategy* 157

Boshoff, WH, 'Why Define Markets in Competition Cases?' (Stellenbosch Economic Working Papers 10/2013)

——, 'Market Definition as a Problem of Statistical Inference' (2014) 10 *Journal of Competition Law & Economics* 861

Bourreau, M and de Streel, A, 'Digital Conglomerates and EU Competition Policy' (CERRE Report, March 2019)

Bower, JL and Christensen, CM, 'Disruptive Technologies: Catching the Wave' (1995) 73 *Harvard Business Review* 43

Brenkers, R and Verboven, F, 'Market Definition with Differentiated Products: Lessons from the Car Market' in JP Choi (ed), *Recent Developments in Antitrust: Theory and Evidence* (Cambridge, MA, MIT Press, 2007) 153–86

Broder, DF, *US Antitrust Law and Enforcement: A Practice Introduction*, 3rd edn (Oxford, Oxford University Press, 2016)

Bruzzone, G and Boccaccio, M, 'Standards under EU Competition Law: The Open Issues' in G Caggiano, G Muscolo and M Tavassi (eds), *Competition Law and Intellectual Property: A European Perspective* (Alphen aan den Rijn, Kluwer Law International, 2012) 85–111

Budzinski, O and Christiansen, A, 'The Oracle/PeopleSoft Case: Unilateral Effects, Simulation Models and Econometrics in Contemporary Merger Control' (2007) 34 *Legal Issues of Economic Integration* 133

Buttigieg, E, *Competition Law: Safeguarding the Consumer Interest: A Comparative Analysis of US Antitrust Law and EC Competition Law* (Alphen aan den Rijn, Kluwer Law International, 2009)

Caffarra, C and Bishop, B, 'Dynamic Competition and Aftermarkets' (1998) 19 *European Competition Law Review* 265

Caffarra, C and Walker, M, 'An Exploration into the Use of Economics before Courts in Europe' (2010) 1 *Journal of European Competition Law & Practice* 158

Caillaud, B and Jullien, B, 'Chicken & Egg: Competition among Intermediation Service Providers' (2003) 34 *RAND Journal of Economics* 309

Calkins, S, 'The New Merger Guidelines and the Herfindahl-Hirschman Index' (1983) 71 *California Law Review* 402

Cameron, D and Glick, MA, 'Market Share and Market Power in Merger and Monopolization Cases' (1996) 17 *Managerial and Decision Economics* 193

Cameron, D, Glick, MA and Mangum, D, 'Comments on Articles in the Kaplow Special Issue' (2012) 57 *Antitrust Bulletin* 957

——, 'Good Riddance to Market Definition?' (2012) 57 *Antitrust Bulletin* 719

Camesasca, PD, 'Mayday or Heyday? Dynamic Competition Meets Media Ownership Rules after Premiere' (2000) 21 *European Competition Law Review* 76

Camesasca, PD and van den Bergh, RJ, 'Achilles Uncovered: Revisiting the European Commission's 1997 Market Definition Notice' (2002) 47 *Antitrust Bulletin* 143

Capobianco, A and Nyeso, A, 'Challenges for Competition Law Enforcement and Policy in the Digital Economy' (2018) 9 *Journal of European Competition Law & Practice* 19

Carlton, DW, 'Market Definition: Use and Abuse' (2007) 3 *Competition Policy International* 3

Carrier, MA, 'Two Puzzles Resolved: Of the Schumpeter–Arrow Stalemate and Pharmaceutical Innovation Markets' (2008) 93 *Iowa Law Review* 393

——, *Innovation for the 21st Century: Harnessing the Power of Intellectual Property and Antitrust Law* (Oxford, Oxford University Press, 2009)

Carstensen, PC, 'Introduction' (2012) 57 *Antitrust Bulletin* 655

Chamberlin, EH, 'Product Heterogeneity and Public Policy' (1950) 40 *American Economic Review* 85

——, *The Theory of Monopolistic Competition: A Re-orientation of the Theory of Value*, 8th edn (Cambridge, MA, Harvard University Press, 1969)

Chin, AR, 'The Misapplication of Innovation Market Analysis to Biotechnology Mergers' (1997) 3 *Boston University Journal of Science and Technology Law* 6

Christensen, CM, *The Innovator's Dilemma: The Revolutionary Book that Will Change the Way You Do Business* (New York, Collins, 2003)

Coate, MB and Fischer, JH, 'Is Market Definition Still Needed after All These Years' (2014) 2 *Journal of Antitrust Enforcement* 422

Coate, MB and Simons, JJ, 'In Defense of Market Definition' (2012) 57 *Antitrust Bulletin* 667

Coates, K, *Competition Law and Regulation of Technology Markets* (Oxford, Oxford University Press, 2011)

Colangelo, G and Maggiolino, M, 'Applying Two-Sided Markets Theory: The *MasterCard* and *American Express* Decisions' (2018) 14 *Journal of Competition Law & Economics* 115

Colangelo, M, 'Parity Clauses and Competition Law in Digital Marketplaces: The Case of Online Hotel Booking' (2017) 8 *Journal of European Competition Law & Practice* 3

Colangelo, M and Zeno-Zencovich, V, 'Online Platforms, Competition Rules and Consumer Protection in Travel Industry' (2016) 2 *Journal of European Consumer and Market Law* 75

Constantinesco, L-J, *Rechtsvergleichung, Band II: Die rechtsvergleichende Methode* (Cologne, Carl Heymanns, 1972)

Cooter, R and Ulen, T, *Law & Economics*, 6th edn (Boston, Pearson, 2012)

Cornish, WR, *Intellectual Property: Patents, Copyright, Trade Marks and Allied Rights* (London, Sweet & Maxwell, 1981)

Cortés, E, Dawson, A and Hatton, C, 'Squaring the Circle: The EU's Quest for Balance between Antitrust and Intellectual Property' [2015] *European Antitrust Review* 16

Coscelli, A and Edwards, G, 'Dominance and Market Power in EU Competition Law Enforcement' in I Lianos and D Geradin (eds), *Handbook on European Competition Law* (Cheltenham, Edward Elgar, 2013) 350–84

Cotter, TF, 'Innovation and Antitrust Policy' in RD Blair and DD Sokol (eds), *The Oxford Handbook of International Antitrust Economics*, vol 2 (Oxford, Oxford University Press, 2015) 132–52

Crane, DA, 'The Economics of Antitrust Enforcement' in KN Hylton (ed), *Antitrust Law and Economics* (Cheltenham, Edward Elgar, 2010) 1–22

——, 'Search Neutrality and Referral Dominance' (2012) 8 *Journal of Competition Law & Economics* 459

——, 'Market Power without Market Definition' (2014) 90 *Notre Dame Law Review* 31

——, 'Rationales for Antitrust: Economics and Other Bases' in RD Blair and DD Sokol (eds), *The Oxford Handbook of International Antitrust Economics*, vol 1 (Oxford, Oxford University Press, 2015) 3–16

Crémer, J, de Montjoye, Y-A and Schweitzer, H, 'Competition Policy for the Digital Era' (Report for DG Competition, 3 April 2019)

Crocioni, P, 'The Hypothetical Monopolist Test: What it Can and Cannot Tell You' (2002) 23 *European Competition Law Review* 354

——, 'Leveraging of Market Power in Emerging Markets: A Review of Cases, Literature, and a Suggested Framework' (2007) 4 *Journal of Competition Law & Economics* 449
Cross, FB and Miller, RL, *The Legal Environment of Business: Text and Cases – Ethical, Regulatory, Global, and E-Commerce Issues*, 7th edn (Mason, Cengage, 2009)
Curzon Price, T and Walker, M, 'Incentives to Innovate v Short-Term Price Effects in Antitrust Analysis' (2016) 7 *Journal of European Competition Law & Practice* 475
Czapracka, K, *Intellectual Property and the Limits of Antitrust: A Comparative Study of US and EU Approaches* (Cheltenham, Edward Elgar, 2009)
Dabbah, MM, *International and Comparative Competition Law* (Cambridge, Cambridge University Press, 2010)
Davis, PJ and Garcés, E, *Quantitative Techniques for Competition and Antitrust Analysis* (Princeton, Princeton University Press, 2009)
De Cruz, P, *Comparative Law in a Changing World*, 3rd edn (New York, Routledge-Cavendish, 2008)
Desai, DR and Waller, SW, 'Brands, Competition, and the Law' [2010] *Brigham Young University Law Review* 1425
DeSanti, S and Cohen, W, 'Competition to Innovate: Strategies for Proper Antitrust Assessments' in R Cooper Dreyfuss, D Leenheer Zimmerman and H First (eds), *Expanding the Boundaries of Intellectual Property: Innovation Policy for the Knowledge Society* (Oxford, Oxford University Press, 2001) 317–41
Deutscher, E and Makris, S, 'Exploring the Ordoliberal Paradigm: The Competition-Democracy Nexus' (2016) 11 *Competition Law Review* 181
Devlin, A, *Fundamental Principles of Law and Economics* (New York, Routledge, 2015)
Devlin, A and Jacobs, M, 'Antitrust Divergence and the Limits of Economics' (2010) 104 *Northwestern University Law Review* 253
——, 'Antitrust Error' (2010) 52 *William & Mary Law Review* 75
Drexl, J, 'The Relationship between the Legal Exclusivity and Economic Market Power: Links and Limits' in I Govaere and H Ullrich (eds), *Intellectual Property, Market Power and the Public Interest* (Brussels, Peter Lang, 2008) 13–33
——, 'Anticompetitive Stumbling Stones on the Way to a Cleaner World: Protecting Competition in Innovation without a Market' (2012) 8 *Journal of Competition Law & Economics* 507
Easterbrook, FH, 'The Limits of Antitrust' (1984) 63 *Texas Law Review* 1
Eblen, CC, 'Defining the Geographic Market in Modern Commerce: The Effect of Globalization and E-Commerce on *Tampa Electric* and its Progeny' (2004) 56 *Baylor Law Review* 49
Eiszner, JR, 'Innovation Markets and Automatic Transmissions: A Shift in the Wrong Direction?' (1998) 43 *Antitrust Bulletin* 297
Ellig, J, 'Industrial Organization' in PJ Boettke (ed), *The Elgar Companion to Austrian Economics* (Cheltenham, Edward Elgar, 1994) 244–48
Emch, E and Thompson, TS, 'Market Definition and Market Power in Payment Card Networks' (2006) 5 *Review of Network Economics* 45
Encaoua, D and Hollander, A, 'Competition Policy and Innovation' (2002) 18 *Oxford Review of Economic Policy* 63
Europe Economics, 'The Development of Analytical Tools for Assessing Market Dynamics in the Knowledge Based Economy' (12 September 2003)
European Commission, 'Intellectual Property Rights and Standardization: Communication from the Commission' (27 October 1992) COM(1992) 445 final
——, 'XXIVth Report on Competition Policy 1994' (Luxembourg, 1995)
——, 'The European Commission Accepts an Undertaking from Digital Concerning its Supply and Pricing Practices in the Field of Computer Maintenance Services' (IP/97/868, 10 October 1997)
——, 'XXVIIth Report on Competition Policy 1997' (Luxembourg, 1998)
——, 'DG Competition Discussion Paper on the Application of Article 82 of the Treaty to Exclusionary Abuses' (December 2005)

——, 'Study on the Coverage, Functioning and Consumer Use of Comparison Tools and Third-Party Verification Schemes for Such Tools: Final Report Prepared by ECME Consortium (in Partnership with Deloitte)' (EAHC/FWC/2013 85 07, 2013)
——, 'Comparison Tools: Report from the Multi-stakeholder Dialogue – Providing Consumers with Transparent and Reliable Information' (March 2013)
——, 'Market Definition in a Globalised World' [March 2015] *Competition Policy Brief* 1
——, 'Antitrust: Commission Sends Two Statements of Objections on Exclusivity Payments and Predatory Pricing to Qualcomm' (IP/15/6271, 8 December 2015)
——, 'Advancing the Internet of Things in Europe' (19 April 2016) SWD(2016) 110 final
——, 'Antitrust: Commission Sends Statement of Objections to Google on Android Operating System and Applications' (IP/16/1492, 20 April 2016)
——, 'Online Platforms and the Digital Single Market – Opportunities and Challenges for Europe' (25 May 2016) COM(2016) 288 final
——, 'Preliminary Report on the E-Commerce Sector Inquiry' (15 September 2016) SWD(2016) 312 final
——, 'Final Report on the E-Commerce Sector Inquiry' (10 May 2017) COM(2017) 229 final
——, 'Merger Statistics: 21 September 1990–31 August 2019' (31 August 2019)
——, 'Antitrust: Commission Fines Qualcomm €997 Million for Abuse of Dominant Market Position' (IP/18/421, 24 January 2018)
——, 'Proposal for a Regulation on Promoting Fairness and Transparency for Business Users of Online Intermediation Services' (26 April 2018) COM(2018) 238 final
——, 'Antitrust: Commission Fines Google €4.34 Billion for Illegal Practices Regarding Android Mobile Devices to Strengthen Dominance of Google's Search Engine' (IP/18/4581, 18 July 2018)
——, 'Mergers: Commission Prohibits Siemens' Proposed Acquisition of Alstom' (IP/19/881, 6 February 2019)
——, 'Antitrust: Commission Sends Statement of Objections to BMW, Daimler and VW for Restricting Competition on Emission Cleaning Technology' (IP/19/2008, 5 April 2019)
——, in OECD (ed), *Policy Roundtable: Competition and Regulation Issues in the Pharmaceutical Industry* (2000) DAFFE/CLP(2000)29, 339–47
——, in OECD (ed), *Policy Roundtable: Two-Sided Markets* (2009) DAF/COMP(2009)20, 157–85
Evans DS, 'The Antitrust Economics of Multi-Sided Platform Markets' (2003) 20 *Yale Journal on Regulation* 325
——, 'The Middle Way on Applying Antitrust to Information Technology' [2009] *Antitrust Chronicle* 1
——, 'Background Note' in OECD (ed), *Policy Roundtable: Two-Sided Markets* (2009) DAF/COMP(2009)20, 23–48
——, 'The Antitrust Economics of Free' (2011) 7 *Competition Policy International* 71
——, 'Lightening up on Market Definition' in E Elhauge (ed), *Research Handbook on the Economics of Antitrust Law* (Cheltenham, Edward Elgar, 2012) 53–89
——, 'Two-Sided Markets' in ABA Section of Antitrust Law (ed), *Market Definition in Antitrust: Theory and Case Studies* (Chicago, ABA Publishing, 2012) 437–70
——, 'Attention Rivalry among Online Platforms' (2013) 9 *Journal of Competition Law & Economics* 313
Evans DS and Hylton, KN, 'The Lawful Acquisition and Exercise of Monopoly Power and its Implications for the Objectives of Antitrust' (2008) 4 *Competition Policy International* 203
Evans DS and Noel, M, 'Defining Antitrust Markets When Firms Operate Two-Sided Platforms' [2005] *Columbia Business Law Review* 667
Evans DS and Schmalensee, R, 'Some Economic Aspects of Antitrust Analysis in Dynamically Competitive Industries' in AB Jaffe, J Lerner and S Stern (eds), *Innovation Policy and the Economy*, vol 2 (Cambridge, MA, MIT Press, 2002) 1–49
——, 'Markets with Two-Sided Platforms' in ABA Section of Antitrust Law (ed), *Issues in Competition Law and Policy* (Chicago, ABA Publishing, 2008) 667–93

——, 'The Antitrust Analysis of Multisided Platform Businesses' in RD Blair and DD Sokol (eds), *The Oxford Handbook of International Antitrust Economics*, vol 1 (Oxford, Oxford University Press, 2015) 404–47

Ezrachi, A, 'The Competitive Effects of Parity Clauses on Online Commerce' (2015) 11 *European Competition Journal* 488

——, 'Sponge' (2017) 5 *Journal of Antitrust Enforcement* 49

——, *EU Competition Law: An Analytical Guide to the Leading Cases*, 6th edn (Oxford, Hart Publishing, 2018)

Ezrachi, A and Maggiolino, M, 'European Competition Law, Compulsory Licensing, and Innovation' (2012) 8 *Journal of Competition Law & Economics* 595

Ezrachi, A and Robertson, VHSE, 'Competition, Market Power and Third-Party Tracking' (2019) 42 *World Competition* 5

Ezrachi, A and Stucke, ME, *Virtual Competition: The Promise and Perils of the Algorithm-Driven Economy* (Cambridge, MA, Harvard University Press, 2016)

Farrell, J and Klemperer, P, 'Coordination and Lock-in: Competition with Switching Costs and Network Effects' in M Armstrong and RH Porter (eds), *Handbook of Industrial Organization*, vol 3 (Amsterdam, North Holland, 2007) 1967–2072

Farrell, J and Shapiro, C, 'Antitrust Evaluation of Horizontal Mergers: An Economic Alternative to Market Definition' (2010) 10 *The BE Journal of Theoretical Economics* 1

Fatur, A, *EU Competition Law and the Information and Communication Technology Network Industries: Economic Versus Legal Concepts in Pursuit of (Consumer) Welfare* (Oxford, Hart Publishing, 2012)

Federal Trade Commission, 'Anticipating the 21st Century – Competition Policy in the New High-Tech, Global Marketplace: Staff Report', vol 1 (May 1996)

——, 'FTC Closes its Investigation of Genzyme Corporation's 2001 Acquisition of Novazyme Pharmaceuticals, Inc' (13 January 2004)

——, 'Statement Concerning *Google/DoubleClick*, Case 071-0170' (20 December 2007)

——, 'Proposed Acquisition of Instagram by Facebook, Case 121-0121' (22 August 2012)

——, 'Google Agrees to Change its Business Practices to Resolve FTC Competition Concerns in the Markets for Devices like Smart Phones, Games and Tablets, and in Online Search' (3 January 2013)

——, 'Data Brokers: A Call for Transparency and Accountability' (May 2014)

Feldman, R, 'Patent and Antitrust: Differing Shades of Meaning' (2008) 13 *Virginia Journal of Law & Technology* 1

Feng, K, 'Patent-Related Mergers and Market Definition Under the 2010 Horizontal Merger Guidelines: The Need to Consider Technology and Innovation Markets' (2012) 34 *Thomas Jefferson Law Review* 197

Filistrucchi, L, 'Market Definition in Multi-Sided Markets' in OECD (ed), *Rethinking Antitrust Tools for Multi-Sided Platforms* (2018) 37–54

Filistrucchi, L, Geradin, D and van Damme, E, 'Identifying Two-Sided Markets' (2013) 36 *World Competition* 33

Filistrucchi, L et al, 'Market Definition in Two-Sided Markets: Theory and Practice' (2014) 10 *Journal of Competition Law & Economics* 293

First, H and Waller, SW, 'Antitrust's Democracy Deficit' (2013) 81 *Fordham Law Review* 2543

Fisher, FM, 'Horizontal Mergers: Triage and Treatment' (1987) 1 *Journal of Economic Perspectives* 23

Franck, J-U and Peitz, M, 'Market Definition and Market Power in the Platform Economy' (CERRE Report, May 2019)

Frändberg, Å, 'An Essay on Legal Concept Formation' in JC Hage and D von der Pfordten (eds), *Concepts in Law* (Berlin, Springer, 2009) 1–16

Früh, A, *Immaterialgüterrechte und der relevante Markt: Eine wettbewerbsrechtliche und schutzrechtliche Würdigung technologischer Innovation* (Cologne, Carl Heymanns, 2012)

Furman, J et al, 'Unlocking Digital Competition' (Report of the Digital Competition Expert Panel, March 2019)

Gal, MS, '3D Challenges: Ensuring Competition and Innovation in 3D Printing' (SSRN abstract n 3356891, 20 March 2019)

Gal, MS and Elkin-Koren, N, 'Algorithmic Consumers' (2017) 70 *Harvard Journal of Law & Technology* 309

Gal, MS and Rubinfeld, DL, 'The Hidden Costs of Free Goods: Implications for Antitrust Enforcement' (2016) 80 *Antitrust Law Journal* 521

Galloway, J, 'Driving Innovation: A Case for Targeted Competition Policy in Dynamic Markets' (2011) 34 *World Competition* 73

Garza, DA, 'Market Definition, the New Horizontal Merger Guidelines, and the Long March Away from Structural Presumptions' [2010] *Antitrust Source* 1

Gebicka, A and Heinemann, A, 'Social Media & Competition Law' (2014) 37 *World Competition* 149

Geiger, C (ed), *Constructing European Intellectual Property: Achievements and New Perspectives* (Cheltenham, Edward Elgar, 2013)

Genevaz, S and Vidal, J, 'Going Digital: How Online Competition Changed Market Definition and Swayed Competition Analysis in *Fnac/Darty*' (2017) 8 *Journal of European Competition Law & Practice* 30

Geradin, D, 'Pricing Abuses by Essential Patent Holders in a Standard-Setting Context: A View from Europe' ('The Remedies for Dominant Firm Misconduct' Conference, University of Virginia, June 2008)

——, 'Competition Law' in JM Smits (ed), *Elgar Encyclopedia of Comparative Law*, 2nd edn (Cheltenham, Edward Elgar, 2012) 208–15

Gerasymenko, A and Afendikova, S, 'The Relevant Temporal Market Definition in Antitrust Analysis' (2018) 4 *Baltic Journal of Economic Studies* 68

Gerber, DJ, 'Competition Law and the Institutional Embeddedness of Economics' in J Drexl, L Idot and J Monéger (eds), *Economic Theory and Competition Law* (Cheltenham, Edward Elgar, 2009) 20–44

——, 'Comparative Law and Global Regulatory Convergence: The Example of Competition Law' in M Adams and J Bomhoff (eds), *Practice and Theory in Comparative Law* (Cambridge, Cambridge University Press, 2012) 120–42

——, 'Global Competition Law Convergence: Potential Roles for Economics' in T Eisenberg and G Ramello (eds), *Comparative Law and Economics* (Cheltenham, Edward Elgar, 2016) 206–35

Geroski, PA, 'Thinking Creatively about Markets' (1998) 16 *International Journal of Industrial Organization* 677

——, 'Competition in Markets and Competition for Markets' (2003) 3 *Journal of Industry, Competition and Trade* 151

Ghidini, G, *Innovation, Competition and Consumer Welfare in Intellectual Property Law* (Cheltenham, Edward Elgar, 2010)

Gifford, DJ and Kudrle, RT, 'Antitrust Approaches to Dynamically Competitive Industries in the United States and the European Union' (2011) 7 *Journal of Competition Law & Economics* 695

——, *The Atlantic Divide in Antitrust: An Examination of US and EU Competition Policy* (Chicago, University of Chicago Press, 2015)

——, 'Antitrust Goals, Procedures, and Policies in the US and the EU' (2017) 62 *Antitrust Bulletin* 239

Gilbert, RJ, 'Competition and Innovation' in ABA Section of Antitrust Law (ed), *Issues in Competition Law and Policy* (Chicago, ABA Publishing, 2008) 577–600

Gilbert, RJ and Sunshine, SC, 'Incorporating Dynamic Efficiency Concerns in Merger Analysis: The Use of Innovation Markets' (1995) 63 *Antitrust Law Journal* 569

——, 'The Use of Innovation Markets: A Reply to Hay, Rapp, and Hoerner' (1995) 63 *Antitrust Law Journal* 75

Gilbert, RJ and Tom, WK, 'Is Innovation King at the Antitrust Agencies? The Intellectual Property Guidelines Five Years Later' (2001) 69 *Antitrust Law Journal* 43

Ginsburg, DH and Wright, JD, 'Dynamic Analysis and the Limits of Antitrust Institutions' (2012) 78 *Antitrust Law Journal* 1

Glader, M, 'Innovation Economics: The Antitrust Guidelines on Horizontal Co-operation' (2001) 24 *World Competition* 513

——, *Innovation Markets and Competition Analysis: EU Competition Law and US Antitrust Law* (Cheltenham, Edward Elgar, 2006)
Glassman, ML, 'Market Definition as a Practical Matter' (1980) 49 *Antitrust Law Journal* 1155
Glick, MA, Cameron, DJ and Mangum, DG, 'Importing the Merger Guidelines Market Test in Section 2 Cases: Potential Benefits and Limitations' (1997) 42 *Antitrust Bulletin* 121
Glick, MA and Campbell, D, 'Market Definition and Concentration: One Size Does Not Fit All' (2007) 52 *Antitrust Bulletin* 229
Goeteyn, G, Smith, P and Ashall, S, 'Away From Market Shares? The Increasing Importance of Contestability in EU Competition Law Cases' (2015) 6 *Journal of European Competition Law & Practice* 197
Gohari, RS, *Verweigerung von Geschäftsbeziehungen - Kartellrechtliche Analyse nach Schweizer, EU- und US-Recht* (Bern, Stämpfli, 2017)
Goldfarb, A and Tucker, C, 'Substitution between Offline and Online Advertising' (2011) 7 *Journal of Competition Law & Economics* 37
Graef, I, 'Market Definition and Market Power in Data: The Case of Online Platforms' (2015) 38 *World Competition* 473
Graham, C and Morton, J, 'Latest EU Developments in Standards, Patents and FRAND Licensing' (2014) 36 *European Intellectual Property Review* 700
Greene, H, 'Guideline Institutionalization: The Role of Merger Guidelines in Antitrust Discourse' (2006) 48 *William & Mary Law Review* 771
Greenhalgh, C and Rogers, M, 'The Value of Intellectual Property Rights to Firms and Society' (2007) 23 *Oxford Review of Economic Policy* 541
Griggs, L, 'A Teleological Approach to Market Definition: Has it Led to Single Product Market Definition?' (2002) 4 *University of Notre Dame Australia Law Review* 77
Grimes, WS, 'Market Definition in Franchise Antitrust Claims: Relational Market Power and the Franchisor's Conflict of Interest' (1999) 67 *Antitrust Law Journal* 243
Grunes, AP and Stucke, ME, 'No Mistake about it: The Important Role of Antitrust in the Era of Big Data' [2015] *Antitrust Source* 1
Gundlach, GT, 'Aftermarkets, Systems, and Antitrust: A Primer' (2007) 52 *Antitrust Bulletin* 17
Gürkaynak, G et al, 'Multisided Markets and the Challenge of Incorporating Multisided Considerations into Competition Law Analysis' (2017) 5 *Journal of Antitrust Enforcement* 100
Gyselen, L, 'Competition in Innovation: A Novel Concept? The Case Law on Pharmaceuticals' in P Lugard and L Hancher (eds), *On the Merits: Current Issues in Competition Law and Policy - Liber Amicorum Peter Plompen* (Cambridge, Intersentia, 2005) 31–49
Hale, GE and Hale, RD, *Market Power: Size and Shape under the Sherman Act* (Boston, Little, Brown & Comp, 1958)
Hall, GR, 'Market Definition and Antitrust Policy' (1963) 20 *Washington & Lee Law Review* 47
Harris, BC and Simons, JJ, 'Focusing Market Definition: How Much Substitution is Necessary?' (1989) 12 *Research in Law and Economics* 207
Harris, RG and Jorde, TM, 'Antitrust Market Definition: An Integrated Approach' (1984) 72 *California Law Review* 3
Hartman, R et al, 'Assessing Market Power in Regimes of Rapid Technological Change' (1993) 2 *Industrial and Corporate Change* 317
Hausman, JA, Leonard, GK and Vellturo, CA, 'Market Definition under Price Discrimination' (1996) 64 *Antitrust Law Journal* 367
Hawker, NW, 'Automotive Aftermarkets: A Case Study in Systems Competition' (2011) 56 *Antitrust Bulletin* 57
Hay, GA, 'Innovations in Antitrust Enforcement' (1995) 64 *Antitrust Law Journal* 7
Hayek, FA, 'The Meaning of Competition' in FA Hayek, *Individualism and Economic Order* (New York, Routledge, 1949)
Heide, T, 'Trade Marks and Competition Law after Davidoff' (2003) 25 *European Intellectual Property Review* 163

Heinemann, A, 'The Contestability of IP-Protected Markets' in J Drexl (ed), *Research Handbook on Intellectual Property and Competition Law* (Cheltenham, Edward Elgar, 2008) 54–79

——, 'Behavioural Antitrust: A "More Realistic Approach" to Competition Law' in K Mathis (ed), *European Perspectives on Behavioural Law and Economics* (Berlin, Springer, 2015) 211–42

Hemphill, TA, 'Role of Competition Policy in the US Innovation System' (2003) 30 *Science and Public Policy* 285

Hesse, RB and Soven, JH, 'Defining Relevant Product Markets in Electronic Payment Network Antitrust Cases' (2006) 73 *Antitrust Law Journal* 709

Hewitt, G, 'Background Note' in OECD (ed), *Policy Roundtable: Merger Review in Emerging High Innovation Markets* (2002) DAFFE/COMP(2002)20, 19–35

Heyer, K, 'A World of Uncertainty: Economics and the Globalization of Antitrust' (2005) 72 *Antitrust Law Journal* 375

Hickman, D, 'Patents: Competition Law a Defence to Patent Infringement Claims?' (2003) 25 *European Intellectual Property Review* N114

Hildebrand, D, *The Role of Economic Analysis in EU Competition Law: The European School*, 4th edn (Alphen aan den Rijn, Kluwer Law International, 2016)

Hiltunen, E and Hiltunen, K, *Technolife 2035: How Will Technology Change Our Future?* (Newcastle upon Tyne, Cambridge Scholars Publishing, 2015)

Hoerner, RJ, 'Innovation Markets: New Wine in Old Bottles?' (1995) 64 *Antitrust Law Journal* 49

Holzweber, S, 'Market Definition for Multi-Sided Platforms: A Legal Reappraisal' (2017) 40 *World Competition* 563

Hoppner, T, 'Defining Markets for Multi-Sided Platforms: The Case of Search Engines' (2015) 38 *World Competition* 349

Horton, JJ and Zeckhauser, RJ, 'Owning, Using and Renting: Some Simple Economics of the "Sharing Economy"' (NBER Working Paper, vol 22029, 2016)

Hovenkamp, HJ, 'Post-Chicago Antitrust: A Review and Critique' [2001] *Columbia Business Law Review* 257

——, 'The Reckoning of Post-Chicago Antitrust' in A Cucinotta, R Pardolesi and RJ van den Bergh (eds), *Post-Chicago Developments in Antitrust Law* (Cheltenham, Edward Elgar, 2002) 1–33

——, 'Innovation and the Domain of Competition Policy' (2008) 60 *Alabama Law Review* 103

——, 'Schumpeterian Competition and Antitrust' (2008) 4 *Competition Policy International* 273

——, 'Antitrust and Innovation: Where We are and Where We Should Be Going' (2011) 77 *Antitrust Law Journal* 749

——, 'Mergers with Dominant Firms: The *Lundbeck* Case' [2011] *Competition Policy International Antitrust Chronicle* 1

——, 'Competition for Innovation' [2012] *Columbia Business Law Review* 799

——, 'Markets in Merger Analysis' (2012) 57 *Antitrust Bulletin* 887

——, 'Response: Markets in IP and Antitrust' (2012) 100 *Georgetown Law Journal* 2133

——, 'Reimagining Antitrust: The Revisionist Work of Richard S Markovits' (2016) 94 *Texas Law Review* 1221

Hovenkamp, HJ et al, *IP and Antitrust: An Analysis of Antitrust Principles Applied to Intellectual Property Law*, 3rd edn, vol I (New York, Wolters Kluwer, 2018)

Hruska, AC, 'A Broad Market Approach to Antitrust Product Market Definition in Innovative Industries' (1992) 102 *Yale Law Journal* 305

Husa, J, *A New Introduction to Comparative Law* (Oxford, Hart Publishing, 2015)

Hylton, KN, '*Brown Shoe* versus the Horizontal Merger Guidelines' (2011) 39 *Review of Industrial Organization* 95

Ibáñez Colomo P, 'Beyond the "More Economics-Based Approach": A Legal Perspective on Article 102 TFEU Case Law' (2016) 53 *CML Rev* 709

——, 'Restrictions on Innovation in EU Competition Law' (2016) 41 *EL Rev* 201

Ibáñez Colomo P and de Stefano, G, 'Protecting the Integrity and Reputation of Legal Research: JECLAP's New Rules on Disclosure' (2017) 8 *Journal of European Competition Law & Practice* 623

International Competition Network, 'ICN Merger Guidelines Workbook' (April 2006)
——, 'Recommended Practices for Merger Analysis' (2010)
——, 'Competition Enforcement and Consumer Welfare – Setting the Agenda' (May 2011)
——, 'Unilateral Conduct Workbook – Chapter 3: Assessment of Dominance' (May 2011)
International Tennis Federation, 'ITF Rules of Tennis' (2012)
Italianer, A, 'Innovation and Competition' in B Hawk (ed), *International Antitrust Law & Policy* (Huntington, Juris Publishing, 2013)
Jaeger, M, 'The Standard of Review in Competition Cases Involving Complex Economic Assessments: Towards the Marginalisation of the Marginal Review?' (2011) 2 *Journal of European Competition Law & Practice* 295
——, 'Perspective of the Judiciary' (12th GCLC Annual Conference, Brussels, 27 January 2017)
Jalonen, H, 'The Uncertainty of Innovation: A Systematic Review of the Literature' (2011) 4 *Journal of Management Research* 1
Janka, SF and Uhsler, SB, 'Antitrust 4.0: The Rise of Artificial Intelligence and Emerging Challenges to Antitrust Law' (2018) 39 *European Competition Law Review* 112
JND, 'Product Market Definition under the Sherman and Clayton Acts' (1962) 110 *University of Pennsylvania Law Review* 861
Jones, A and Sufrin, BE, *EU Competition Law: Text, Cases, and Materials*, 6th edn (Oxford, Oxford University Press, 2016)
Jones Harbour, P, 'Dissenting Statement, in the Matter of Google/DoubleClick' (20 December 2007)
Jones Harbour, P and Koslov, TI, 'Section 2 in a Web 2.0 World: An Expanded Vision of Relevant Product Markets' (2010) 76 *Antitrust Law Journal* 769
Jorde, TM and Teece, DJ, 'Competing through Innovation: Implications for Market Definition' (1988) 64 *Chicago-Kent Law Review* 741
Kagan, J, 'Bricks, Mortar, and Google: Defining the Relevant Antitrust Market for Internet-Based Companies' (2011) 55 *New York Law School Law Review* 271
Kahwaty, HJ and Tyler, CB, 'Market Definition: Achieving an Integrated Analysis' (2014) 59 *Antitrust Bulletin* 667
Kalintiri, A, 'What's in a Name? The Marginal Standard of Review of "Complex Economic Assessments" in EU Competition Enforcement' (2016) 53 *CML Rev* 1283
Kaplow, L, 'The Accuracy of Traditional Market Power Analysis and a Direct Adjustment Alternative' (1982) 95 *Harvard Law Review* 1817
——, 'Why (Ever) Define Markets?' (2010) 124 *Harvard Law Review* 438
——, 'Market Definition and the Merger Guidelines' (2011) 39 *Review of Industrial Organization* 107
——, 'Market Definition Alchemy' (2012) 57 *Antitrust Bulletin* 915
——, 'Market Definition' in RD Blair and DD Sokol (eds), *The Oxford Handbook of International Antitrust Economics*, vol 1 (Oxford, Oxford University Press, 2015) 345–63
Kathuria, V, 'A Conceptual Framework to Identify Dynamic Efficiency' (2015) 11 *European Competition Journal* 319
Katsoulacos, Y, Avdasheva, S and Golovanova, S, 'Legal Standards and the Role of Economics in Competition Law Enforcement' (2017) 12 *European Competition Journal* 277
Katsoulacos, Y and Ulph, D, 'On Optimal Legal Standards for Competition Policy: A General Welfare-Based Analysis' (2009) 57 *Journal of Industrial Economics* 410
Katz, A, 'Making Sense of Nonsense: Intellectual Property, Antitrust, and Market Power' (2007) 49 *Arizona Law Review* 837
Katz, ML and Shapiro, C, 'Systems Competition and Network Effects' (1994) 8 *Journal of Economic Perspectives* 93
Katz, ML and Shelanski, HA, '"Schumpeterian" Competition and Antitrust Policy in High-Tech Markets' (2005) 14 *Competition* 47
——, 'Mergers and Innovation' (2007) 74 *Antitrust Law Journal* 1
Kauper, TE, 'The Problem of Market Definition under EC Competition Law' (1997) 20 *Fordham International Law Journal* 1682

Keeling, DT, *Intellectual Property Rights in EU Law – Volume I: Free Movement and Competition Law* (Oxford, Oxford University Press, 2003)

Kehder, C, *Konzepte und Methoden der Marktabgrenzung und ihre Anwendung auf zweiseitige Märkte* (Baden-Baden, Nomos, 2013)

Kelly, K, *The Inevitable: Understanding the 12 Technological Forces That Will Shape Our Future* (New York, Viking, 2016)

Kern, BR, 'Innovation Markets, Future Markets, or Potential Competition: How Should Competition Authorities Account for Innovation Competition in Merger Reviews?' (2014) 37 *World Competition* 173

Kern, BR, Dewenter, R and Kerber, W, 'Empirical Analysis of the Assessment of Innovation Effects in US Merger Cases' (2016) 16 *Journal of Industry, Competition and Trade* 373

Keyte, JA, 'Market Definition and Differentiated Products: The Need for a Workable Standard' (1995) 63 *Antitrust Law Journal* 697

Keyte, JA and Schwartz, KB, '"Tally-Ho!": UPP and the 2010 Horizontal Merger Guidelines' (2011) 77 *Antitrust Law Journal* 587

Keyte, JA and Stoll, NR, 'Markets? We Don't Need No Stinking Markets! The FTC and Market Definition' (2004) 49 *Antitrust Bulletin* 593

Killick, J, Jourdan, J and Pêcheux, P, 'The *Servier* Judgment: The General Court Annuls the Commission's Market Definition But Confirms the Illegality of Certain Patent Settlement Agreements' (2019) 10 *Journal of European Competition Law & Practice* 25

Kingsbury, AF, 'Market Definition in Intellectual Property Law: Should Intellectual Property Courts Use an Antitrust Approach to Market Definition?' (2004) 8 *Marquette Intellectual Property Law Review* 63

Kischel, U, *Comparative Law* (A Hammel (trans), Oxford, Oxford University Press, 2019)

Klein, B, 'Market Power in Franchise Cases in the Wake of *Kodak*: Applying Post-contract Hold-up Analysis to Vertical Relationships' (1999) 67 *Antitrust Law Journal* 283

Kolstad, O, 'Competition Law and Intellectual Property Rights: Outline of an Economics-Based Approach' in J Drexl (ed), *Research Handbook on Intellectual Property and Competition Law* (Cheltenham, Edward Elgar, 2008) 3–26

Korah, V, *Intellectual Property Rights and the EC Competition Rules* (Oxford, Hart Publishing, 2006)

Körber, T, 'Internet Search Engines and Competition Law' (2014) 9 *Journal of Intellectual Property Law & Practice* 517

Kovacic, WE, 'The Influence of Economics on Antitrust Law' (1992) 30 *Economic Inquiry* 294

——, 'The Intellectual DNA of Modern US Competition Law for Dominant Firm Conduct: The Chicago/Harvard Double Helix' [2007] *Columbia Business Law Review* 1

——, 'Competition Policy in the European Union and the United States: Convergence or Divergence?' in X Vives (ed), *Competition Policy in the EU: Fifty Years on from the Treaty of Rome* (Oxford, Oxford University Press, 2009) 314–43

Kwoka, J, 'The Structural Presumption and the Safe Harbor in Merger Review: False Positives or Unwarranted Concerns?' (2018) 81 *Antitrust Law Journal* 837

Lacy, GS, 'Standardizing Warhol: Antitrust Liability for Denying the Authenticity of Artwork' (2011) 6 *Washington Journal of Law, Technology & Arts* 185

Lamadrid de Pablo, A, 'The Double Duality of Two-Sided Markets' (2015) 64 *Competition Law Journal* 5

Landes, WM and Posner, RA, 'Market Power in Antitrust Cases' (1981) 94 *Harvard Law Review* 937

Landman, LB, 'The Economics of Future Goods Markets' (1997) 21 *World Competition* 63

——, 'Competitiveness, Innovation Policy, and the Innovation Market Myth: A Reply to Tom and Newberg on Innovation Markets as the "Centerpiece" of "New Thinking" on Innovation' (1998) 13 *St John's Journal of Legal Commentary* 223

——, 'Innovation Markets in Europe' (1998) 19 *European Competition Law Review* 21

Larouche, P, 'Relevant Market Definition in Network Industries: Air Transport and Telecommunications' (2000) 1 *Journal of Network Industries* 407

Larrey, P, *Connected World - From Automated Work to Virtual Wars: The Future, by Those Who are Shaping it* (London, Portfolio Penguin, 2017)
Lasserre, B, 'Market Definition: A Resilient Feature of Competition Enforcement?' in B Hawk (ed), *International Antitrust Law & Policy* (Huntington, Juris Publishing, 2013) 191–214
Legrand, P, 'The Impossibility of "Legal Transplants"' (1997) 4 *Maastricht Journal of European and Comparative Law* 111
——, 'Negative Comparative Law' (2015) 10 *Journal of Comparative Law* 405
Lemley, MA, 'Industry-Specific Antitrust Policy for Innovation' [2011] *Columbia Business Law Review* 637
Lemley, MA and McKenna, MP, 'Is Pepsi Really a Substitute for Coke? Market Definition in Antitrust and IP' (2012) 100 *Georgetown Law Journal* 2055
Lerner, AP, 'The Concept of Monopoly and the Measurement of Monopoly Power' (1934) 1 *Review of Economic Studies* 157
Li, Z, 'New Developments in Civil Antitrust Litigation in China' (2012) 2 *Competition Policy International Antitrust Chronicle* 1
Liebowitz, SJ and Margolis, SE, 'Network Effects and the *Microsoft* Case' in J Ellig (ed), *Dynamic Competition and Public Policy: Technology, Innovation, and Antitrust Issues* (Cambridge, Cambridge University Press, 2001) 160–92
Lind, RC and Muysert, P, 'Innovation and Competition Policy: Challenges for the New Millennium' (2003) 24 *European Competition Law Review* 87
Lind, RC, Muysert, P and Walker, M, 'Innovation and Competition Policy: Part I - Conceptual Issues' (OFT377, Economic Discussion Paper 3, March 2002)
Lind, RC et al, 'Report on Multiparty Licensing' (22 April 2003)
Lindsay, A and McCarthy, E, 'Do We Need to Prevent Pricing Algorithms Cooking up Markets?' (2017) 38 *European Competition Law Review* 533
Lindsay, MA, 'Safeguarding the Standard: Standards Organizations, Patent Hold-up, and Other Forms of Capture' (2012) 57 *Antitrust Bulletin* 17
Lopatka, JE, 'Market Definition?' (2011) 39 *Review of Industrial Organization* 69
Luchetta, G, 'Is the Google Platform a Two-Sided Market?' (2014) 10 *Journal of Competition Law & Economics* 185
Lundqvist, B, *Standardization under EU Competition Rules and US Antitrust Laws: The Rise and Limits of Self-Regulation* (Cheltenham, Edward Elgar, 2014)
Machlup, F, *The Economics of Sellers' Competition: Model Analysis of Sellers' Conduct*, 5th print (Baltimore, Johns Hopkins Press, 1952)
Mackaay, E, *Law and Economics for Civil Law Systems* (Cheltenham, Edward Elgar, 2013)
Madero Villarejo, C, 'Abuses of a Dominant Position in Information Technology Industries (IT)' in C-D Ehlermann and I Atanasiu (eds), *What is an Abuse of a Dominant Position?* (Oxford, Hart Publishing, 2006) 523–56
Madiéga, T, 'Innovation and Market Definition under the EU Regulatory Framework for Electronic Communications' (2006) 29 *World Competition* 55
Maggiolino, M, 'The Economics of Antitrust and Intellectual Property Rights' in SD Anderman and A Ezrachi (eds), *Intellectual Property and Competition Law: New Frontiers* (Oxford, Oxford University Press, 2011) 73–92
Maggiolino, M and Montagnani, ML, 'AstraZeneca's Abuse of IPR-Related Procedures: A Hypothesis of Anti-trust Offence, Abuse of Rights, and IPR Misuse' (2011) 34 *World Competition* 245
Mäihäniemi, B, 'The Role of Innovation in the Analysis of Abuse of Dominance in Digital Markets: The Analysis of Chosen Practices of Google Search' (2017) 1 *Market and Competition Law Review* 111
Maisel, L, 'Submarkets in Merger and Monopolization Cases' (1983) 72 *Georgetown Law Journal* 39
Mandrescu, D, 'Applying (EU) Competition Law to Online Platforms: Reflections on the Definition of the Relevant Market(s)' (2018) 41 *World Competition* 453
Manne, GA, 'In Defence of the Supreme Court's "Single Market" Definition in *Ohio v American Express*' (2019) 7 *Journal of Antitrust Enforcement* 104

Manne, GA and Rinehart, W, 'The US E-Books Case against Apple: The Procompetitive Story' [2012(3)] *Concurrences* 18

Manne, GA and Wright, JD, 'Innovation and the Limits of Antitrust' (2010) 6 *Journal of Competition Law & Economics* 153

——, 'Google and the Limits of Antitrust: The Case against the Case against Google' (2011) 34 *Harvard Journal of Law & Public Policy* 171

Marco Colino, S, *Competition Law of the EU and UK*, 8th edn (Oxford, Oxford University Press, 2019)

Marco Colino, S et al, 'The *Lundbeck* Case and the Concept of Potential Competition' [2017(2)] *Concurrences* 24

Markesinis, BS, *Comparative Law in the Courtroom and the Classroom* (Oxford, Hart Publishing, 2003)

Markham, JW, 'The Joint Effect of Antitrust and Patent Laws upon Innovation' (1966) 56 *American Economic Review* 291

Markovits, RS, 'Predicting the Competitive Impact of Horizontal Mergers in a Monopolistically Competitive World: A Non-Market-Oriented Proposal and Critique of the Market Definition-Market Share-Market Concentration Approach' (1978) 56 *Texas Law Review* 587

——, 'International Competition, Market Definition, and the Appropriate Way to Analyze the Legality of Horizontal Mergers under the Clayton Act: A Positive Analysis and Critique of Both the Traditional Market-Oriented Approach and the Justice Department's Horizontal Merger Guidelines' (1988) 64 *Chicago-Kent Law Review* 745

——, *Economics and the Interpretation and Application of US and EU Antitrust Law*, vols I and II (Berlin, Springer, 2014)

Martin, DD, 'The Brown Shoe Case and the New Antimerger Policy' (1963) 53 *American Economic Review* 340

McChesney, FS, 'Easterbrook on Errors' (2010) 6 *Journal of Competition Law & Economics* 11

McDavid, JL and Steuer, RM, 'The Revival of Franchise Antitrust Claims' (1999) 67 *Antitrust Law Journal* 209

McElroy, FW, 'Alternatives to the US Antitrust Agency Approach to Market Definition' (1996) 11 *Review of Industrial Organization* 511

Michaels, R, 'The Functional Method of Comparative Law' in M Reimann and R Zimmermann (eds), *The Oxford Handbook of Comparative Law*, 2nd edn (Oxford, Oxford University Press, 2019) 345–89

Monopolkommission, 'Wettbewerbspolitik: Herausforderung digitale Märkte' (Sondergutachten 68, 2015)

Monti, G, 'Article 82 EC and New Economy Markets' in C Graham and F Smith (eds), *Competition, Regulation, and the New Economy* (Oxford, Hart Publishing, 2004) 2–11

Morse, MH, 'Product Market Definition in the Pharmaceutical Industry' (2003) 71 *Antitrust Law Journal* 633

Nazzini, R and Nikpay, A, 'Object Restrictions and Two-Sided Markets in EU Competition Law after Cartes Bancaires' (2014) 10 *Competition Policy International* 157

Negrinotti, M, 'Abuse of Regulatory Procedures in the Intellectual Property Context: The AstraZeneca Case' (2008) 29 *European Competition Law Review* 446

Nelken, D, 'Using the Concept of Legal Culture' (2004) 29 *Australian Journal of Legal Philosophy* 1

——, 'Using Legal Culture: Purposes and Problems' (2010) 5 *Journal of Comparative Law* 1

Nelson, RR and Winter, SG, 'The Schumpeterian Tradeoff Revisited' (1982) 16 *American Economic Review* 114

Nevo, H, *Definition of the Relevant Market: (Lack of) Harmony between Industrial Economics and Competition Law* (Cambridge, Intersentia, 2015)

Newberg, JA, 'Antitrust for the Economy of Ideas: The Logic of Technology Markets' (2000) 14 *Harvard Journal of Law & Technology* 83

Newborn, SA and Snider, VL, 'The Growing Judicial Acceptance of the Merger Guidelines' (1992) 60 *Antitrust Law Journal* 849

Newman, N, 'Search, Antitrust, and the Economics of the Control of User Data' (2014) 31 *Yale Journal on Regulation* 401

Nicholas, T, 'What Drives Innovation?' (2011) 77 *Antitrust Law Journal* 787

Niels, G, Jenkins, H and Kavanagh, J, *Economics for Competition Lawyers*, 2nd edn (Oxford, Oxford University Press, 2016)

OECD (ed), *Policy Roundtable: Market Definition* (2012) DAF/COMP(2012)19

——, *Policy Roundtable: The Role and Measurement of Quality in Competition Analysis* (2013) DAF/COMP(2013)17

OECD and Eurostat (eds), 'Oslo Manual: Guidelines for Collecting and Interpreting Innovation Data' (2005)

OECD Secretariat, 'Executive Summary' in OECD (ed), *Policy Roundtable: Merger Review in Emerging High Innovation Markets* (2002) DAFFE/COMP(2002)20, 7–11

Office of Fair Trading, 'Market Definition' (OFT 403, December 2004)

Ordover, JA, Sykes, AO and Willig, RD, 'Herfindahl Concentration, Rivalry, and Mergers' (1982) 95 *Harvard Law Review* 1857

Ortiz Blanco, L, *Market Power in EU Antitrust Law* (Oxford, Hart Publishing, 2012)

Padilla, AJ, 'The Role of Supply-Side Substitution in the Definition of the Relevant Market in Merger Control' (Madrid, June 2001)

Pardolesi, R and Renda, A, 'The European Commission's Case against Microsoft: Kill Bill?' (2004) 27 *World Competition* 513

Pate, RH, 'The Common Law Approach and Improving Standards for Analyzing Single Firm Conduct' (Thirtieth Annual Conference on International Antitrust Law and Policy, New York, 23 October 2003)

Patterson, MR, *Antitrust Law in the New Economy: Google, Yelp, LIBOR, and the Control of Information* (Cambridge, MA, Harvard University Press, 2017)

Peeperkorn, L and Paulis, E, 'Competition and Innovation: Two Horses Pulling the Same Cart' in P Lugard and L Hancher (eds), *On the Merits: Current Issues in Competition Law and Policy – Liber Amicorum Peter Plompen* (Cambridge, Intersentia, 2005) 17–29

Peeperkorn, L and Verouden, V, 'Market Definition' in J Faull and A Nikpay (eds), *The EU Law of Competition*, 3rd edn (Oxford, Oxford University Press, 2014) 42–56

Petit, N, 'Significant Impediment to Industry Innovation: A Novel Theory of Harm in EU Merger Control?' (ICLE White Paper 2017-1, 2017)

Petit, N and Neyrinck, N, 'Behavioral Economics and Abuse of Dominance: A Fresh Look at the Article 102 TFEU Case-Law' [2010] *Österreichische Zeitschrift für Kartellrecht* 203

Petrovčič, U, *Competition Law and Standard Essential Patents: A Transatlantic Perspective* (Alphen aan den Rijn, Kluwer Law International, 2014)

Pickering, V and Dolmans, M, 'The 1997 Digital Undertaking' (1998) 19 *European Competition Law Review* 108

Pitofsky, R, 'New Definitions of Relevant Market and the Assault on Antitrust' (1990) 90 *Columbia Law Review* 1805

Pleatsikas, C and Teece, DJ, 'The Analysis of Market Definition and Market Power in the Context of Rapid Innovation' (2001) 19 *International Journal of Industrial Organization* 665

Podszun, R, 'The Arbitrariness of Market Definition and an Evolutionary Concept of Markets' (2016) 61 *Antitrust Bulletin* 121

Podszun, R and Franz, B, 'Was ist ein Markt? – Unentgeltliche Leistungsbeziehungen im Kartellrecht' (2015) 3 *Neue Zeitschrift für Kartellrecht* 121

Pohlmeier, J, *Netzwerkeffekte und Kartellrecht* (Baden-Baden, Nomos, 2004)

Posner, RA, 'Antitrust in the New Economy' (2001) 68 *Antitrust Law Journal* 925

——, *Antitrust Law*, 2nd edn (Chicago, University of Chicago Press, 2001)

——, *Economic Analysis of Law*, 9th edn (New York, Wolters Kluwer, 2014)

Pradelles, F and Scordamaglia-Tousis, A, 'The Two Sides of the Cartes Bancaires Ruling: Assessment of the Two-Sided Nature of Card Payment Systems under Article 101(1) TFEU and Full Judicial Scrutiny of Underlying Economic Analysis' (2014) 10 *Competition Policy International* 139

Prassl, J, *Humans as a Service: The Promise and Perils of Work in the Gig Economy* (Oxford, Oxford University Press, 2018)
Priest, GL, 'Networks and Antitrust Analysis' in ABA Section of Antitrust Law (ed), *Issues in Competition Law and Policy* (Chicago, ABA Publishing, 2008) 641–66
Rabel, E, 'Die Fachgebiete des Kaiser-Wilhelm-Instituts für ausländisches und internationales Privatrecht (gegründet 1926) (1900–1935)' in M Planck (ed), *25 Jahre Kaiser Wilhelm-Gesellschaft zur Förderung der Wissenschaften*, vol III (Berlin, Springer, 1937) 77–190
Ranchordas, S, 'Innovation Experimentalism in the Age of the Sharing Economy' (2015) 19 *Lewis & Clark Law Review* 871
Rapp, RT, 'The Misapplication of the Innovation Market Approach to Merger Analysis' (1995) 64 *Antitrust Law Journal* 19
Ratliff, JD and Rubinfeld, DL, 'Online Advertising: Defining Relevant Markets' (2010) 6 *Journal of Competition Law & Economics* 653
——, 'Is There a Market for Organic Search Engine Results and Can Their Manipulation Give Rise to Antitrust Liability?' (2014) 10 *Journal of Competition Law & Economics* 517
Reitzes, J and Moss, D, 'Airline Alliances and Systems Competition' (2008) 45 *Houston Law Review* 293
Remer, M and Warren-Boulton, FR, '*United States v. H&R Block*: Market Definition in Court since the 2010 Merger Guidelines' (2014) 59 *Antitrust Bulletin* 599
Rheinstein, M, 'Teaching Comparative Law' (1938) 5 *University of Chicago Law Review* 615
Richards, JD, 'Is Market Definition Necessary in Sherman Act Cases When Anticompetitive Effects Can Be Shown with Direct Evidence?' (2012) 26 *Antitrust* 53
Ritter, C, 'Antitrust in Two-Sided Markets: Looking at the US Supreme Court's *Amex* Case from an EU Perspective' (2019) 10 *Journal of European Competition Law & Practice* 172
Robertson, VHSE, 'Delineating Digital Markets under EU Competition Law: Challenging or Futile?' (2017) 12 *Competition Law Review* 131
——, 'A Brief Comment on the 2017 Update of the US Intellectual Property Licensing Guidelines' (2018) 39 *European Competition Law Review* 461
——, 'The Relevant Market in Competition Law: A Legal Concept' (2019) 7 *Journal of Antitrust Enforcement* 158
Robinson, J, *The Economics of Imperfect Competition* (London, Macmillan, 1954)
Rochet, J-C and Tirole, J, 'Platform Competition in Two-Sided Markets' (2003) 1 *Journal of the European Economic Association* 990
Röller, L-H, 'Antitrust Economics: Catalyst for Convergence' (George Mason Law Review Symposium, Washington, DC, 6 October 2004)
Roman, VD, 'Digital Markets and Pricing Algorithms: A Dynamic Approach towards Horizontal Competition' (2018) 39 *European Competition Law Review* 37
Rosati, E, *Originality in EU Copyright: Full Harmonization through Case Law* (Cheltenham, Edward Elgar, 2013)
Rosch, JT, 'Antitrust Regulation of Innovation Markets' (ABA Antitrust Intellectual Property Conference, Berkeley, 5 February 2009)
Ross, A, *The Industries of the Future* (New York, Simon & Schuster, 2017)
Rubinfeld, DL, 'Market Definition with Differentiated Products: The Post/Nabisco Cereal Merger' (2000) 68 *Antitrust Law Journal* 163
Rubinfeld, DL and Hoven, J, 'Innovation and Antitrust Enforcement' in J Ellig (ed), *Dynamic Competition and Public Policy: Technology, Innovation, and Antitrust Issues* (Cambridge, Cambridge University Press, 2001) 65–94
Russo, F and Stasi, ML, 'Defining the Relevant Market in the Sharing Economy' (2016) 5 *Internet Policy Review* 1
Rysman, M, 'The Economics of Two-Sided Markets' (2009) 23 *Journal of Economic Perspectives* 125
Säcker, FJ, *The Concept of the Relevant Product Market: between Demand-Side Substitutability and Supply-Side Substitutability in Competition Law* (Frankfurt am Main, Peter Lang, 2008)
Sagers, CL, *Antitrust*, 2nd edn (New York, Wolters Kluwer, 2014)

Samuel, G, *An Introduction to Comparative Law Theory and Method* (Oxford, Hart Publishing, 2014)
Scheffman, DT and Spiller, PT, 'Econometric Market Delineation' (1996) 17 *Managerial and Decision Economics* 165
Scherer, FM, *Industrial Market Structure and Economic Performance* (Chicago, Rand McNally, 1970)
——, 'Technological Innovation and Monopolization' in ABA Section of Antitrust Law (ed), *Issues in Competition Law and Policy* (Chicago, ABA Publishing, 2008) 1033–068
Schilling, MA, 'Towards Dynamic Efficiency: Innovation and its Implications for Antitrust' (2015) 60 *Antitrust Bulletin* 191
Schmalensee, R, 'Another Look at Market Power' (1982) 95 *Harvard Law Review* 1789
——, 'Antitrust Issues in Schumpeterian Industries' (2000) 90 *AEA Papers and Proceedings* 192
Schmidt, HKS, 'Article 82's "Exceptional Circumstances" that Restrict Intellectual Property Rights' (2002) 23 *European Competition Law Review* 210
——, 'The Influence of IP Rights on Product Definition in Competition Law: The Curious Case of Tying' (2010) 26 *International Company and Commercial Law Review* 224
Schmidt, ILO, 'The Suitability of the More Economic Approach for Competition Policy: Dynamic vs. Static Efficiency' (2007) 28 *European Competition Law Review* 408
Schumpeter, JA, *Business Cycles: A Theoretical, Historical, and Statistical Analysis of the Capitalist Process*, vol I (New York, McGraw-Hill, 1939)
——, *Capitalism, Socialism and Democracy*, 5th edn (London, Allen & Unwin, 1976)
Schwalbe, U and Maier-Rigaud, F, 'Background Note' in OECD (ed), *Policy Roundtable: Market Definition* (2012) DAF/COMP(2012)19, 21–103
Schwalbe, U and Zimmer, D, *Kartellrecht und Ökonomie: Moderne ökonomische Ansätze in der europäischen und deutschen Zusammenschlusskontrolle*, 2nd edn (Frankfurt am Main, Verlag Recht und Wirtschaft, 2011)
Schweitzer, H, 'The European Competition Law Enforcement System and the Evolution of Judicial Review' in C-D Ehlermann and M Marquis (eds), *European Competition Law Annual 2009: The Evaluation of Evidence and its Judicial Review in Competition Cases* (Oxford, Hart Publishing, 2011) 79–146
Schweitzer, H et al, 'Modernisierung der Missbrauchsaufsicht für marktmächtige Unternehmen: Endbericht' (Report for the Federal Ministry for Economic Affairs and Energy, 29 August 2018)
Sellers, JM, 'The Black Market and Intellectual Property: A Potential Sherman Act Section Two Antitrust Defense?' (2004) 14 *Albany Law Journal of Science & Technology* 583
Shapiro, C, 'The 2010 Horizontal Merger Guidelines: From Hedgehog to Fox in Forty Years' (2010) 77 *Antitrust Law Journal* 49
——, 'Competition and Innovation: Did Arrow Hit the Bull's Eye?' in J Lerner and S Stern (eds), *The Rate and Direction of Inventive Activity Revisited* (Chicago, University of Chicago Press 2012) 361–404
Shelanski, HA, 'Information, Innovation, and Competition Policy for the Internet' (2013) 161 *University of Pennsylvania Law Review* 1663
Sherry, EF and Teece, DJ, 'Royalties, Evolving Patent Rights, and the Value of Innovation' (2004) 33 *Research Policy* 179
Shy, O, *The Economics of Network Industries* (Cambridge, Cambridge University Press, 2001)
Sidak, JG and Teece, DJ, 'Dynamic Competition in Antitrust Law' (2009) 5 *Journal of Competition Law & Economics* 581
——, 'Rewriting the Horizontal Merger Guidelines in the Name of Dynamic Competition' (2009) 16 *George Mason Law Review* 885
Silberman, AH, 'The Myths of Franchise "Market Power"' (1996) 65 *Antitrust Law Journal* 181
Sleuwaegen, L, de Voldere, I and Pennings, E, 'The Implications of Globalization for the Definition of the Relevant Geographic Market in Competition and Competitiveness Analysis' (Ref Ares(2014)77066, January 2001)
Smith, BJ, 'Vertical vs. Core Search: Defining Google's Market in a Monopolization Case' (2012) 9 *NYU Journal of Law & Business* 331

Smith, K, 'Measuring Innovation' in J Fagerberg, DC Mowery and RR Nelson (eds), *The Oxford Handbook of Innovation* (Oxford, Oxford University Press, 2006) 148–77
Smith, RL, 'Defining and Proving Markets and Market Power' in J Duns, A Duke and BJ Sweeney (eds), *Comparative Competition Law* (Cheltenham, Edward Elgar, 2015) 27–55
Solano Díaz, P, 'EU Competition Law Needs to Install a Plug-in' (2017) 40 *World Competition* 393
Soma, JT and Davis, KB, 'Network Effects in Technology Markets: Applying the Lessons of *Intel* and *Microsoft* to Future Clashes between Antitrust and Intellectual Property' (2000) 8 *Journal of Intellectual Property Law* 1
Sousa Ferro, M, '"Ceci n'est pas un marché": Gratuity and Competition Law' (2015) 1 *Concurrences* 1
——, 'Judicial Review: Do European Courts Care about Market Definition?' (2015) 6 *Journal of European Competition Law & Practice* 400
——, *Market Definition in EU Competition Law* (Cheltenham, Edward Elgar, 2019)
Spaak, T, 'Explicating the Concept of Legal Competence' in JC Hage and D von der Pfordten (eds), *Concepts in Law* (Berlin, Springer, 2009) 67–80
Spulber, DF and Yoo, CS, 'Antitrust, the Internet, and the Economics of Networks' in RD Blair and DD Sokol (eds), *The Oxford Handbook of International Antitrust Economics*, vol 1 (Oxford, Oxford University Press, 2015) 380–403
Stein, AM and Brett, BJ, 'Market Definition and Market Power in Antitrust Cases: An Empirical Primer on When, Why and How' (1979) 24 *New York Law School Law Review* 639
Stewart, IB, 'Mergers and Competition: An Analysis of Section 50 of the Trade Practices Act' (2000) 74 *Australian Law Journal* 533
Stigler, GJ and Sherwin, RA, 'The Extent of the Market' (1985) 28 *Journal of Law and Economics* 555
Stigler, Committee on Digital Platforms, 'Final Report' (September 2019)
Stucke, ME, 'Behavioral Antitrust and Monopolization' (2012) 8 *Journal of Competition Law & Economics* 545
Stucke, ME and Grunes, AP, *Big Data and Competition Policy* (Oxford, Oxford University Press, 2016)
Subcommittee on Antitrust, Competition Policy and Consumer Rights, 'The Power of Google: Serving Consumers or Threatening Competition?' (J–112–43, Washington DC, 21 September 2011)
Sullivan, LA, *Handbook of the Law of Antitrust* (St Paul, West Academic Publishing, 1977)
Sundararajan, A, *The Sharing Economy: The End of Employment and the Rise of Crowd-Based Capitalism* (Cambridge, MA, MIT Press, 2016)
Surblytė, G, 'Competition Law at the Crossroads in the Digital Economy: Is it All about Google?' (2015) 1 *Journal of European Consumer and Market Law* 170
Susskind, RE and Susskind, D, *The Future of the Professions: How Technology Will Transform the Work of Human Experts* (Oxford, Oxford University Press, 2015)
Teece, DJ, *Dynamic Capabilities and Strategic Management: Organizing for Innovation and Growth* (Oxford, Oxford University Press, 2009)
——, 'Favoring Dynamic over Static Competition: Implications for Antitrust Analysis and Policy' in GA Manne and JA Wright (eds), *Competition Policy and Patent Law under Uncertainty: Regulating Innovation* (Cambridge, Cambridge University Press, 2011) 203–27
Teece, DJ and Coleman, M, 'The Meaning of Monopoly: Antitrust Analysis in High-Technology Industries' (1998) 43 *Antitrust Bulletin* 801
Telyas, D, *The Interface between Competition Law, Patents and Technical Standards* (Alphen aan den Rijn, Kluwer Law International, 2014)
Temple Lang, J, 'European Community Antitrust Law: Innovation Markets and High Technology Industries' (1997) 20 *Fordham International Law Journal* 717
——, '"Potential" Downstream Markets in European Antitrust Law: A Concept in Need of Limiting Principles' (2011) 7 *Competition Policy International* 106
——, 'Practical Aspects of Aftermarkets in European Competition Law' (2011) 7 *Competition Policy International* 199
Ten Kate, A and Niels, G, 'The Relevant Market: A Concept Still in Search of a Definition' (2009) 5 *Journal of Competition Law & Economics* 297

Thépot, F, 'Market Power in Online Search and Social Networking: A Matter of Two-Sided Markets' (2013) 36 *World Competition* 195
Tom, WK and Newberg, JA, 'Antitrust and Intellectual Property: From Separate Spheres to Unified Field' (1997) 66 *Antitrust Law Journal* 167
Tor, A, 'The Market, the Firm, and Behavioral Antitrust' in E Zamîr and D Teichman (eds), *The Oxford Handbook of Behavioral Economics and the Law* (Oxford, Oxford University Press, 2014) 539–67
Tschentscher, A, 'Dialektische Rechtsvergleichung – Zur Methode der Komparistik im öffentlichen Recht' (2007) 62 *JuristenZeitung* 807
Turner, DF, 'The Role of the "Market Concept" in Antitrust Law' (1980) 49 *Antitrust Law Journal* 1145
Turner, JDC, *Intellectual Property and EU Competition Law*, 2nd edn (Oxford, Oxford University Press, 2015)
Ullrich, H, 'IP-Antitrust in Context: Approaches to International Rules on Restrictive Uses of Intellectual Property Rights' (2003) 48 *Antitrust Bulletin* 837
United States, 'Geographic Market Definition' in OECD (ed), *Working Party No 3 on Co-operation and Enforcement* (8 November 2016) DAF/COMP/WP3/WD(2016)49
——, in OECD (ed), *Policy Roundtable: Market Definition* (2012) DAF/COMP(2012)19, 321–31
——, in OECD (ed), *Policy Roundtable: Merger Review in Emerging High Innovation Markets* (2002) DAFFE/COMP(2002)20, 141–59
US Department of Justice, 'Department of Justice and Microsoft Corporation Reach Effective Settlement on Antitrust Lawsuit' (2 November 2001)
——, 'Antitrust Enforcement and Intellectual Property Rights: Promoting Innovation and Competition' (April 2007)
——, 'Competition and Monopoly: Single-Firm Conduct under Section 2 of the Sherman Act' (September 2008)
——, 'Yahoo! Inc. and Google Inc. Abandon Their Advertising Agreement: Resolves Justice Department's Antitrust Concerns, Competition is Preserved in Markets for Internet Search Advertising' (5 November 2008)
——, 'Statement of the Department of Justice Antitrust Division on its Decision to Close its Investigation of the Internet Search and Paid Search Advertising Agreement between Microsoft Corporation and Yahoo! Inc: Investigation Shows that Agreement Not Likely to Reduce Competition' (18 February 2010)
——, 'Justice Department Secures Largest Negotiated Merger Divestiture Ever to Preserve Competition Threatened by Bayer's Acquisition of Monsanto' (29 May 2018)
US Department of Justice and US Patent & Trademark Office, 'Policy Statement on Remedies for Standards-Essential Patents Subject to Voluntary F/RAND Commitments' (8 January 2013)
Väisänen, TA, *Enforcement of FRAND Commitments under Article 102 TFEU: The Nature of FRAND Defence in Patent Litigation* (Baden-Baden, Nomos, 2011)
Vaishnav, A, 'Product Market Definition in Pharmaceutical Antitrust Cases: Evaluating Cross-Price Elasticity of Demand' [2011] *Columbia Business Law Review* 586
Van den Bergh, RJ, 'Modern Industrial Organization versus Old-Fashioned European Competition Law' (1996) 17 *European Competition Law Review* 75
——, 'The More Economic Approach in European Competition Law: Is More Too Much or Not Enough?' in M Kovač and A-S Vandenberghe (eds), *Economic Evidence in EU Competition Law* (Cambridge, Intersentia, 2016) 13–42
Van den Bergh, RJ and Giannaccari, A, 'L'approcio più economico nel diritto comunitario della concorrenza: Il più è troppo o non (ancora) abbastanza?' (2014) XVI *Mercato concorrenza regole* 393
Vassallo, AP, 'Can One (Ever) Accurately Define Markets?' (2017) 13 *Journal of Competition Law & Economics* 261
Veljanovski, C, 'EC Merger Policy after *GE/Honeywell* and *Airtours*' (2004) 49 *Antitrust Bulletin* 153
Verhaert, J, 'The Challenges Involved with the Application of Article 102 TFEU to the New Economy: A Case Study of Google' (2014) 35 *European Competition Law Review* 265

Vickers, J, 'Competition Law and Economics: A Mid-Atlantic Viewpoint' (2007) 3 *European Competition Journal* 1

Von Hippel, E, *Democratizing Innovation* (Cambridge, MA, MIT Press, 2005)

Von Kalinowski, JO, 'Market Definition under Section 2: The Applicability of Clayton Act Section 7 Analysis' (1978) 10 *Southwestern University Law Review* 95

Vrins, O, 'Intellectual Property Licensing and Competition Law: Some News from the Front – the Role of Market Power and Double Jeopardy in the EC Commission's New Deal' (2001) 23 *European Intellectual Property Review* 576

Waelbroeck, D, 'Vertical Agreements: 4 Years of Liberalisation by Regulation No 2790/99 after 40 Years of Legal (Block) Regulation' in H Ullrich (ed), *The Evolution of European Competition Law: Whose Regulation, Which Competition?* (Cheltenham, Edward Elgar, 2006) 85–110

Waller, SW, 'Antitrust and Social Networking' (2012) 90 *North Carolina Law Review* 1771

Watson, A, *Legal Transplants: An Approach to Comparative Law*, 2nd edn (Athens, GA, University of Georgia Press, 1993)

Weber, M, 'Liability for the Acquisition of Faked or Wrongly Attributed Works of Art in US Law' in K Odendahl and PJ Weber (eds), *Kulturgüterschutz – Kunstrecht – Kulturrecht* (Baden-Baden, Nomos, 2010) 409–33

Werden, GJ, 'The History of Antitrust Market Delineation' (1992) 76 *Marquette Law Review* 123

——, 'Demand Elasticities in Antitrust Analysis' (1998) 66 *Antitrust Law Journal* 363

——, 'Market Delineation under the Merger Guidelines: Monopoly Cases and Alternative Approaches' (2000) 16 *Review of Industrial Organization* 211

——, 'The 1982 Merger Guidelines and the Ascent of the Hypothetical Monopolist Paradigm' (2003) 71 *Antitrust Law Journal* 253

——, 'Why (Ever) Define Markets? An Answer to Professor Kaplow' (2013) 78 *Antitrust Law Journal* 729

White, LJ, 'Antitrust and Merger Policy: A Review and Critique' (1987) 1 *Journal of Economic Perspectives* 13

——, 'Market Definition and Market Power in Payment Card Networks: Some Comments and Considerations' (2006) 5 *Review of Network Economics* 61

——, 'Market Power and Market Definition in Monopolization Cases' in ABA Section of Antitrust Law (ed), *Issues in Competition Law and Policy* (Chicago, ABA Publishing, 2008) 913–24

Whitener, MD, 'Potential Competition Theory: Forgotten But Not Gone' (1991) 5 *Antitrust* 17

Widnell, NA, 'The Crystal Ball of Innovation Market Analysis in Merger Review: An Appropriate Means of Predicting the Future?' (1996) 4 *George Mason Law Review* 369

Winfree, J, 'Fan Substitution and Market Definition in Professional Sports Leagues' (2009) 54 *Antitrust Bulletin* 801

Wirtz, MM and Holzhäuser, M, 'Die kartellrechtliche Zwangslizenz' (2004) 50 *Wettbewerb in Recht und Praxis* 683

Wismer, S, Bongard, C and Rasek, A, 'Multi-Sided Market Economics in Competition Law Enforcement' (2017) 8 *Journal of European Competition Law & Practice* 257

Wismer, S and Rasek, A, 'Market Definition in Multi-Sided Markets' in OECD (ed), *Rethinking Antitrust Tools for Multi-Sided Platforms* (2018) 55–67

Witt, AC, *The More Economic Approach to EU Antitrust Law* (Oxford, Hart Publishing, 2016)

Wright, J, 'One-Sided Logic in Two-Sided Markets' (2004) 3 *Review of Network Economics* 44

Wright, JD, 'Antitrust, Multi-dimensional Competition, and Innovation: Do We Have an Antitrust-Relevant Theory of Competition Now?' in GA Manne and JD Wright (eds), *Competition Policy and Patent Law Uncertainty: Regulating Innovation* (Cambridge, Cambridge University Press, 2011) 228

Wrobel, GG, 'Connecting Antitrust Standards to the Internet of Things' (2014) 29 *Antitrust* 62

Wu, L and Baker, S, 'Applying the Market Definition Guidelines of the European Commission' (1998) 19 *European Competition Law Review* 273

Wu, T, 'The *American Express* Opinion, the Rule of Reason, and Tech Platforms' (2019) 7 *Journal of Antitrust Enforcement* 117

Yaşar, AG, 'Achieving Symbiosis between Disruptive Innovation and Merger Control: Challenges and Remedies' (SSRN abstract n 3015007, 1 June 2017)

Yoo, CS, 'Copyright and Product Differentiation' (2004) 79 *New York University Law Review* 212

Zhang, AH, 'Using a Sledgehammer to Crack a Nut: Why China's Anti-monopoly Law was Inapproriate for *Renren v Baidu*' (2011) 7 *Competition Policy International* 277

Zhang, L, 'Refusal to License Intellectual Property Rights under Article 82 EC in Light of Standardisation Context' (2010) 32 *European Intellectual Property Review* 402

Zimmer, D, 'The Emancipation of Antitrust from Market-Share-Based Approaches' (2016) 61 *Antitrust Bulletin* 133

Zingales, N, 'Product Market Definition in Online Search and Advertising' (2013) 9 *Competition Law Review* 29

Zweigert, K and Kötz, H, *Einführung in die Rechtsvergleichung auf dem Gebiete des Privatrechts*, 3rd edn (Tübingen, Mohr Siebeck, 1996)

INDEX